Sustainable Development Goals Series

The **Sustainable Development Goals Series** is Springer Nature's inaugural cross-imprint book series that addresses and supports the United Nations' seventeen Sustainable Development Goals. The series fosters comprehensive research focused on these global targets and endeavours to address some of society's greatest grand challenges. The SDGs are inherently multidisciplinary, and they bring people working across different fields together and working towards a common goal. In this spirit, the Sustainable Development Goals series is the first at Springer Nature to publish books under both the Springer and Palgrave Macmillan imprints, bringing the strengths of our imprints together.

The Sustainable Development Goals Series is organized into eighteen subseries: one subseries based around each of the seventeen respective Sustainable Development Goals, and an eighteenth subseries, "Connecting the Goals," which serves as a home for volumes addressing multiple goals or studying the SDGs as a whole. Each subseries is guided by an expert Subseries Advisor with years or decades of experience studying and addressing core components of their respective Goal.

The SDG Series has a remit as broad as the SDGs themselves, and contributions are welcome from scientists, academics, policymakers, and researchers working in fields related to any of the seventeen goals. If you are interested in contributing a monograph or curated volume to the series, please contact the Publishers: Zachary Romano [Springer; zachary.romano@springer.com] and Rachael Ballard [Palgrave Macmillan; rachael.ballard@palgrave.com].

Alfredo Pereira Jr • Francisco Sousa
Editors

Principles for Governance

Strategies for Reducing Inequality and Promoting Human Development

Springer

Editors
Alfredo Pereira Jr
State University of São Paulo
Botucatu and Marília, SP, Brazil

Francisco Sousa
MFA Business Office
George Washington University
Arlington, VA, USA

SAIS John Hopkins University
Washington DC – Senior Staff
Johns Hopkins University
Baltimore, MD, USA

ISSN 2523-3084 ISSN 2523-3092 (electronic)
Sustainable Development Goals Series
ISBN 978-3-031-40024-7 ISBN 978-3-031-40025-4 (eBook)
https://doi.org/10.1007/978-3-031-40025-4

This Springer imprint is published by the registered company Springer Nature Switzerland AG
The registered company address is: Gewerbestrasse 11, 6330 Cham, Switzerland

Paper in this product is recyclable.

Introduction

The current economic, military, sanitary and environmental crisis urges for new political initiatives towards the route that leads to qualified economic development and ecosystem stability, making possible the long-term continuity of human life in the planet.

Thomas Piketty's diagnosis on the increase of economic inequality in most countries (Piketty, 2014, 2020, 2021) points to a scenario of chronic social crisis, requiring a transformation of the system, affording to the majority of people to constructively co-participate with the governments in the definition of strategic goals and their implementation, and to access collectively generated benefits guaranteed by institutional democratic mechanisms. Approaching a 'reduction of inequality' program, Piketty makes suggestions of political economic actions to which this book is directed – but not limited, since they seem to be not sufficient to solve the complex network of problems that we are facing in the twenty-first century.

Standing (2017) moves further, in a defense of his Basic Income strategy, proposed to be the best way to fight both poverty in developing countries and other social problems in the whole planet. He refers to the 'monsters' of our time:

1. Inequality (leading to an increase of hunger in several populations)
2. Insecurity in several domains, from nutritional to criminal
3. People's increasing debt and insolvency
4. Chronic stress, leading to suicide, abuse of opioids and other substances
5. Artificial intelligence and robots impacting on human work and causing unemployment
6. Precarity of working conditions (lack of time for living, chasing temporary jobs), as treated in his previous book (Standing, 2013)
7. Threat of extinction deriving from a multifactorial ecological crisis
8. Rise of populism, a type of government based on fear, that turns out to be a pseudo-solution to the problems

The fight against the monsters requires a systemic approach in several sectors of social activity. What are the available resources to make this transformation happen? In this book, we approach the crisis from a multidisciplinary perspective, to propose and discuss a series of actions to be carried by public administration teams, working together with groups of scientific, technological and academic professionals, and social activists.

The chapters that compose the book depart from some common premises and envisage possible ways of public administration actions. The focus of discussion in several chapters is related to issues present in the Brazilian context, familiar for most authors, and we also include collaborations from other countries. Brazil is at the crossroads of problems, which are essential to the assessment of the current global ordeal: environmental deplection, neo-fascist populism and economic recession resulting from decades of economic neoliberal policies. In addition, we expect Brazil to adopt new approaches to social inclusion for overcoming these growing problems, likely using information technologies such as the pioneer official website (*GOV.BR*).

We recognize the necessity for more effective usage of Information Technology (IT) on the basis of humanitarian principles, to cope with the ever-growing complexity of the social system. This effort aims to direct state policies towards advancing social goals that benefit a majority of the population and establish closer accountability of elected officials and public administrators with greater online transparency. The usage of IT in E-government brings the possibility of public policies to reduce economic inequality, avoiding corruption (in the sense of misuse of public resources for private interests) and promoting the activities of production and consumption of goods necessary for human development, in an ecologically sustainable model.

The 'reducing inequality' in the title refers to the need of providing ways of offering a better quality of life to the people, which requires providing income to the increasing number of people who do not find employment, for several reasons, which include the historical processes of colonization, racial and gender discrimination and, more recently, the sanitary crisis and the progressive automation of production methods.

The 'human development' concept in the title encompasses proper human needs and potentialities, and also the need to support efforts towards increasing the methods of ecologically friendly production, both as a strategy to reduce deteriorating environmental conditions and improve quality of life. Destroying the stability and reducing the biodiversity of the ecosystems that generate the resources our descendants need for survival and well-being lead to the destruction and possible extinction of the human species.

Thomas Piketty's description and analysis of the process of increasing social inequality in the last 40 years is very accurate (Piketty, 2014, 2020, 2021). The acceptance of these facts inspired the United Nations' programme of reduction of inequality in the planet. However, the propositional section of Piketty's work is not so inspiring, but reads as another attempt to revive the social democracy of the 1950–1960s 'welfare state'. Below is a shortlist of his suggestions to reduce inequality (Piketty, 2020):

1. Redistribution of property, by means of tributary reforms (progressive taxation of capital, promoting fiscal justice) and giving more power to employees in corporations
2. Distribution of income, by means of basic income programmes
3. Climatic transition, by means of carbon uptake, cleaner (non-fossil) energy and similar projects

4. Creation of an international taxation system and vigilance against illegal financial operations in 'fiscal paradises'
5. Equalitarian rights to education

'Reducing inequality', or course, does not imply total equality – the abolition of social classes, or equalizing existing differences of economic status within the people, but refers to a systemic change that aims to abolish severe poverty and promote human development. However, implicit in Piketty's reasoning is the concept of a 'zero sum game', in which the state intervenes to transfer resources from the rich to the poor, thus reducing the social differences between them. There are two insufficiencies in this view:

(a) Piketty does not account for the mechanisms of money generation and distribution by sovereign states, and how it favours the accumulation process in the current stage of capitalism. He focuses on the inequality that derives from the 'absolute right to property', but not on the inequality that derives from how the money issued by banks under state authorization is distributed in the society. Exclusion from the economic system generates poverty; for instance, Brazil has some 5575 municipalities and over 1000 of them do not have banks or correspondent stores. But this issue can be largely resolved with affordable web technology supported by government. According to *Modern Cartalism* (Knapp, 1905), the money owned (or not) by the people is determined by the state, following legal and political conventions. How is this money distributed in society, such that the minority of citizens takes it all, and the others have to work and be explored by the minority, to get a minimal fraction necessary for their survival? Without understanding this process, largely responsible for the increase of inequality in the twenty-first century, attempts to reconstruct the 'welfare state' are possibly condemned to failure;
(b) The reduction of inequality does not necessarily, in principle, depend on taking money from the riches to give to the poor. If the social system enters a 'positive sum game', the reduction of inequality can be made by means of changes in the state's economic policy, targeting the way in which state-issued money (as the digital currencies already used in some countries; see Chap. 4) is distributed. More precisely, if the federal state selectively (by means of 'qualitative easing', as proposed by Brown, 2019) finances cooperative and sustainable productive actions, and at the same time democratizes money for the people's consumption of the produced goods, a human development cycle can be generated.

From what we have learned from the success of Chinese economy, one cannot do without the market economy and entrepreneurial meritocracy, while establishing a process for eradicating poverty, by utilizing public banks for financing what is socially relevant. The connection between socially relevant productive work and the reward by merit has been very effective in the Chinese 'Market Socialism', contrasting with the apparent limitations of 'Real Socialism' in other countries. This is not a Chinese idiosyncratic policy. For instance, the New Deal in the 1930s in the USA was financed by the

Federal State, with ample resources to make things happen, even the securitization of public debt. 'Qualitative easing' (Brown, 2019), the *selective* issue and distribution of (digital) money by the federal state, may be a strategy useful for most countries to induce human development. This approach can use information technology to direct 'Money to the People' (title of Chap. 4, which paraphrases John Lennon's song 'Power to the People') and manage the flux of money in society, since these operations are not possible with traditional material substrates used for money (paper, metal) that can flow without much control.

A new human development model will not emerge spontaneously, since the articulation for the emergence of novel supply chains require the creation of new types of enterprises, for which it is necessary to have the support of the state. Nobody can work miracles for creating new types of enterprise entirely on their own. In terms of production incentives, if we have a New Deal, it cannot simply be giving money to large companies, which are already well established, or to existing businesses that are not environmentally friendly. There must be criteria, standards and compliance for the private sector to qualify for public financing. In a 'New Green Deal', funding has to be for those who meet certain requirements. Moreover, there is a glaring need for project proposals, as well as qualified personnel to evaluate these submissions.

The adjustment of production and consumption can be better achieved by means of control of the flux of state-issued money. Now, with the Central Bank Digital Currency (CBDC; Lagarde, 2018), all transactions can be tracked, as this money resides in the state's computer network; when a transaction takes place, a tracking code is generated which is then validated and follows on to the recipient account where it is posted as a credit or debt. For each region, public administrators, along with usage of artificial intelligences, could track, for instance, which products the beneficiaries of basic income cash transfers are spending their money on, and then, the state could provide funding to the producers of those consumer goods. This would expand the scale and improve production methods, so that supply is increased in the region markets according to consumer demand. Thus, the increase in demand would not necessarily result in higher prices, keeping goods and services affordable of the products.

These concerns lead to the task of rethinking the state's means of investing in popular production and consumption, and the relation between the concept of human development and theories of complexity and self-organization. In the environmental sphere, we need to re-discuss how to combine conservation and production, to support and (re-)direct the public universities to work in synergy with local communities, how to reshape fundamental education to raise social consciousness, and, ultimately, to redefine the role of science, in public and private researches, in a new economic model that has an ecosystem perspective. In the social area, we should also re-conceptualize how to achieve well-being and well-living, coping with information technology issues, quality of life and satisfaction indicators.

This new model would be parallel to 'agribusiness', the main economic force in Brazil. It would be reckless to fight agribusiness head on, but we also don't need to be complacent, because it has many negative aspects. It does not serve as a general model. If applied to the whole environment, it entails the risk of destroying biodiversity and ecosystem stability. We need two parallel systems: at the same time that agribusiness occupies a certain part of the territory, we should have other parts of the territory in which alternative models can be implemented. And what would be their characteristics? An important trait is to incentive businesses that generate employment and income for people, in order to include people in the economic system. So, employability is an important issue valuing the knowledge that people already have. Within a model of ecological economics, the knowledge that the natives have of plants is recognized as much greater than, for instance, the database of the 'Big Pharma'. So, in a solidary and sustainable model of the social system, we need to value the people's knowledge and social technologies.

Another characteristic is that there must be projects aimed at generating popular consumer goods. It makes no sense, in an ecological economy, the state investing in the development of a very expensive non-strategic product that only a small fraction of the population can buy. This will not have a production scale, that is, it will not enable the generation of many jobs, nor a great transformation in the relationship with nature, because it operates in a very restricted market. The main target of the state's investment should be popular consumption, and this also implies having to generate a distribution network for these products. Organic supermarkets, like *Whole Foods* in the USA or *Coop* in Europe, could be financed with public money in developing countries.

What research shows, in the areas of Sociology and History, is that economic and human development is qualitatively induced by the federal state, financing the branches of the economy considered to be more strategic, and also creating infrastructure conditions so that people can have a better quality of life and enjoy the products that are generated. When the state settles this adjusted production and consumption cycle, human development is more likely to happen. As this adjustment is made by the state, the field of Public Administration is central to the programs of reducing inequality and promoting human development. There is also the participation of state and municipal governments, but the power to issue money to promote what is considered strategic belongs to the federal state.

The book is structured around a series of proposals to fight the crisis and promote human development. While some of these proposals build on already existing formulations, others are original and form a broad framework to inspire politicians and public administrators. These proposals can be adapted to particular circumstances and be applied to different countries, being carried by political organizations with different ideological directions. They are presented in the chapters of the book, including the development of a new branch of Bioethics, the Bioethics of Governance (BG), concerning the responsibility of the state managers of caring for the life of the populations under their leadership. The BG is based on the four principles of Medical Bioethics (Beneficience, Non-Maleficiency, Justice and Autonomy; see Beauchamp & Childress, 2013), the principles of the Ethics of Sustainability (Precaution, Sustainable Economics, Distribution, Viability; see Kibert et al., 2011) and other principles needed for effective democratic governance. Bioethics concerns in public policies are required to avoid distractions, as in the case of mistaken COVID treatments, and false claims about the need of austerity in public accounting (Lara Resende, 2020) – when state financial support to the society is necessary to avoid human suffering and death.

We also envisage the development of a social technology, the *Popular Online Forum* (Pereira Jr. et al., 2014), implemented in the state's computers, using the Internet, having the function of operating as a public non-profit social network, for the interaction of the state staff and representatives with the population, to monitor political initiatives and make sure the principles of the Bioethics of Governance are respected. In this regard, the use of IT in Public Administration would allow better targeting within an institutional framework to promote the interaction of state managers and the people. In Brazil, the site Gov.br can be developed in this direction.

Suggestions of impact investments targeting social and environmental goals are part of a systemic effort to democratize money. These efforts should allow citizens to receive funding for economic innovation, allowing them, among other facilities, to have an account in non-profit public banks and financial institutions, and using it for online transactions, addressing one of the putative principles of the BG: social justice in the emission and distribution of money.

Other initiatives in the direction of human development are the endorsement of proposals of basic income, universal or conditional (Barrowclough, 2018; Standing & Orton, 2018), and job guarantee programs, to include every citizen in the development process, and allowing people to dedicate themselves to socially relevant activities, which are not well recognized and remunerated in the market, such as taking care of children, taking care of the elderly, of people with disabilities and illnesses, planting trees, horticulture, etc.; the implementation of this type of programme locally (e.g. in a chosen city) can be used as preliminary approach to more substantial federal Basic Income programs.

The promotion of 'Green Deals', parallel to the current 'agrobusiness' model, can promote the progressive reformatting of the primary, secondary and tertiary sectors of the economy, according to scientific principles of ecosystem sustainability; incentive to solidary initiatives, such as Cooperativism, the 'Commons' model, Circular and Creative Economy, Recycling and, last but not least, creating Local Currencies supported by community non-profit financial institutions (Brown, 2019).

Complementary efforts towards achieving human development would also include changes in the educational curriculum and methods, allowing teachers and students to actively participate in the discussion of national problems and their solutions in the classroom; and, last but not least, improving the actually existing mechanisms of social assistance to better cope with the principles of the BG and make the innovations accessible to everyone. Cooperative relations in the production system connected to effective practices in efficient social assistance systems can help to promote the reduction of economic inequality and provide new opportunities of work and access to the consumption of necessary goods for the whole population of the planet. All these efforts can elicit the emergence of new modalities of social consciousness within the population, to promote benevolence and actions towards a solidary and sustainable society, and proactive efforts against violence and war. This rise of consciousness can be achieved by means of scientific and non-scientific mediations.

The main goal of this book is to provide insight and stimulation for new government initiatives using information technology. We also intend that the book will become a reference for courses in Public Administration, Cognitive Technology, E-Learning, Finance, Philosophy of Economy, Agronomics, Forest Engineering, Bioethics and Education, among others, given the strong interdisciplinary profile we adopted.

This theoretical effort expresses the need to construct the non-totalitarian societies' institutional security architecture, in order to consider non-military aspects to guarantee the integrity and continuity of sovereign nations, especially the protection of social infrastructure. The concept of 'bio-security' serves all systems related to fauna, flora and human development. The greatest threat to the environment today is poverty, the result of exclusion. The productivity gains produced by the new technologies need to be better distributed in models of 'Sociocracy', in more horizontal structures, reducing the differences of social classes. All of this can be made possible in collaboration with the private sector and state sponsorship. This 'social economy' model can easily be made compatible with existing economic structures, or eventually replacing those that do not fulfil the needs of contemporary societies.

The intent of this book is to promote human development, as a process carried out by the people and for the people. The role of the state is, therefore, comparable to an enzyme catalyzing a reaction (this role was also attributed to public universities, see Pereira Jr., 2005); the state induces the process of human development by the selective emission and distribution of money 'enzymes' (as qualitative easing) for popular activities of production and consumption, creating coherent cycles within the people's social attitudes and behaviours. In this view, the state is not a malign 'Leviathan', but the only available tool for the realization of the interests of the majority of the population in each country. In a final remark, it should be stressed that the financial help programmes here proposed should be directed to the self-organization of the people, not to give them a free lunch – except, of course, if they are in a condition of hunger!

Botucatu and Marília, SP, Brazil — Alfredo Pereira Jr
Arlington, VA, USA — Francisco Sousa

References

Barrowclough, D. (2018). Starting with the poor. In United Nations (Org.), *The ins and outs of inclusive finance: Some lessons from microfinance and basic income*. UNCTAD (United Nations Conference on Trade and Development).

Beauchamp, T., & Childress, J. (2013). *Principles of biomedical ethics* (7th ed.). Oxford University Press.

Brown, E. (2019). *Banking on the people: Democratizing money in the digital age*. The Democracy Collaborative.

Kibert, C., Monroe, M., Plate, R., Peterson, A., & Thiele, L. (2011). Working toward sustainability: Ethical decision-making in a technological world (Wiley series in sustainable design). Wiley. ISBN: 978-0-470-53972-9 First version "The Ethics of Sustainability" available at: http://citeseerx.ist.psu.edu/viewdoc/summary?doi=10.1.1.472.1559

Knapp, G. F. (1905). *The state theory of money*. Simon Publications.

Lagarde, C. (2018). *Winds of change: The case for new digital currency*. International Monetary Fund. Available at: https://www.imf.org/en/News/Articles/2018/11/13/sp111418-winds-of-change-the-case-for-new-digital-currency

Lara Resende, A. (2020). *Consenso e Contrassenso: Por uma economia não dogmática*. Portfolio Editora.

Pereira Jr., A. (2005). *A Universidade Pública e os Desafios do Desenvolvimento* (Vol. 22, pp. 18–23). Inteligência Empresarial (UFRJ).

Pereira Jr., A., Ilario, E., Paixão, V. G., Chinalli, L. A., & Monserrat Neto, J. (2014). Tecnologia, Democracia e Socialismo: o Encontro do Século? *Perspectivas em Ciências Tecnológicas, 3*, 88–108.

Piketty, T. (2014). *Capital in the twenty-first century*. Belknap Press.

Piketty, T. (2020). *Capital and ideology*. Harvard University Press.

Piketty, T. (2021). *Time for socialism: Dispatches from a world on fire, 2016–2021*. Yale University Press.

Standing, G. (2013). *The Precariat: The new dangerous class*. Bloomsbury Publishing.

Standing, G. (2017). *Basic income: A guide for the open-minded*. Yale University Press.

Standing, G., & Orton, I. (2018). Development and basic income: An emerging economic model. In United Nations (Org.), The ins and outs of inclusive finance: Some lessons from microfinance and basic income. UNCTAD (United Nations Conference on Trade and Development).

Contents

Part I

Principles and Information Technology for a Democratic Government

1 Bioethics of Public Governance: A Strategy Towards the "Reduction of Inequality"

Alfredo Pereira Jr, Francisco Sousa, and Enidio Ilario

Abstract

Currently, the planet experiences a multifactorial economic, sanitary, military, epistemological, and ecological crisis that demands new efforts to overcome structural problems of the society. The proposal of a Bioethics of Governance (BG) addresses the challenge, from all countries and ideological colors, to cope with the crisis and its problems, by means of formulating principles that express the obligations of those who are in a position to guide the society towards solutions, minimizing human suffering, disease and death. The elaboration of the principles of BG would be similar to Medical Bioethics, recently extended to the UNESCO Universal Declaration of Bioethics and Human Rights. The set of BG principles should correspond to the minimal, basic obligation of people's representatives at all levels of social organization (from local to national and global) to promote human development. In this chapter, we propose a method to democratically establish a set of BG principles, in a series of representative interdisciplinary meetings. Once the principles are established, they can be used for the people to control the actions of governments, by using public information technology systems and personal devices.

1 Introduction

In participative democracy, initiatives to benefit the majority of the people should, ideally, be led by the people themselves. However, contemporary society is far from being a popular self-organizing system; since at least a thousand years, most collective actions in urban societies are financed – and by this means organized – by state money, and, more recently, by 'lobbies' mounted by powerful corporations. Should the people let these entities make the decisions that determine the trajectory of the society, and the (in)stability of the environment that makes possible human life on the planet?

Considering the leading role of states and public representatives, the second article of UNESCO's *Universal Declaration of Bioethics and Human*

A. Pereira Jr (✉)
State University of São Paulo, Botucatu and Marília, SP, Brazil
e-mail: alfredo.pereira@unesp.br

F. Sousa
MFA Business Office, George Washington University, Washington, DC, USA

SAIS John Hopkins University Washington DC – Senior Staff, Johns Hopkins University, Baltimore, MD, USA

E. Ilario
University of Campinas – Center for Logic Epistemology and History of Sciences, Campinas, Brazil

A. Pereira Jr, F. Sousa (eds.), *Principles for Governance*, Sustainable Development Goals Series,
https://doi.org/10.1007/978-3-031-40025-4_1

Rights (UDBHR) says: "This Declaration is addressed to states…to provide a universal framework of principles and procedures to guide states in the formulation of their legislation, policies or other instruments in the field of bioethics" (see https://en.unesco.org/themes/ethics-science-and-technology/bioethics-and-human-rights). In this chapter, we relate these bioethical concerns with the functions of public governance. We aim to propose a formulation of universal principles of Bioethics of Governance (BG), intended to be applied to all democratic countries, in superposed and/or complementary ways to their own laws and ethical rules. The initial task would be to formulate the principles of BG, letting to a second phase the adaptation to the legal and institutional framework of each country.

Why would it be the BG, not just the Ethics of Governance? This is because the most relevant concerns address issues of humans and other forms of *life* in the planet, not only in the area of health, but involving survival in general: having money to buy food, availability of water, having a job with regular income, and satisfaction of human basic needs, in synergy with the other forms of life in the planetary environment. Van Rensselaer Potter, a biochemist and researcher in the field of oncology, created the neologism 'Bioethics' (Potter, 1971), as a much broader notion than biomedical ethics, addressing health problems resulting from the degradation of the environment and the habitat of other species.

The human economy works within a wider ecosystem, but we often forget our dependence on eco-systemic conditions for the continuity of human life in the planet. So, when mentioning Bioethics, it is not only the human life that is under the responsibility of the governors, but also the responsibility for the co-evolution of life in the ecosystem in which we are inserted, because the continuity of our own existence will depend on how this ecosystem will evolve over time, and how human action influences the evolution.

In each country, the proposal addresses public governance at the federal, state, and municipal levels. We have elected officials who are not properly monitored by the people, and their behaviors often go against the interests of the continuity of life. Monitoring them is a task that cannot be made personally, with the presence of the people in the physical dependences of the state, but becomes possible with the use of Internet communication and related computer/information technologies. Evidently, not only because of the pandemics but taking into consideration the size of the population, we cannot have a Greek-style participatory democracy or, for example, assemblies, as those used by union and student movements for discussion and decision-making, where people participate directly.

Article 18 of UNESCO's UDBHR makes two statements: "Persons and professionals concerned and society as a whole should be engaged in dialogue on a regular basis", and "Opportunities for informed pluralistic public debate, seeking the expression of all relevant opinions, should be promoted". These are of course very constructive propositions, but they need principles from which to evaluate what is good and what is bad in a given field of activity. Besides general bioethical principles, which would be the most relevant principles to be obeyed by the state managers and politicians who aim to become state managers?

The monitoring and control of professionals is similar to what happened in the medical field. BG has in common with Medical Bioethics the idea that there must be surveillance and punishment, so that the established principles are followed. The professional who flees from the principles may have the permit revoked, the professional action restricted or even reach the point of no longer being able to exercise medical activities. So, it is extremely relevant, beyond establishing purely philosophical directions, to promote public practices of monitoring and punishment to avoid that the agreed principles are violated. It is not enough to establish principles if we do not indicate how, in practice, society will do, so that these principles are effectively fulfilled. So the idea is to establish principles and also means for these principles to become effective; but this cannot just be "top down social engineering", from the rulers to the people; there must also be a form of action from the bottom up, by the people organizing themselves and controlling the state managers. So it's a two-way street.

The lack of popular participation and control of state governance is harmful in the face of situations such as an economic crisis, with the increase of chronic unemployment and hunger; health and environmental issues, and last but not the least, the 'global warming' problems. In the absence of popular participation, politicians implement solutions of interest to the powerful groups that finance and control them. Overcoming the limitations of these wrong solutions requires more time and human suffering, for the undue and unethical actions of the state managers to be corrected.

For a social self-organizing process to happen, it is necessary to have ways for people to interact with the state's managers. Considering the size of human populations, today this task requires not only information technology to interconnect the people, as suggested before, but also the involvement of the scientific community to provide qualified information for decision processes. The participatory democracy of a country or a state, or even a municipality, needs to have a proper communication apparatus that uses information technology resources. Today we have several information technology resources, such as social networks, which have not been properly used for constructive popular self-organizing processes. The 'new digital public sphere' that emerged with widespread fast internet in the last decades is badly used by society, being vitiated by fake news, slander, and orchestrated movements to manipulate people so that they buy certain products or vote for certain candidates. Without qualified information and communication, malign "Fake News" may find an open field to propagate and contaminate public affairs.

In this chapter, after discussing the problems we are facing today, we elaborate on some principles supplementary to Medical Bioethics and Sustainability Ethics, to guide public governance, composing the BG. The proposal is to establish the principles and a system of connections between the three powers of the Democratic Republic and the people, which would enable a dynamic monitoring of the actions of the representatives, as these actions are in accordance or not with these established principles.

2 Proposal of a Method for the Creation of a Democratically Agreed Set of Principles for BG

How to democratically establish agreed principles for public governance, to be used by peoples of different countries to fight the planetary crisis? We propose a strategy that begins with the recognition of fundamental ethical values, as those expressed in the well-known four principles of Medical Bioethics (Beneficience, Non-Maleficiency, Justice and Autonomy; see Beauchamp & Childress, 2013), and the principles of the Ethics of Sustainability (Precaution, Sustainable Economics, Distribution, Viability; see Kibert et al., 2011). The latter can be summarized in three concepts (Smookler, 2010); the public government's actions to fight the planetary crisis should be based on what is:

1. Bearable – which means "possibly sustainable", that is, that can be continued over time
2. Equitable – which implies a "reduction of inequality" (as proposed by Piketty, 2020)
3. Viable – which means that there are means (environmental and human resources, science and technology) to achieve the results

These values are the basis for the process of elaboration of the BG, but they cannot be the proper principles intended for BG because those should be more concrete, empirical, and practical propositions, to be followed by the state managers (and politicians who intend to become state managers), and aptly monitored by the people, using information technology and personal devices (today, mostly the smartphone).

Between the abstract principles that express fundamental values and the concrete principles of BG, there is a planetary crisis. The crisis puts the demands on the governments. Therefore, the principles of BG should be adequate responses of the democratic society to the crisis, on the basis of fundamental values.

Who will make the connections between fundamental values and aspects of the crisis, and establish the principles of BG? We should begin with a plan for the elaboration and implementation of BG, making a proposal about the process of discussion and elaboration of the BG. It is to be continued by means of interdisciplinary meetings with scholars and representatives of all sectors of contemporary society, reproducing in a larger scale the type of scientific and political movement that originated Medical Bioethics, in response to tragic events such as the genocides that occurred in Nazifascist and Tuksgee experiments.

The UNESCO *Declaration of Bioethics and Human Rights* (DBHR) calls for "reflection on societal changes and even on global balances brought about by scientific and technological developments" (see https://en.unesco.org/themes/ethics-science-and-technology/bioethics-and-human-rights). *The scientific community* has a leading role in the process of creation and implementation of the BG, departing from the existing agreed bioethical documents. After the scientific community establishes a consensus about the fundamental principles of BG, the effective implementation of the principles in each country will depend on initiatives of the representatives in the three powers, under pressure of organized sectors of the populations. For all these actions to happen, the approximation and productive interaction of the academic system with politicians and local communities is suggested.

An event of this nature was a milestone in the recognition of the responsibility of nations for the survival of life on the planet, the United Nations Conference on Environment and Development, also known as *Eco-92*, *Earth Summit* and *Rio 92*, held in June 1992 in the city of Rio de Janeiro, Brazil. Unlike the meeting that took place 20 years earlier in Stockholm, Sweden, it was attended by a massive number of heads of state, and it is also worth mentioning the participation of non-governmental organizations, which held a parallel meeting called the 'Global Forum', giving rise to the so-called 'Earth Charter' from this parallel event. The idea of sustainable development resulted from Rio 92, in addition to the recognition of the Precautionary Principle:

> In order to protect the environment, the precautionary approach shall be widely applied by states according to their capabilities. Where there are threats of serious or irreversible damage, lack of full scientific certainty shall not be used as a reason for postponing cost-effective measures to prevent environmental degradation. (https://www.scielo.br/j/ea/a/szzGBPjxPqnTsHsnMSxFWPL/?format=pdf&lang=pt)

It was also at Rio 92 that the UNFCCC (United Nations Framework Convention on Climate Change) was born, an international treaty that implies the periodic meeting of the member countries of the Convention, and the first Conference of the Parties (COP) took place in Berlin in Germany in 1995, followed by in Kyoto in Japan in 1997, resulting in the famous Kyoto Protocol. The COP-30 conference will probably take place in the city of Belém, Brazil, in 2025, three decades after the first Earth Summit was also held in Brazil and there is hope that it will become as important a milestone as Rio 92 was.

3 The Facets of the Crisis: Environmental, Economic, Social, Epistemological, and Sanitary

The current crisis is the result of several factors:

1. Environmental: In the *Antropoceno* era, human interference on the climate – measured by the increasing CO_2 levels in the atmosphere, a causal factor for global warming – became critical. Invasion of native biomes increased the risk of epidemics caused by viruses, bacteria, and fungi from other animal species and their mutations, when hosted in the human environment; in the long run, there is also the risk of a drastic reduction of biodiversity, exponentially increasing the instability of ecosystems. These threats are combined with cyclical economical crises of capitalism – and political limitations of both the democratic representative and totalitarian socialist systems, challenging the resilience of human societies to overcome the obstacles to human survival and possible development.

2. Economic: During the last 50 years, in most democratic countries, public managers lent their faith to economical neoliberalism, the concept of a self-organizing competing market that could distill public virtues out of private egotism, thus implementing a maximally reduced federal state that would regulate the competition without competing with private businesses. The results were not as expected, because the lack of public investment led the economy to a chronic recessive period, and the abandoning of the promotion of public welfare increased inequality, in which the rich accumulates and the poor does not have resources to cope with survival and ordinary expenses. In a way similar to the ecosystem's interactive networks, the decrease in the number of preys, and their route to extinction, is not good for the predators. Economic development requires a continuous and dynamic balance of production and consumption, which obviously – but, unfortunately, not evident to public managers – implies the existence of proportional levels of production and consumption. If consumption is high and production levels are low, there will be a tendency to inflationary processes, while if production is high and consumption is low, the capitalists will cut their investments and dismiss part of the workers. John Maynard Keynes elaborated on partially successful solutions for the crisis of capitalism (Keynes, 1936) that led several developed countries to the welfare state in a process that began in the post-war period and extended to the 1970s, when the increase in the public debt – considered as necessary to finance the public assistance systems – reached an alarming level, for the macroeconomic standards of that time – the proportion of the debt relative to the country's Gross Domestic Product. The neoliberalism of Margaret Thatcher and Ronald Reagan was the reaction to increasing public debt, implementing a series of actions to reduce it. Until recently, when the defenders of Modern Monetary Theory arose to the center of the political stage – thanks to the left wing of the USA Democratic Party – neoliberalism was thought as politically correct, even for politicians inclined to promote social justice.
3. Social: As a result of the neoliberal era, the most serious challenge for the economical system became the increasing social inequality that progressively pushes a large part of the population to poverty, and, consequently, to the outside of the market, an event that exerts negative feedback on the productive system. Thomas Piketty's analysis of social inequality (Piketty, 2020) became a classical work that inspires the United Nations and research funding organizations to seek real solutions to the crisis. In this situation, many corporations preferred to reduce investment in the productive system and bet on financial applications. In the political arena, opportunists wave their hands to magical solutions, and/or induce the people to destructive behaviors against themselves, such as the war, targeting on the more vulnerable sectors of the population.
4. Epistemological: Scientific knowledge plays a fundamental role in education and supporting economic development. In any country that is developed, the university works as an enzyme of social actions and reactions, leading to progress. The university is also where we think about the different conceptions of 'progress'. The scientific community, considered as composed of intellectual workers, generates the knowledge that makes possible the creation of new productive enterprises and transformation processes. This statement may give a certain impression that scientists are the owners of the truth, but scientists have divergences and no guarantee of having achieved an absolute truth. There are good scientists skeptical about the role of science in human development and/or deniers of the consequences of human actions on the planetary crisis, advancing systematic doubts about global warming and medical issues. The epistemological crisis arises when skepticism in science, combined with belief in intentionally fraudulent non-scientific doctrines, reaches large parts of the populations. Things become complicated when each person and group say

they know and tell the truth in social networks, where there is no referee system to evaluate the merit of what is said. How to decide? What we have to defend, in fact, is the scientific method, that is, how each result was obtained, because that's what supports it. For instance, when there was a debate on therapeutic effects of Chloroquine and Ivermectin at the USA Congress, a first-rate scientist appeared, defending the efficacy of these drugs for the treatment of COVID-19. However, these drugs were poorly tested for COVID, because a person can recover from it without taking any medicine. In those cases that were presented as evidence for the effectiveness of these drugs, it was not really a causal effect of them. It's the same as if you have hepatitis A and you take a drug, then in a month you'll be cured. But it wasn't the drug. Hepatitis A cures in a month, even without taking medicine. For this reason, we should recognize the value of Article 18.1 of the UNESCO DBHR to fight the epistemological crisis, stating: "professionalism, honesty, integrity, and transparency in decision-making should be promoted, in particular declarations of all conflicts of interest and appropriate sharing of knowledge. Every endeavour should be made to use the best available scientific knowledge and methodology in addressing and periodically reviewing bioethical issues".

5. Sanitary: The expansion of vectors of infectious and contagious diseases, formerly known as "tropical diseases", as an effect of global warming, has impacted on regions previously free of such scourges. Contact with previously unknown microorganisms and mutations induced by chemical agents manipulated by human agribusinesses and industries increase the risk of new pandemics, as in the COVID-19 case. Previewing this situation and the importance of vaccination, Article 21.3 of the UNESCO DBHR says: "Transnational health research should be responsive to the needs of host countries, and the importance of research contributing to the alleviation of urgent global health problems should be recognized".

4 Outline of Putative Principles of the BG: Departing from the UNESCO Declaration

Departing from the agreed UNESCO documents, we can make a preliminary list of the practical principles that may compose BG, such as:

1. Human Dignity and Quality of Life: The DBHR introduction states that "ethical issues raised by the rapid advances in science and their technological applications should be examined with due respect to the dignity of the human person…such developments should always seek to promote the welfare of individuals, families, groups or communities and humankind as a whole in the recognition of the dignity of the human person and universal respect for, and observance of, human rights and fundamental freedoms". At the end of the twentieth century, Brazilian bioethicist Garrafa (1999) observed that issues like these were not included in the traditional agenda of bioethics. On October 19, 2005, UNESCO guided the theme, not without great debates and clashes. The Bioethics for the twenty-first-century document was approved by acclamation by the Organization's 191 Member States at its 33rd General Conference. In practice, considering e.g. the movements for Human Rights in the Third World (including 30 years of *Action for Citizenship* in Brazil; www.acaodacidadania.org.br), the main index of dignity and quality of life has been the *absence of hunger*. Governments have the duty to ensure that no one goes hungry in the country in which they are responsible for providing conditions for feeding and nutritional security, especially the financing of the production of healthy human food in advance of the population's needs for consumption.

2. Peace and Personal Safety (Non-Violence): The issue of personal safety is understood as the right of people not being involuntarily the target of violent actions that put their life and health at risk. State managers should carry public actions to avoid events like lost bullets in police interventions, personal attacks of various types, in which the person is constrained by violence. Special attention should be directed to efforts to prevent war, of all types: conventional, chemical, biological, and nuclear. Unfortunately, some international organizations dedicated to promote peace were built during the 'cold war' period and for this reason, may have not been really concerned in avoiding *any* type of war – they seem to behave as if some types are legitimate. However, considering that all types of war are 'negative sum games' (see next section, about this concept) in which everyone that is involved is at risk of losing life, physical and mental health, if the BG is concerned with the generation of 'positive sum' games, then it should be against all types of war.
3. Availability of Employment and Income: The issue treated by Keynes (1936) is becoming clearer nowadays: the federal state, the only institution that can emit money (by means of public banks or similar state organizations) or guarantee the money generated by operations of private banks, has the obligation to put the money into circulation in a selective way, that promotes the creation of jobs and income for people. For this, companies, whether state or private, that provide jobs, are activities of interest within the country's strategic plan. The federal state has the responsibility of creating mechanisms for people to have a source of income, or directly providing a basic income for every citizen.
4. Non-Discrimination: Governments must be against any racial, gender, appearance, or status discrimination of persons or groups. As stated in article 11 of the DBHR, "No individual or group should be discriminated against or stigmatized on any grounds, in violation of human dignity, human rights and fundamental freedoms". Candidates to public elections who defend discriminatory flags should be rejected by the people, and in more radical cases be blocked by instances of BG before the elections happen;
5. Non-Pollution: Today there are several companies that get their profit at the cost of polluting the environment and causing intoxication of the people who live around their installations, e.g. in Brazil, the cases of legal mining in the state of Minas Gerais, and illegal mining in indigenous territories in the Amazon.
6. Access to Water: Water is increasingly becoming a critical factor for human survival, human health. People in many regions of the planet have difficulty accessing drinking water. Article 14 (b) of the DBHR refers to the access to adequate nutrition and water;
7. Right to Education: This is a universal claim that is recognized in Article 14 (e) of the DBHR, referring to the reduction of poverty and illiteracy.
8. Right to Health: Some countries have a Public Health System, e.g. Brazil and UK. Article 14 (a) of the DBHR refers to "access to quality health care and essential medicines, especially for the health of women and children, because health is essential to life itself and must be considered to be a social and human good". In the current post-COVID situation, there are many persons going hungry, a shortage of certain products, and also a lot of low-quality food being consumed, which, because it is cheap, people with a low income end up consuming. This creates a public health nutritional problem that should be treated by the governments.
9. Availability of Energy for Domestic Purposes. The state should have a plan to generate and distribute the energy necessary for the productive system and utility needs of all the population, attacking the existing bottlenecks existent at each geographic region. Nowadays, we have wind and solar energy and the so-called 'green hydrogen', as renewable alternatives to conventional fossil sources.

10. Mobility for People and Commerce. There are logistic problems affecting the circulation of people and commercial goods that can be solved only with state planning and intervention. For example, the production flow in Brazil is complicated, since the railways were discouraged. We have urban traffic problems, raising the need to encourage the use of bicycles and electric vehicles.
11. Preservation of Biodiversity of Biomes and Stability of Ecosystems. In Brazil, there are serious issues, like the devastation of the Amazon and Atlantic forests, and desertification of the 'Cerrado' (the most common biome in central regions of the country). Article 17 of the DBHR states: "Due regard is to be given to the interconnection between human beings and other forms of life, to the importance of appropriate access and utilization of biological and genetic resources, to respect for traditional knowledge and to the role of human beings in the protection of the environment, the biosphere and biodiversity".
12. Respect and Support for Practices (Social Technologies) and Knowledge of Communities. Sustainable projects of development must respect and support indigenous knowledge and practices in relation to the forests. Article 14 (d) of the DBHR claims against the "elimination of the marginalization and the exclusion of persons on the basis of any grounds".

5 The Roles of the State and the Digital Public Arena in the Implementation of BG

Since the advent of modern science and the industrial revolution, there is a central role of science and the scientific method in the formation of the people's mentality, and also in the search for solutions to each country's problems. This means that the scientific community has a degree of responsibility in the formation of citizens and supporting satisfactory public governance. However, most of the population, mainly in undeveloped countries, does not have university degrees and/or respective educational formation to understand the complexity of the current planetary crisis. Very few have the wisdom of the Swedish teenager Greta Thumberg, being able to grasp the consequences of the current crisis at an early age. So, we have to think of strategies to ensure that science and the university have a greater influence on this whole process, having political economic issues at the center.

The coordination of economic development is made by the federal state (Mazzucato, 2013). The role of the state is not directly participating in economic activities (e.g. substituting private by public enterprises), but indirectly, by means of directive actions mediated by the sovereign money that only the state can issue or legitimate (in the case of being issued by private banks). Issuing sovereign money is a prerogative and duty of the state, central to national security systems. The state has to provide money for the economy to function, fueling the market; otherwise, there will be no sufficient production and consumption.

Contemporary mass societies have large urban populations depending on having money to buy food and other goods necessary for survival. The production-consumption cycle is hampered by the lack of money, as it happens, not only in Brazil, but also in other countries, which have adopted political economic paradigms of *monetary scarcity*. 'Thatcherian Neoliberalism' is a name given to this model of public administration, and with that people get progressively into debt; there is a chronic recession of the productive economy because the people do not have money to buy, to sell, to create companies, and a part of the population does not even have the money to satisfy basic survival needs. The economy tends to a deeper recession and increasing unemployment, and COVID-19 pandemic has made this much worse, leading to measures such as, in the United States, the issuance of trillions of dollars to reactivate the economy. However, the use of this "quantitative easing" method, without increasing the production and offer of goods for popular consumption, is likely to increase inflation.

For the purpose of changing the political economic model, it is very important to promote games with 'positive sum'. The concepts of 'negative' and 'zero sum' games were created by physicist John von Neumann, exponent of several areas of Physics, together with Oskar Morgenstern (von Neumann & Morgenstern, 1944), which were used in political economy. We have the famous Nash equilibrium, which demonstrated the possibility of positive-sum game in the economy, when the players keep their strategies stable for some time and then collaborate with each other. This was a major paradigm shift, because in the more traditional view, both of Neoliberalism and Marxism, it was believed that games would be, at best, zero-sum, that is, when one player wins, the other(s) loses. In the positive-sum game everyone wins, while in the negative sum game everyone loses – as in war, when people on both sides die.

The idea of participatory democracy can be grounded on positive-sum games and here we refer to the work of Fehr and Gachter (2002), in which they seek to demonstrate, with human games, that for the positive-sum game to really work, there has to be what they called *altruistic punishment*. That is, people who are collaborating, cooperating, need to implement a system of monitoring, surveillance, and punishment of those – the *freeloaders* – who take advantage of cooperation to exploit others, or to obtain undue advantages at the expense of the efforts of the people who are collaborating. If not identified and punished, the freeloaders destroy the positive-sum game, which becomes a zero-sum or negative-sum game.

So, for the positive-sum game to really be successful, there has to be the surveillance and punishment of the profiteers, which is something very different from the surveillance and punishment that Foucault (1995) dealt with when talking about psychiatric institutions. Philosophically speaking, it would be a surveillance and punishment system diametrically opposed to what happens in psychiatric hospitals and other 'totalitarian institutions'. In these institutions – prisons, asylums – the individual is subjected to unilateral rules and repressed in his freedom. They implement a system of surveillance and punishment that represses freedom. Surveillance and punishment, for us to have a 'positive sum' game in Politics, would be the opposite. It would be the exercise of freedom by individuals, autonomously, to monitor and punish those who take advantage of cooperation to have their own undue gains. This proposal entails a democratic refusal of laissez-faire (Classical Liberalism) and also of Neoliberalism, in which the state fails to complete its function of inducing development, to favor a pseudo-free competition that supposedly would raise an 'invisible hand' that promotes collective benefit from private egotism and vices. However, this is not what happens, because this competition system leads a few ones to become increasingly rich, the majority of people becoming poorer, and others increasingly excluded from the market system. The goal of BG would be to reverse this process, leading to the inclusion of people in the economic political system.

This is something that more radically liberal (or libertarian) authors would cringe a little bit with. It is the idea that for the 'positive sum' game to be effective, it is necessary to identify people who take advantage of cooperation to gain undue advantage. In Neoliberalism, the competition for profit happens as if people had total freedom (inside the Law) to do whatever they want, but a society that succeeds in cooperation must have mechanisms to avoid the action of freeloaders. If they can act freely, they make cooperation unfeasible. What would be a collective gain becomes a private gain. Unfortunately, the digital public sphere has become a battleground in which the absence of altruistic punishment makes the game a 'null' or 'negative sum' one.

In the current situation of neoliberalism, the strongest competitors who take advantage of social cooperation have almost total freedom, even above the law, to do actions that harm the majority. They have gains and are even praised by a part of the population as being smart people, who managed to get rich at the expense of others, but they are not watched to see how they achieved these gains and they are not even punished for having had illicit activities, or activities that, if they were in a position of governance, would be activities that would go against the BG. When, for example, freeloaders use privileged information to carry out real estate speculation and thereby achieve economic and financial gains, as happens in many cities in Brazil, information about areas that can be used for land subdivisions is improperly made; there is leakage of privileged information from public managers, addressed to the private area, involving their own relatives, or the business groups connected the person who is in the governance position. This is a very common example of conflict of interest in the interior of Brazil, where privileged information is used to enrich these people to the detriment of the interests of the majority.

Assuming that the role of the state is to finance production and consumption, the goal of human development programs is to generate sustainable cycles of production and consumption within the people, affording complementary social programs of education, health, arts, leisure, sports, etc.... People are the producers and consumers. The state and the governors cannot make things in the place of the people, and should not give them a 'free lunch' (except, of course, if there is a condition of hunger); their role is to help the people to self-organize, according to a strategic program and using the state's capacity of issuing money and using it to implement desired social functions. This presupposes the participation of the people, not only in the economy but also in politics, culture, and all sectors of social life.

BG looks for principles to anchor the expansion of popular qualified participation in all areas of social activity, using information technology and the digital media of the Internet. However, current uses of social media are geared towards 'negative sum' games, such as attacks on personal reputations, slander, fake news, and manipulation of public opinion. Is there the need to create another system to generate a positive-sum game? Then we get into a problem, so that to have the other system that would allow qualified popular participation, with a positive-sum game, we would already have to have the conditions that this system itself would generate. There is a problem there of how to get out of the current situation to build this system that will enable popular participation and then we will be able to implement a process of human development.

The solution to this bootstrapping problem is the rise of the scientific community to a protagonist position, together with allies in the government, politicians, intellectuals, artists, international organizations, and other thinkers interested in rebuilding the social system to overcome the increase of social inequality and the other challenges of our time. In order to transform the current digital public sphere into a 'positive sum' game, there must be initially an apparatus that promotes the propagation of truthful information, operating like Archimedes' lever support point. This may have begun with UNESCO's GEObs: "GEObs is a system of databases with worldwide coverage in bioethics and other areas of applied ethics in science and technology such as environmental ethics, science ethics, and technology ethics. This UNESCO initiative is freely accessible online to all Member States and the general public. It is designed to serve as a valuable reference, collaborative, consultative, and comparative resource hub of ethics activities around the world". https://en.unesco.org/themes/ethics-science-and-technology/geobs

For the implementation of BG, allowing better participation of people in the democratic system, the next step – following the formation of reliable observatories providing ethically correct information – is to establish desirable principles for the conduct of governors in relation to human life, and to life in the planet in general. This needs to be done through congresses, based on the negative experiences we are having. These principles would be references for the evaluation of the governments, as discussed in the following chapters of the book.

6 Controversial Bioethical Issues Regarding Ecological Economy

Ecological Economy issues should be at the center of the BG proposal in Brazil, both for conservation and production projects. What would be a 'synergy with nature'? The question is meaningful for those who value and intend to use the resources of the ecosystem for economic activities in a sustainable paradigm. For instance, in Brazil, in the *Cerrado* biome, synergy with nature would mean to interact constructively with the resources of this biome. The bioethical issue is: Should we cultivate these native species, or destroy them, replacing with cultures of exotic species – currently more valued in the market (e.g. soy, corn, rice, cotton, etc.)?

In Brazil, many exotic species were introduced by the colonizers because they did not know the native plants and how to use them, and because they came to the new continent with previously formed feeding habits. They were not interested in asking: What resources do the native people themselves already use? Should we cultivate them for our consumption and also on a commercial scale?

In a program of Ecological Economy, we pay attention to the resources that the ecosystem provides: plants, mainly fruits, but also seeds, roots, all kinds of resources that the ecosystem offers, to think about how to use them economically in a way that doesn't destroy the biodiversity of the ecosystem. Using science – Forestry Engineering, Agronomy, Biology; using the knowledge that we currently have in the universities combined with the knowledge of the people who live in these places, we can find methods to use the resources, valuing them, making products of commercial interest, even for export. They may be preferentially directed to the internal market, producing popular consumer goods. This has to come together with a program of Basic Income, as discussed in other chapters of this book, so that the people will have the money to purchase the products. The government cannot encourage production without demand for consumption by the people; nor encourage consumption without having production.

That's where the universities come in, training qualified people to elaborate and evaluate projects, checking if they are in line with our advances in the environmental, nutritional, agronomic, and related scientific areas. Production of food is strategic for Brazil, but still undervalued. There are indications that Brazil may be the largest producer of human food for the entire planet, while other countries suffer from low production of food. Brazil is already big in production of food for animal feed; agribusiness supplies a lot of this part of feed and meat production; however, the main interest of agribusiness is not to produce quality human food. Agribusiness producers are not so interested in supplying the domestic market, because it is much more profitable to export and receive in dollars than to sell here and receive in Brazilian currency, which is sometimes very devalued.

For these reasons, projects of Ecological Economy would be parallel to agribusiness. They are not *against* agribusiness, but it's not in favor either. It is another model that we could implement given the vastness of our territory and our great wealth of ecosystems. With state support, we can implement a new model parallel to agribusiness, avoiding the even greater expansion of agribusiness into the areas of native forests, leaving agribusiness in the area it already occupies. They already know very well about more *intensive* agriculture, which does not need to expand the area they occupy; they can increase their productivity using the same areas. In degraded areas, we can do restoration projects, planting trees. This needs to be a national program with support from the Federal state. In areas around large cities, we can also develop family farming projects, sustainable agriculture, agro-ecological models, aimed at producing human healthy food. It is also essential to preserve native vegetation and also reforest areas of mountains, hills, and slopes, for the production of sources of drinking water. This is where the issue of public health also comes in, because the quality of the food and water greatly interferes with public health.

The above type of proposal of an Ecological Economy is not just because of the issues of global warming and environmental degradation resulting from the destruction of biodiversity; movements against these phenomena are very fair, trying to reduce carbon emissions and many other initiatives. The most sensitive minds in the planet are worrying about the environmental issues; because of global warming alone, many cities will be flooded, many regions will become desert. Such questions refer very directly to the reflections of Hans Jonas in the book 'The Principle of Responsibility' (Jonas, 2006), proposing a new categorical imperative: "Act in such a way that the effects of your action are compatible with the permanence of a truly human life on Earth."

A *New Green Deal* in Brazil is not just a response to those ethical issues, but needs strong state guidance and financing attractiveness. If it were just for morals, we would hardly be able to succeed, because it's like asking people to give up the most immediate, easier profits, to projects that have a return in the long run – to save the life of the human species in this planet. There may be half a dozen people who are going to be altruistic, who want to save humanity despite their own private interests, but there's going to be a large majority who will continue behaving in exactly the same way, looking for their immediate private interests.

For this reason, the main claim in favor of a Brazilian 'New Green Deal' is strategic. What is Brazil's place in the world economy? Is it only to supply raw materials? To sell iron and then buy steel at a much higher price, missing the opportunity to add value? If Brazil continues in this position at full speed, allowing mining to invade forested areas and indigenous territories, the problems here discussed will increase. This type of positioning of the Brazilian economy within the international production system does not give us conditions to have a better quality of life for the people. Brazil needs to find another economic model, in which we use our resources in a sustainable way, adding value to these resources, and this model does not need very sophisticated, cutting-edge technology. We can use *social technologies*, which are those knowledge and practices that the people themselves have already developed.

For example, the production of *manioc* (cassava) in the Northeast, using the social technology of rudimentary flour mills moved by human energy, does not need to be drastically changed to provide employment and income for the people of the region, while at the same time generating a product that is commercially attractive. The people already have the technology to extract the cassava, which is difficult, because it is a root buried in the ground that you have to dry, grind, make the *tapioca*, make the cassava four, which is used as human food in the whole country. With federal support, in a broad program to support Ecological Economy, we can improve the methods without brainwashing the people, expanding the scale of commercialization, making this type of enterprise go beyond the region where it already happens, to have a national scale and maybe even an international one.

A type of company that we still don't have in Brazil is the organic supermarket. Currently, if you want to have healthy, organic food, you really pay a little more, because you will be dealing with producers and traders who work on a very small scale. So, they can't keep the business offering a low price. Now, if large chains of organic supermarkets are formed, as, in the United States, the *Whole Foods*, and in Europe the *Coop*, then there would be a much greater demand for human food, organic, healthy products, produced in a fair way, without the exploitation of work, no slave or child labor. In Europe and United States, there are very few products from South America in food markets. Organic supermarkets usually meet a series of criteria, such as paying workers a fair wage, not using certain types of pesticides, chemical fertilizers, and hormones. Among the products we find, there is the Colombian coffee and the Brazilian nut. Like these products, there are many others that we could also be developing, using social technologies and exporting.

There is a Brazilian fruit, the *Pequi*, from which we can extract an oil to be used in culinary, that is an antioxidant, a recommended functional food according with research made in public universities. Some people, some groups, may manage to do this, as happened with *Açaí*, but this case is not the rule, it is the exception, because the Açaí berry was accidentally known by some Americans who passed through here; they decided to invest in it, and it worked very well. Today it is found all over the world.

For this to become a new model – taking Açaí and Brazil nuts as examples and transforming this into the new model – it needs a heavy funding program from the Federal state, which is the player in the game that has the money to make it real. This support has to be combined with popular income programs, for the people to have money to buy the products. So it's a matter of governance. The governments have to manage this situation, for this to happen and work, it won't happen by accident. A public initiative that uses scientific knowledge, that generates a situation of well-being for the people, an increase in production, the people consuming the products they produced, having money to buy them…. These goals need the public administration, they need competent people in the public administration to act, through the tools they have, to put the money in the right way, in the right ventures, in the right people for this to happen. There is no use for the state to direct the money to non-strategic spending, favoring private interests. That doesn't make the economy develop. A correct government policy – or rather, a state policy, because policies permeate various governments – is not linked to candidates, but to the national strategic goals. These policies have to be well targeted to encourage and support the activities of the majority of the people, within a development program that uses scientific knowledge. Hence the university plays a fundamental role: to bring the necessary knowledge for things to happen in an adequate pathway.

For instance, the research carried out by *Embrapa* (*Brazilian Agricultural Research Corporation*, a public corporation belonging to the federal government) has made it possible to successfully cultivate several exotic plant species in the tropical climate. Researchers at *Embrapa* adapted, through artificial selection, biotechnology, and genetic engineering, several plants for large-scale production. The producers couldn't make these crops yield by themselves. In addition to Embrapa, we have a number of universities that also carry out research that brings important results for a development project within the ecological economy model.

Ecology is no longer just about preserving nature, but about making a series of managements so that the ecosystem remains stable while used by people to produce goods, or finding stability for a degraded ecosystem. The idea of static equilibrium is a poor one, because if the ecosystem depends on it we can't do anything, we can't produce, we can't use nature's resources, because it will take nature out of equilibrium. This is an old view; nowadays, we know that this is not the case. The important factor is if there is biodiversity, because biodiversity regulates itself, producing stability in ecosystems. With biodiversity, problems such as climate change, water scarcity, problems with rivers, floods in certain places, and drought in others may be overcome. The deleterious effects of the destruction of forests, pollution, use of fossil fuels, fires, mining without proper care, etc. can be treated with investment on biodiversity.

We're not going to preserve the ecosystem so that it stays at a certain ideal point of balance that we can't do anything about. We are going to use the ecosystem, we are going to use the cycle of nature at the same time that we allow nature to recover. We create forms of regeneration, replacement of these resources, so that there is stability in the system while maintaining its biodiversity. So, this is what current science tells us, this is a very important tool for us to act. And within that, we have the possibility to choose which of these ecosystem resources we will use in the production process. Products of interest should be of popular interest, because it is necessary to produce what the people will consume, to avoid inflation of their prices. So, we have to think about producing goods that can provide a better quality of life for the vast majority of the population. The elite, the rich are already doing very well, they don't need the state to invest in luxury products like bags that cost thousands of dollars. The state must basically invest in the production of healthy food for the people, availability of water, clean energy, sanitation, means of transport, what the people need. Then we will have a healthy society and, if we achieve an ecological economy, a healthy society within a healthy ecosystem. Because human society is a part of the ecosystem, so we cannot separate one from the other. This is the basics of the BG.

7 Concluding Remarks

In this chapter, we describe a methodological proposal for the creation of BG and putative principles to be discussed internationally, in an adequate epistemological framework. The main message is that governments should care for human and animal life on Earth, promoting 'positive sum games' instead of destructive agendas and/or assuming religious directions to save human souls in a transcendental domain, while their bodies and other forms of life are destroyed in the only habitable planet we know of. State managers, while occupying a position of responsibility for social security, must take care and *promote life* in all of its dimensions, and fight *against human suffering, disease, and death.*

The implementation of BG can improve the democratic system, creating a philosophical reference for the implementation of digital systems of mediation between state managers (at the federal, state, and municipal levels) and the people. Human relationships are currently mediated by Internet communication and digital mechanisms. We need to create a new system that can work in a constructive way, with the possibility of popular control of the three powers, and also a change in monetary policy, which would be 'qualitative' easing (Brown, 2019), that is, selective public spending, investing in what matters for the majority of the population, generating employment and providing better living conditions, in accord with the principles of BG.

References

Beauchamp, T., & Childress, J. (2013). *Principles of biomedical ethics* (7th ed.). Oxford University Press.

Brown, H. (2019). *Banking on the people: Democratizing money in the digital age*. The Democracy Collaborative.

Fehr, E., & Gachter, S. (2002). Altruistic punishment in humans. *Nature, 415*(6868), 137–140.

Foucault, M. (1995). *Discipline & punish: The birth of the prison*. Vintage Books.

Garrafa, V. (1999). Bioética, Saùde e Ciudadanía. *Sao Paulo: O Mundo Da Saude, 23*(5).

Jonas, H. (2006). *O Princípio Responsabilidade: ensaio de uma ética para uma civilização tecnológica*. PUC RIO.

Keynes, J. M. (1936). *The general theory of employment, interest, and money*. Palgrave Macmillan.

Kibert, C., Monroe, M., Plate, R., Peterson, A., & Thiele, L. (2011). *Working toward sustainability: Ethical decision-making in a technological world*. Wiley. ISBN: 978-0-470-53972-9.

Mazzucato, M. (2013). *The entrepreneurial state: Debunking public vs. private sector myths*. Anthem Press.

Piketty, T. (2020). *Capital and ideology*. Harvard University Press.

Potter, V. R. (1971). *Bioethics: Bridge to the future*. Prentice Hall.

Smookler, H. (2010). *Sustainability ethics*. Seminar presented at the UCLA Extension Public Policy Program. Available at https://cupdf.com/document/sustainability-ethics.html

von Neumann, J., & Morgenstern, O. (1944). *Theory of games and economic behavior*. Princeton University Press.

Alfredo Pereira Jr is a Philosopher of Science, Professor at the Universidade Estadual Paulista Julio de Mesquita Filho (UNESP), Brazil, retired, with degrees in Philosophy and Administration. Currently working in the UNESP Graduate Program in Philosophy. Alfredo has around 300 publications in several areas of knowledge, mostly in the Philosophy of Neuroscience and Theory of Consciousness. Orcid: https://orcid.org/0000-0002-5960-041XAD, Scientific Index: 56th place in Top 100 Latin America Philosophers/Scientists.

Francisco Sousa is an Economist for social enterprises: structures, operations management, and market development. Mr. Sousa's work experience includes risk management, trading markets, and consulting services with client assignments in the United States and overseas. Mr. Sousa has worked with IBM, Canada's Manufacturers Life Insurance, CACI International, George Washington and Johns Hopkins Universities. In recent years, Mr. Sousa served on the 'Community Housing Finance Commission' in Arlington, VA appointed by the County Board.

Enidio Ilario, Associate member of the Brazilian Society of Bioethics, Expert Member of UNESCO's Global Ethics Observatory (GEObs), Co-coordinator of the Interdisciplinary Seminars on Economic and Social Inequality, and member of the Self-Organization Group, both at the Center for Logic, Epistemology and History da Ciência (CLE) at Universidade Estadual de Campinas (UNICAMP). Doctor in Psychology, Master in Philosophy from Pontificial Catholic University of Campians (PUCCAMP), Specialist in Collective Health from FCM-UNICAMP, and in Internal Medicine from Sociedade Brasileira de Ciências Médicas (SBCM).

2 From Ideology to Practice: Political Currents and the Democratic Convergence

Marcus Vinicius Pestana and Alfredo Pereira Jr

Abstract

How do ideologies contribute to the guidance of political practice today? To answer this question, first, it is necessary to identify the main ideological fields that shape current forms of governments and the consequences of their adoption. In this chapter, we briefly review the ideas and theories that moved the contemporary political arena (Conservatism, Liberalism, Communism, Fascism, Social Democracy, Eurocommunism and Authoritarian Populism) and argue for the need of a democratic convergence. With this approach, we do not intend to defend a particular ideology, or to propose a new one, but to evaluate which components of the existing currents are more adequate to the practice of democracy. The term "democracy", in this context, refers to a type of political-economic system (formed by the people, conducted by the people and serving the people) that has been assumed, at least partially, in the discourse of contemporary political currents – even when, in practice, they act against the democratic principles of governance.

M. V. Pestana (✉)
Federal University of Juiz de Fora,
Juiz de Fora, Brazil

A. Pereira Jr
State University of São Paulo, Botucatu and Marília,
SP, Brazil
e-mail: alfredo.pereira@unesp.br

1 Introduction

The disconnection between theoretical production and political practice, and the era of "alternative truths" and torrential production of statements in social networks have left a certain intellectual indigence and unprecedented sectarianism as a legacy, interdicting the debate.

The very role of ideas in transforming life and the world is a controversial topic. The revolutionary perspective has always believed that a new set of ideas and values can dynamite the old order and build a new world. The conservative perspective is anchored in the conviction that traditions, essential values and customs should guide us in the necessary prudence for the production of reforms and changes.

The evolution of human civilization is determined by a series of objective material conditions that limit or enhance changes. The degree of freedom we have is not unlimited. Utopias and dreams collide with concrete limits imposed by reality. But, there is no doubt that ideas have an enormous transforming force. Objective and subjective elements intersect. The intensity of each one also arouses enormous controversies in the

A. Pereira Jr, F. Sousa (eds.), *Principles for Governance*, Sustainable Development Goals Series,
https://doi.org/10.1007/978-3-031-40025-4_2

evolution of modern thought. The political struggle is the result of the confrontation of divergent ideas that assume the characteristics of power projects in search of hegemony in society.

In one of his main songs, Cazuza,[1] a Brazilian rock star in the 1980s, cried out: "Ideology, I want one to live by". Human beings are not given to ruminating on a drowsy daily life and passively accepting injustices and distortions that jump out at the eyes. Critical capacity, the drive for innovation, existential restlessness, and the vocation for change, thoughtful or disruptive, are part of human nature.

In Brazil, there was an extreme radicalization of the political debate. But not only here, just look at the examples of the USA, France or Italy, among other countries. The central issue is that the current debate is as deep as a saucer. The goal is not to listen, understand, dialogue, debate, build consensus. The search is for the annihilation of the enemy and the mobilization of the respective ideological bubbles, with sparse theoretical and solid historical references. Just trigger the rhetorical machine gun: "fascists", "communists", "conservatives", "neoliberals", without the slightest knowledge about these ideological formulations, and often with a poverty of spirit and intellect.

Einstein ironically warned: "Two things are infinite: the universe and human stupidity, but as far as the universe is concerned, I'm still not entirely sure". Exaggeration aside, it is obvious that a productive debate has to be respectful, recognizing the legitimacy of opponents, with open hearts and minds, reasoned, and intelligent. In our country, sometimes things get worse because of what Brazilian humorist and writer Millôr Fernandes identified: "When an ideology gets very old, it comes to live in Brazil" (Fernandes, 2002). Fascism and communism are two ideologies already overcome by history, dated, but still remain as a source of inspiration for many.

[1]Brazilian singer and lyricist (1958–1990) known for his political and social compositions. The quoted verse belongs to the song "Ideologia" (Cazuza, "Ideologia" album. Rio de Janeiro: Philips Records. 1988, Available at: https://open.spotify.com/track/4s6qyTWg5r4EyOi2QIwM9g).

With the open spirit of John Maynard Keynes (1883–1946) – "when the facts change, I change my opinion" – we will review and discuss a little about current ideologies and their historical context that have borne fruits in the modern world: conservatism, liberalism, communism, fascism, social democracy, Eurocommunism and authoritarian populism.

2 Conservatism: Permanencies and Changes

Let's start with conservative thinking. Why? Currently, many citizens in several countries, for various reasons, define themselves as conservative.

The emergence of conservatism is the result, in the political field, of three great revolutions: the Glorious Revolution, in England, at the end of the seventeenth century; the American Revolution, in the USA, from 1776; and the French Revolution, which began in 1789. From the economic point of view, the English Industrial Revolution, in the second half of the eighteenth century, with the emergence of industrial capitalism, is an influence. In terms of religion, the Protestant Reformation and the Catholic Counter-Reformation. In the cultural field, the Enlightenment, the age of reason, and the displacement of theocentrism place the human being at the center of human history. In short, a long transition from feudal society, absolute monarchy, the agrarian and rural world, the condemnation of usury, the symbiosis between State and Church, to the birth of capitalism, modern democracy, the urban-industrial universe, the new productive technologies and new production relations, the secular State, the legitimation of individualism and profit as engines of development.

Initially, the transition resulted in the polarization between conservatism and liberalism, until both converged, keeping the differences in conception, in the fight against the nascent socialist movement. The conservative defends permanent things, the primeval contract of eternal society, the pact between God and generations past, present and future. He advocates preserving fundamental principles and promoting prudent reforms. It is not to be confused with reactionary and anti-liberal traditionalism. They see reactionaries as those who want to sacrifice the future in the name of restoring an idyllic vision of the past. But they condemn the revolutionary impetus of the progressives, who want to break radically with the past and the present, in the name of the utopian creation of another possible world.

Conservatives endorse democratic principles, constitutional government, checks and balances, and elections, but not the idea of a social contract that favors freedom over order and responsibilities, according to them, to qualify liberal individualism. Conservatives are better identified with the processes of transition to capitalism in England and the USA, where traces of permanence preserve some traditions and customs, and have enormous reservations in relation to the French Revolution, its Jacobinism and tendency towards the centralization of power. The process of transformation of the State and society in the United Kingdom has preserved the monarchy until today, in its constitutional parliamentary form. The emphasis on the various equations between the concepts of freedom, equality, tradition, order and responsibility, divided conservatives and liberals for two centuries.

In an expanded version of the thought of a contemporary English conservative thinker, Roger Scruton, we could say that modern conservatism begins with the defense of tradition against claims of popular sovereignty, and has become an appeal in the name of religion, family, customs, the "Common Law", of high culture, against the materialist doctrines of progress. They remain convinced that good things are more easily destroyed than created, and determined to preserve them in the face of politically engineered changes. Conservatives believe that there is an enduring moral order, adhere to custom, convention, and continuity, and value the principle of pre-establishment. Several governments and political parties in the modern world, including Brazil, were guided by conservative principles, combined with some liberal guidelines. But it is impossible to associate in a single category of analysis concrete experiences such as the trajectory of the UK Conservative Party, from Churchill to Thatcher, the German Christian Democratic Union, Angela Merkel at the head, religious fundamentalism, and its current authoritarian versions. Conservatism reappears, in a bastard and heretical version, in the authoritarian populism of the twenty-first century.

For those who want to be introduced to conservative thought, we recommend reading two books by the English thinker Roger Scruton, "Conservatism, an invitation to the great tradition" and "How to be a conservative" (Scruton, 2018, 2019), and by the American intellectual, Russel Kirk, "Edmund Burke, rediscovering a genius" (Kirk, 2009).

3 Liberalism: Market and Individualism

Liberalism emerges as the set of ideas and theories that animated the political and economic transformation that gave rise to the capitalist system and modern democracy, from the rupture with the Old Regime, the absolute monarchy of divine origin, the remaining feudal traits and the commercialism. The tradition of liberal thought, in its individualist or contractualist aspects, is present in the works of Locke, Hobbes, Voltaire, Montesquieu, Rousseau, Adam Smith, Tocqueville, John Stuart Mill, Von Mises, Hayek, Karl Popper and Milton Friedman, among others.

This trajectory was born intertwined with the four great revolutions of the seventeenth and eighteenth centuries: the Glorious Revolution, the American Revolution, the Industrial Revolution and the French Revolution. Conservative thinking was more of a reaction to what they saw as the excesses of liberalism in breaking with the past, customs, traditions and institutions. Conservatives, roughly speaking, identified more with the transformation processes in England and the USA, which ensured, according to them, traces of continuity, than with the disruptive traits of the French Revolution. In liberal thought, individualists and contractualists lead, through different paths, to liberalism, the defense of the market economy and the rule of law.

For contractarians, in the line of Locke and Rousseau, before the social contract all human beings were free and equal and, after the social pact, individuals give up a portion of individual freedom to build a society that guarantees them the minimum individual freedoms, the right to property and the protection of the Rule of Law, which is legitimized by the social contract and ensures the validity of the rules of the democratic game. The nascent capitalist society and the Rule of Law would be the result, therefore, of the social contract politically constructed by individuals and by the various social segments. The new institutions, according to the contractarians, were not derived from the simple extension of the individualist pursuit of profit and survival, but this view is not necessarily in conflict with individualism, since individuals may adhere to the contract using their "free will".

Individualism, a direct child of the Enlightenment, has a conceptual framework based on the combination of individualism, liberalism and the philosophy of history, more linked to Anglo-Saxon thought, with roots in David Hume (1711–1776), and his vision based on natural morality and the impulses of human beings moved by selfishness aimed at survival vis-à-vis altruism. In the work of Adam Smith, although many analysts see that the reading of "The Wealth of Nations", his classic work (Smith, 2018), has to be done in light of his theory of morality. The synthesis would be in the perception that the economy moves driven by private individual interests in search of profit and survival. Hence his most famous metaphor, the "invisible hand", which he himself defines about individual actions: "By seeking his own interest, he (the individual) often promotes that of society more efficiently than when he really intends to promote it". This would translate into the classic expression "laissez faire, laissez passer" of physiocrat origin, letting the market and society freely conduct reality, far from State intervention, but without reducing conviction to the formula popularized by Bernard de Mandeville (1670–1773), "private vices, public benefits" (subtitle of his book "Fable of the Bees", published in 1714), being, in this case, all passion identified as vice and all virtue as personal renunciation, a view contested by him, for whom self-love is seen as much broader than simple selfishness. The tensions between individualism and liberalism, and its own seminal connection with the Enlightenment, will feed the polemics involving several philosophers, economists and thinkers.

Regardless of the divergences in the field of moral philosophy, liberal thought as an ideological engine of political actions starts from the "Natural Rights" that human beings would have to life, freedom and property. The free initiative of each individual would lead to the development of society as a whole. The greatest translation of this principle would be Adam Smith's "invisible hand". For this, the State should be minimal, taking care only of tasks that are impossible to be given to society and the market, such as national defense, guaranteeing compliance with the Constitution and Laws, defending the stability of the currency and promoting free competition. Governments should have extremely limited powers not to interfere with the political and economic freedom of individuals.

Classical liberalism was recycled in the twentieth century by Frederich Hayek and Ludwig Von Mises, exponents of the Austrian School, and later by Milton Friedman, leader of the Chicago School. Monetary policy should act according to the needs of free market play, fiscal policy should be austere, restrictive and prudent, the tax burden as low as possible and the most uniform in its incidence, protectionism extinguished with its barriers and tariffs, the exchange fluctuating, and all sorts of state interventionism avoided (minimum wage, sectoral subsidies, compulsory social insurance, customs barriers, professional licensing, nationalization of activities). Public regulation should be discrete in sectors of natural monopoly (energy, telecommunications, railroads, sanitation). That is, more society of free citizens, more market, less State – the Minimum State and decentralized and deconcentrated power. Even social education, health and social security policies should ideally seek voluntary market solutions with subsidies for poor families. Even a Minimum Income Program, like the Negative Income Tax, put money directly in the hands of poor free citizens, not of the state bureaucracy.

Neoliberalism appeared in flesh and blood, already in the 1980s, after the fiscal crisis of the welfare state and social democracy, in the governments led by Margaret Thatcher, First Minister (from 1979 to 1990) in the United Kingdom, and Ronald Reagan, (President from 1981 to 1988) in the USA, guided by the ideas of reduction of the State, privatization of state-owned companies, liberalization of the international economy with free flow of goods and capital, the end of protectionism and reduction of taxes. There is a flagrant contradiction in the liberal conception: capital and goods should have free circulation without any imposed obstacles, but the workforce should not. The vigorous impact of large immigration from poor countries in Africa, the Middle East and Asia to Europe has become one of the biggest dilemmas in the contemporary Western world and the basis for the revival of xenophobic nationalisms and being a cornerstone to explain the strength of the extreme right European and worldwide and its global expression in the twenty-first century, authoritarian populism, rescuing traits present in Nazi-fascism.

For those who want to get started in liberal thought, we recommend reading "Theory of Moral Sentiments" and "Wealth of Nations" by Adam Smith (2010, 2018), "On the Social Contract" by Jean-Jacques Rousseau (2012), "Idea of a Universal History from a Cosmopolitan point of view" by Immanuel Kant (2012), "Democracy in America" by Alexis de Tocqueville (2002), the recent book by Mario Vargas Llosa, "The Call of the Tribe" (Llosa, 2018), and by Friedrich August Von Hayek, "The Road to Serfdom" (Hayek, 2005) and by Milton Friedman, "Capitalism and Freedom" (Friedman, 2020).

4 Communism: Proletariat and Revolution

"A specter haunts Europe – the specter of communism". With these words, Marx and Engels open "The Communist Manifesto" of 1948, and conclude in triumphant proclamation: "Proletarians of all countries, unite" (Marx & Engels, 2014). The programmatic text was the founding document of the Communist League. At the beginning of the nineteenth century, the ills and injustices produced by the new industrial capitalist system generated the socialist movement, as the initial political expression of the nascent working class.

The 1830s were marked by several workers' insurrections in France. In England, in 1831, the Chartist Party, the first workers' party, was founded. In 1871, the Paris Commune broke out. The trade union movement was strengthening in Western Europe. All of this was a reaction to the poor working and living conditions of workers at the time, 16-h workdays with only half an hour of break; child and female labor under inhumane conditions. The numerous contingent of workers displaced from the countryside to the city, in the new urban-industrial scenario, subjected to terrible housing and sanitation conditions, boosted the insurrection mood. Crime and prostitution growing, wild competition between companies and countries, causing inequalities and wars, the reserve army swelled with unemployment (caused by the evolution of machinery and steam power), all of this added fuel to the fire of the class struggle and frightened the new hegemonic class, the bourgeoisie, which was still struggling to extinguish the feudal remnants and the absolutist monarchy.

Initially, the struggles had a union gain perspective and reformist character. In the field of ideas, the so-called "Utopian Socialism" of Sant Simon, Fourier, Robert Owen and Proudhon emerged, and, later, already in the context of the international socialist movement, the reformism of Eduard Bernstein and Karl Kautsky. Some achievements happened, but with their differences, in ideas and time, they did not pose a revolutionary perspective of taking power. It was about humanizing capitalism and valuing democracy.

Marx, Engels and Lenin broke with this tradition. The founders of Marxism start from the principle that the entire history of humanity is the history of class struggles, and that the struggles are the result of economic relations expressed in the mode of production and in the social relations of production, which ultimately determine political, legal, philosophical, religious and cultural institutions. They founded what they call "Scientific Socialism", based on the philosophical principles of Historical and Dialectical Materialism, a totalizing vision –an integrated vision of philosophical, moral, intellectual, natural, historical and political economy ideas – with a teleological dimension – a final destination to conquer. In this, they also differ from conservative, democratic, republican and liberal thinking, which did not consider a finished project previously conceived to be built from its fundamental ideas. It was about letting the story flow freely and move forward.

Workers, for Marxism-Leninism, are exploited by capitalists, the owners of the factories and companies extract a value produced by the workers, the 'surplus value', beyond the cost of production. However, the amount paid to the workers, who compose the large majority of the population, would not be sufficient to sell them all the products, since the total mass of their salaries is lower than the total price of the products, because the system operates on the extraction of the 'surplus value' for the capitalists to get the profit that is necessary for the success of their enterprises. Capitalism in its advanced stages would imply the anarchy of production, producing cyclical crises of overproduction and underconsumption, and chronic and cyclic mismatch between supply and effective demand.

Only the proletarian revolution with the seizure of power by the workers, the dictatorship of the proletariat, the preparation of the extinction of the State and classes, and the abolition of private property would harmonize the mode of production and the social character of production, and would promote the fair distribution of the 'surplus' and the end of the anarchy of production, replacing market laws with centralized planning. This would prepare the advent of a communist society, without the State and without classes.

Marx and Engels believed that the Revolution would break out in countries with advanced and mature capitalism, but the 1917 Revolution took place in the USSR, a predominantly agrarian society, with incipient capitalism and nonexistent democratic experience, and then in 1949 in China, a country even more backward in the construction of capitalism. The Russian communists came to power with the motto "peace, bread and freedom", taking advantage of the breaches opened by the First World War, widespread hunger and czarist authoritarianism. What followed did not correspond to the communist utopia. Lenin dies early. The revolution in the USSR resulted in the Stalinist period, the cold war, the enlargement of the state apparatus (Maximum State) and the birth of a new elite linked to the governmental apparatus and the single party.

The dissolution of the USSR and the fall of the Berlin Wall marked the rise and fall, in about a century and a half, of the experience of power created by the Communist movement. In China, a hybrid arrangement between one-party Marxism-Leninism-Maoism survives, with a market economy, a consequence of liberalizing reforms and radical political authoritarianism of Deng Xiaoping (1904–1997) set from 1976 to 1997. There still remains the exotic experience of North Korea, with its militaristic and aggressive traits, epidemic poverty and absolute closure of borders; the decadent experience of Bolivarian socialism in Venezuela; the terminal sunset of Cuba; the Vietnam inspired by the Chinese trajectory, and the Ortega dictatorship in Nicaragua.

In this sense, the political experiences that were born from the matrix of Marxism-Leninist thought are dated and no longer exert influence in the contemporary world, except for China's global protagonism. Unlike the extinct 3rd. Communist International, Stalin at the head, China wants to export goods and investments, not revolutions and ideology. Pragmatism is the guiding principle of the Chinese government and Communist Party.

For those who want to know the bases of this current of thought, we suggest reading the "Communist Manifesto" (Marx & Engels, 2014), the text by Engels, "From Utopian Socialism to Scientific Socialism" (Engels, 2005), "The 18th Brumaire of Louis Bonaparte", by Karl Marx (2020), "The Civil War in France" by Karl Marx (2014), and "Imperialism, the highest stage of capitalism" and "The State and the Revolution" by V.I. Lenin (2011a, b).

5 Fascism: Totalitarianism and War

Among the currents of thought that have animated political struggles in recent centuries, now we address one of the most complex and intriguing historical phenomena of modern civilization: Nazi-fascism.

A major earthquake hit Europe in the 1920s, 1930s and early 1940s. It is not possible to attribute such tragic events that resulted in the Second World War, the Holocaust, and the economic and social crisis experienced in the period, only to individual psychological factors and to the personal attributes of Mussolini and Hitler. The leader, whether the "Duce" or the "Führer", can only be understood within the historical context that took shape after the First World War and the monumental crisis that befell Europe, Germany and Italy, aggravated by the Great Depression of 1929. It is intriguing to understand how two unbalanced leaders, without empathy for human beings, charismatic, narcissistic, egotistical and violent, managed to come to power through parliamentary democracy and consolidate the most terrible totalitarian experience in all of modern history (Stalinism has already been the object of previous analysis) with an impact on every corner of the Earth, in the most shocking war experienced by Humanity. The leader, let alone vicious dictators like Mussolini and Hitler, undoubtedly has a crucial role in events, but they would not arrive and remain in power unless under certain historical conditions for obtaining minimal consensus and social support in their countries.

Care must be taken not to generalize the use of fascism to characterize any and all authoritarian experience of power. In Brazil today, it is common to see people exchanging pleasantries on social networks, in parliament and on the streets, calling each other "fascists" and "communists", without any rigor with the use of terms. There was "Japanese fascism", a subject of controversy, being yet another militaristic ultranationalism with different characteristics. 'Generalissimo' Franco, who governed Spain from 1936 to 1975, had undisguised admiration for Mussolini, and received support from Germany and Italy, with bombing and troops, in the Spanish Civil War, which consolidated him in power and is considered the experimental laboratory for the Second Great War. However, like Portugal and Salazar, Franco maintained a position of neutrality in the war, in the face of Nazi, French and English pressure. Salazar, in Portugal, shared his admiration for Mussolini and his ideas, but not for Hitler, with whom he differed due to anti-Catholic elements in the Nazi leader's thinking and the persecution of the Jews, which resulted in the dramatic Holocaust. Integralists in Brazil in the 1930s and 1940s, under the guidance of Plinio Salgado (1895–1975) prayed for the fascist primer. Other experiences in Europe were aborted by the success of the democratic process. Some experiences had undeniable common traits of fascist inspiration: among others, the use of the symbolic language of an authoritarian national and social order translated into the uniforms of its militants – black shirts for the Italian fascists, khakis for the German Nazis and green ones for the Brazilian Integralists – and their symbols – the Nazi swastika, the Italian fascio (a bundle of birch sticks with an ax in the middle) and the sigma for Brazilian fascists.

Other seeds were planted by Nazi-fascist ideology around the world and survive to this day to threaten democracy, but the original conception has its own peculiarities and characteristics, which should not be confused with simple dictatorial authoritarianism. In this sense, it is a mistake to use the term "fascism" to characterize, for example, dictatorships in Latin America or Africa, fundamentalist totalitarianism in the Middle East and the recent wave of right-wing and left-wing authoritarian populism involving leaders such as Trump, Erdogan, Orbán, Putin, Chavez, Maduro, Ortega, Berlusconi, Salvini, Le Pen, Bolsonaro, and the current Italian Prime Minister, Giorgia Meloni. There are elements similar to Fascism and Nazism, but nothing comparable to the experience of Hitler and Mussolini in power, although many of these leaders admire the Duce and the Führer in their hearts, and there is an attempt at historical revisionism, carried forward by identified conservative historians with the extreme right.

Nazi-fascism can only be understood as a consequence of the economic and social crisis derived from the Treaty of Versailles, which fixed the indemnities and sanctions for the losers of the First World War, and which resulted in recession, unemployment, hyperinflation, particularly in Germany and Italy, and triggered serious popular dissatisfaction and the demoralization of parliamentary democracy. The greatest economist of the twentieth century and adviser to the British government in the negotiation, J. M. Keynes, premonitorily warned in his classic "The Economic Consequences of Peace", that Europe would not recover with the humiliating conditions imposed on the defeated and that a serious crisis would set in Italy, although this country was a part of the victorious coalition, but sidelined and not included in the division of the spoils of war. This situation had a fundamental impact on Mussolini's position.

Keynes diverged from the positions of the victorious European countries and the USA, which led to his removal from public office for many years. If that wasn't enough, the crash of the New York Stock Exchange triggered, as an additional spice, the Great Depression of 1929, and let's not forget the Soviet Revolution of 1917, raising the great common enemy: the threat of the communist program of the 3rd. International. This environment was decisive for Mussolini and Hitler to gain social and political support in their countries, and became the basis for their totalitarian adventures, triggering off a new conflict on a global scale.

It is important to note that Mussolini and Hitler reached the post of prime minister in their countries not through coups, but within the rules of parliamentary democracy, with the complacency of conservative elites who believed they would be able to manipulate the two future dictators. It is true that Hitler attempted a coup in 1923, which resulted in his arrest, where he wrote his book "Mein Kampf". Faced with the crisis installed in Italy and the March on Rome, in 1922, King Vittorio Emanuele invited the then deputy Benito Mussolini to assume power and form a new cabinet, after several frustrated attempts to stabilize Italian politics with conservative prime ministers and liberals. For his part, after the failure of the Weimar Republic – a coalition of Catholics, liberals and German social democrats – to overcome the economic and social crisis, the president of Germany, Paul von Hindenburg, appointed, in 1933, the Nazi leader Adolf Hitler for the post of German Chancellor. In other words, democracy generated the "Egg of the Serpent" (1977) – a great film by the Swedish director Ingmar Bergman – opening the doors of power that would later become the tragic experience of Nazi-fascism.

Mussolini and Hitler had identities and differences. Hitler admired the success of Mussolini, who came to power 10 years earlier, and his ideas. Afterwards, he assumed the leading role, given Germany's economic and military superiority. Both had unhealthy, narcissistic, violent, authoritarian and suspicious minds. They were charismatic leaders and excellent orators using demagoguery and popular dissatisfaction to charm the masses. Mussolini was more cultured, read voraciously, spoke three languages in addition to Italian and even became editor of the newspaper "Avanti!" of the Italian Socialist Party, before converting to fascism. Hitler was more limited intellectually, according to his contemporaries. Both were men of action, pragmatists, focused on power, not theorists. Mussolini distrusted Hitler and received the same feeling in return. Mussolini had an active personal life, a wife, lovers, pleasures and five legitimate children. Hitler did not have an active personal life, he was reclusive, his world was the exercise of power. Mussolini worked harder and liked to be involved in managing government affairs. Hitler was more isolated and delegated day-to-day management to his close circle. The relationship was one of political and military alliance, but also one of personal competition. Mussolini was slow to declare war on the side of Germany. Hitler was irritated with the Italian's delay and with his disastrous unilateral and unannounced attempt to invade Greece. The strand of racist nationalism was most pronounced in Hitler's obsession with Aryan superiority and the project of racial cleansing of Germany and its surroundings. The competition between the two is brilliantly reviewed in art form by Charles Chaplin in his masterpiece "The Great Dictator" (1940), in the barbershop scene, where each of the two dictators tries to place his chair higher than that of his ally.

Here we can already return to the essential question: After all, what is fascism? What are its essential traits, even considering the differences between the German and Italian experience?

Fascism and Nazism were totalitarian regimes, bellicose, terrorist dictatorships, moved by radical chauvinist nationalism. Faced with the profound economic and social crisis in Europe and the resulting social dissatisfaction, they came forward in the name of order, to end the class struggle and unite everyone around the Nation, embodied in the State, the party and its greatest leader. They exterminated liberal and parliamentary democracy and cultivated the ghost of the communist threat from the USSR. Anti-Semitism generated the most tragic page of World War II, the Holocaust. Nazi-fascism had a mass social base. Ideologically, the Nazi project was more finished and closed around military expansionism to impose the superiority of the Aryan race, building an empire of vast territorial dimensions. Mussolini was more volatile in the construction of the ideas of Italian fascism, and, as Togliatti characterized him, like a chameleon inspired by eclectic and confusing sources, and adapted to concrete circumstances in search of legitimation, consent and social consensus. Nazi-fascism was sexist, assigning an inferior role to women, and built a corporate regime that subordinated the unions to the State and the Party, a symbol of which is the "Carta del Lavoro", which served as inspiration for the Brazilian laws. They proposed centralized planning to face the anarchy of capitalist production. They consolidated, as essential tools for the success of their project, strong, disciplined, centralized governments and parties, with little internal democracy. They raised their own and parallel armed forces to come to power, the 'militias' that threatened and intimidated political, social, trade union and intellectual leaders with violence. Aiming at consolidating political hegemony after coming to power and turning to dictatorial totalitarianism, they undertook enormous efforts, through various social organizations and institutions, to root Nazi-fascist politics and ideology, co-opting and dominating parts of the Armed Forces, police, intelligentsia, trade unions, the artistic and cultural environment, youth, the leisure and sports segment and organizations in rural areas and social assistance.

For those who want to delve into the study of fascism, particularly the German and Italian experiences, we recommend reading the books: "Lessons on Fascism", by Palmiro Togliatti (2017); "The Economic Consequences of Peace"; by J.M. Keynes (2019); "Listen, Little Man!", by Wilhelm Reich (1974); "Fascism, an alert", by Madeleine Albright (2019); the biography "Hitler", by Ian Kershaw (1999); "Mussolini, the definitive biography", by R. J. B. Bosworth (2014); and "Brazilian Fascism", by Pedro Dória (2020).

6 Social Democracy: Reform and Democracy

Social democracy was born from the split within the international socialist movement, organized in the Second International, founded in 1890, notably in Germany. Some striking facts were decisive: The conquest of the right to universal and secret suffrage and laws limiting working hours and improving the living conditions of workers in some advanced countries. The outbreak of World War I and the Russian Revolution in 1917 were watershed. A series of leaders of the Second International began to realize the impracticability of revolution in advanced countries and to value the mutations of capitalism and the advances of parliamentary democracy.

The most relevant theoretical framework was the publications by Eduard Bernstein, German parliamentarian and leader of the Social Democracy Party, in 1899. He adopted a revisionist, reformist critical view of Marxism, valuing the democratic and parliamentary struggle. The vision began to gain numerous adherents in European social democratic parties.

The First World War was the landmark of the rupture. Russian leaders broke with the Social Democrats and founded the Third International, under the command of the Bolshevik Party. From then on, social democrats focused on creating mass parties with electoral strength, accumulating government experience, fighting for the evolution of legislation that promoted workers' rights, and defending reforms and state intervention to advance and humanize capitalism, without breaking with the market economy and democracy.

The world crisis of 1929 dynamited the canons of classical liberal thought. The thought of the greatest economist of the twentieth century, John Maynard Keynes, appears, offering the theoretical foundation for state intervention in the economy, in certain circumstances, to guarantee full employment. The 'cold war' resulting from the world configuration after World War II – a bipolar geopolitical scenario opposing the two blocs led by the USA and the USSR – was the background in which this approach was adopted by governments and social democratic parties to find a way of evolution of the political-economic system, overcoming the geopolitical polarization.

After World War II, social democracy became hegemonic, giving rise to the "Welfare State". Significant leaders would emerge in Europe and the Americas in later decades: François Mitterrand, in France; Helmut Schmidt, in Germany; Olof Palme, Sweden; Mário Soares, in Portugal; Felipe González, in Spain; Bill Clinton and Barack Obama, in the USA; Tony Blair in the United Kingdom; Fernando Henrique Cardoso, in Brazil; Ricardo Lagos in Chile, among others. At the same time, the Italian Communist Party unleashed its "aggiornamento" process, revising Marxism and adopting democracy as a universal and permanent value, without a class character: neither bourgeois nor working class, preparing the future convergence with social democracy. An important role in this movement was the public dialogue between the Italian communists and the Italian philosopher and senator, Norberto Bobbio, who defined himself as 'social-liberal'. Each of the social-democratic leaders lived with their moment and their circumstances, but they were united by the social-democratic conviction that the market is the best organizer of economic factors, but that it produces crises and inequalities, and that the State must act. And that democracy is the only way to build a better society.

The fiscal crisis in the 1970s, the neoliberal reaction led by Thatcher and Regan in the 1980s, the end of the USSR, the fall of the Berlin Wall, the advance of globalization and the technological revolution, changed the scenario. The new leaders of social democracy face the challenge of reinventing concepts and strategies in the face of the new world.

Permanent dialogue is essential to build a horizon that combines, in the concrete historical conditions of the twenty-first century, sustainable development, social reforms, macroeconomic balance, growth and democracy – the indispensable universal and permanent value.

For those who want to get started in the history of social democracy and its content, we recommend reading the texts "Evolutionary Socialism" by Eduard Bernstein (1988), "What Is a Social Revolution" by Karl Kautsky (2021), "The End of Laissez-faire" by J. M. Keynes (2009), "The Era of Rights" and "Right and Left" by Norberto Bobbio (1991, 1996), "The Social-Democratic Proposal" organized by Hélio Jaguaribe (1998), "The Third Way" by Anthony Giddens (1999) and "A Soma e o Resto" by Fernando Henrique Cardoso (Cardoso & Oliveira, 2011).

7 Eurocommunism: Democracy and New Reformism

As previously discussed, the industrial revolution, the birth of capitalism, the urbanization of society and liberal democracy produced the most radical change that had ever occurred in the history of civilization so far, leaving the traces of feudalism and absolute monarchy behind, and propelling an unprecedented advance of the productive forces and the process of technological innovation. The working and living conditions of large sections of the population subjected to misery and exploitation triggered the trade union and socialist movement in the nineteenth century. The thought of Marx and Engels appears, which intended to offer a broad critical diagnosis of the functioning of capitalist society and build a theory with a view to guiding the workers' struggle towards socialism.

Contrary to what Marx and Engels imagined, the proletarian revolution did not take place in countries of advanced industrialization and mature capitalism, but in rural Russia, with incipient capitalism, by means of an "assault on power" by the Bolsheviks, taking advantage of the concrete circumstances of the First World War and the popular revolt against famine and czarist authoritarianism.

Lenin died early and one of the most controversial figures in the modern world took power: Josef Stalin. Far from an idyllic and humanist social construction, the USSR became the scene of a great tragedy, the Stalinist tragedy.

Throughout this period, several events had an impact on the international communist movement: the leak, in 1956, of the secret report by Nikita Khrushchev, Stalin's successor after his death, denouncing the atrocities of the Stalinist dictatorship; the Cuban Missile Crisis, the extreme point of the Cold War, in 1962; the invasion of Czechoslovakia and the Soviet liquidation of the Prague Spring in 1968; the ambiguous and complex relationship between the Soviet Union and China, led by another luminary of communism, Mao Zedong; the Vietnam War, the internationalization of the economy, changes in the world of work, the rise of Gorbachev, with *Perestroika* and the *Glasnost*; the fall of the Berlin Wall; the dissolution of the USSR and the collapse of communism.

Throughout the entire trajectory of the international communist movement, led by the Soviets from 1919 until its dissolution, an original, differentiated, heterodox and creative experience developed from the historical evolution of the Italian Communist Party, the PCI.

The theoretical and ideological axis of the Italians, a tortuous path, full of contradictions, comes from the progressive elaboration of its leaders Antonio Gramsci, Palmiro Togliatti, Enrico Berlinguer and a fertile group of intellectuals, in search of the Italian path, and more than that, the European pathway to socialism.

The PCI started to defend political democracy as a universal and permanent value, without a class nature, the "passive revolution", a path of progressive development through continuous economic and social reforms, the dispute for hegemony through the "war of positions" in all fields of social life, and permanent dialogue with other democratic political and social forces. As we have already pointed out, the public and theoretical dialogue, from 1950 on, with the social-liberal thinker Norberto Bobbio was an important milestone that deserves to be recorded in this ideological evolution.

Faced with the creation of the European Community, the embryo of the European Union and the Euro, the PCI advanced in its search for a European path, and not just an Italian one, towards socialism, pointing to the expansion of its party construction, first with the creation of the Partito Democrático della Sinistra (PDS), then to Democratici di Sinistra (DS) and, in sequence, merging with Margarida, a group of progressive Catholics, giving rise to the current Democratic Party (PD). The leaders and intellectuals of Eurocommunism do not deny Marxism in its fundamental concepts and method of analysis, although they have thrown overboard the unedifying legacy of Leninism, Stalinism and Maoism.

For those who want to be introduced to the study of Eurocommunism, we recommend reading the books "For a new reformism" by Giuseppe Vacca (2001), "From Stalinism to Democracy" by Marco Mondaini (2011), "A modernization without the modern" by Luiz Werneck Vianna (2011), "Reformismo de Esquerda and Political Democracy" by Luiz Sérgio Henriques (2018), "Itineraries for a Democratic Left" by Alberto Aggio and "Gramsci in his time" organized by Alberto Aggio, Luiz Sérgio Henriques and Giuseppe Vacca (Aggio, 2019).

8 Authoritarian Populism: Digital World and Rupture

"Evil winds blow on the blue planet"; with this sentence, the Spanish thinker Manuel Castells described the whirlwind of crises that embraced the contemporary world in the twenty-first century. One of the visible faces is the emergence of experiences of power that are imprecisely defined as "authoritarian populism". From the right wing, Donald Trump in the US; Silvio Berlusconi, Matteo Salvini and Giorgia Meloni in Italy; Marine Le Pen in France; Viktor Orbán in Hungary; Vladimir Putin in Russia; Benjamin Netanyahu in Israel; Erdogan in Türkiye and Jair Bolsonaro in Brazil. From the left: Chavez and Maduro in Venezuela and Daniel Ortega in Nicaragua. All would be leaders who came to power through the democratic path and once there, they seek to destabilize the institutions, the Constitution and the rules of the game, subverting democracy itself.

They use the broad space of tolerance, proper to democracy, to install a culture of intolerance. They take advantage of the freedom of the press, opinion and manifestation to build a world of "alternative truths", fake News, of the manipulation of information. They instrumentalize modern digital tools, the algorithms of the "chaos engineers", to feed their "ideological bubbles" of support. They stimulate insane polarization, reduce spaces for dialogue and consensus building and stigmatize opponents as enemies to be eliminated, that is, they use the tools of democracy against democracy itself. And, obviously, they only make their projects viable because they have social support based on accumulated frustrations, hatreds and resentments in relation to democracy, globalization, the traditional political system and the social exclusion of large portions of the population.

The currents of thought that moved political struggles in the modern world are organized around five essential vectors: (i) moral and religious values, (ii) commitment or not to freedom and democracy, (iii) role of the state in society and the economy, (iv) pursuit of social equity and (iv) role of nations on the world stage. But the great watershed is the democratic issue. Of the currents analyzed so far, conservatives, liberals, social democrats and Eurocommunists are committed to democracy as a universal and permanent value. Reactionaries, real socialist communism and nazi-fascism do not. The 'authoritarian populists' of the twenty-first century, neither. Issues related to moral, philosophical and religious values must belong to the individual, family and civil society spheres, but the 'authoritarian populists' manipulate this discussion in the political sphere from a conservative perspective, against identity movements mostly linked to the lefts.

The relations between State and market, the degree of government intervention and the organization of the economic system assume varied formats and formulations. There is no experience anywhere in the world that resembles the 'laissez-faire' of classical liberalism, but there is reasonable convergence around the vision of the socially necessary State with a democratically regulated market economy, in line with the concrete historical conditions and peculiar characteristics of each country. Even China, a totalitarian state inheriting the communist tradition incorporated the market into its logic, in what they call "Market Socialism", and others call "State Capitalism". In the search for a more egalitarian society, based on greater social equity, there are disagreements about the format and intensity of public policies, but there is no one who defends the maintenance of the unacceptable and scandalous social inequalities that still prevail in the contemporary world. On the national issue, there are divergent positions, but conservatives, liberals, social democrats and modern socialists converge on the perception that globalization is an irreversible trend and that each country must discover the best way of insertion and fight for the growing institutionalization of global governance in issues such as the environment and climate, public health, human development and employment, international trade and financial flows.

In contrast to all these constructive tendencies, 'authoritarian populists' instrumentalize the negative effects of globalization, such as unemployment or immigration, with xenophobic nationalist and anti-globalization discourses. The British today suffer the economic and social consequences of the *Brexit*, whose campaign was led by the conservative populist Boris Johnson. As exceptions, the far-right Italian prime minister did not adopt a confrontational stance against the European Union when she came to power, and Putin is seeing his popularity plummet in Russia, although the government-induced media try to deceive the population, due to his imperialist nationalism embodied in the invasion of Ukraine.

'Authoritarian populists' are not to be confused with the Nazi-fascist totalitarianism of the first half of the twentieth century, although Mussolini and Hitler came to power through democracy and eroded institutions from within to prepare for a dictatorial turn. Times are different. There are common traits, but they do not allow a common theoretical framework. They have in common the xenophobic nationalist discourse, the anti-democratic character and the social support in its support, but Nazi-fascism had a more closed and finished ideology, while the 'authoritarian populist' are more pragmatic. Like chameleons, they disguise themselves as liberals in the economy at times, at others they pose as liberal democrats vociferating against "censorship" of their transgressions on social networks, but in essence, they seek power to stifle democratic freedoms and impose their projects and objectives. Today, historical conditions impede military imperialist adventures and limit their ambitions, except (again) for Putin's in Ukraine. It is evident that right and left 'authoritarian populist' experiences have differences in narrative and content. However, both strands pursue totalitarian goals and must deserve the revulsion of all democrats.

The instigating question that remains is: How in the twenty-first century, in the society of knowledge, globalization and digital communication, visibly authoritarian and reactionary populist leaders manage to gather electoral support and win elections? Once again, the phenomenon cannot be explained by the leader's charisma, intelligence and political astuteness alone. It is evident that a leader with sufficient characteristics is needed to galvanize social sentiment in a given conjuncture, but reality is more complex. The 'authoritarian populists' only manage to rise to power because frustrations and social and cultural changes have accumulated in the social and economic fabric, creating the culture broth that allows them to flourish.

The roots certainly lie in the structural traits of postmodern capitalism and contemporary society. Globalization causes unemployment in the short term. The fast and frightening scientific and technological revolution – of the Internet age, robotics, genetic engineering, artificial intelligence – impacts unevenly on the various segments of society, producing exclusion and resentment. Society is no longer a simple struggle between capitalists and workers, with a middle class swinging in the middle. The flexibilization of the world of work and the multiplicity of interests produces a fragmented and diverse society. Political parties, central institutional tools in the functioning of democracy, lose their capacity of leadership. Politicians are seen as a "new aristocracy" detached from the interests of the population, with its own internal logic.

It is not for nothing that in general the 'authoritarian populists' are outsiders who mobilize their bubbles and voters against "everything that is there", "the old politics" and "the traditional political system". The masses, relegated to a situation of low access to quality public services, in terms of education, health, culture and existence, seeing a hostile and non-inclusive State, not recognizing themselves in postmodern society, are easily captured by charismatic leaders, which exploit the accumulated frustrations and resentments, waved with salvationist and messianic projects, promising radical changes. The situation is aggravated by the growing protagonism of social networks and digital communication platforms, where each citizen can play the role of an "individual political party", express their opinions, create ties and identity groups and suffer the manipulative action of algorithms.

Outside of democracy, we believe, there is no salvation. Democracy is, by its human nature, imperfect, but it is the only political regime that carries in itself mechanisms of checks and balances, of learning and self-correction and of permanence of stable institutions and rules for the functioning of the democratic game. What is the course correction capability in North Korea, Russia, Venezuela or Nicaragua? Low, certainly. However, the world turns and no transformation is impossible. Democratic forces, of all stripes, have shown themselves to be resilient and responsive in countries such as the USA, France, Spain, Germany, Brazil, Portugal, among many others, but the ghost of 'authoritarian populism' hovers around the world threatening democracy. It is enough, for example, to cite the current example of France, where far-right and far-left populism unite and mobilize the population against Macron's necessary and inevitable pension reform. "The price of freedom is eternal vigilance". What will come of the next American or French elections? The struggle for democracy must unite, on a global scale, conservatives, liberals, social democrats, progressives and democratic socialists against any and all forms of authoritarianism, but the fight will only be successful if the forces that defend democracy restore their ties with society and manage to raise concrete answers to the anxieties and dreams of common citizens.

For those who want to go deeper into the study of the phenomenon of 'authoritarian populism' at the beginning of the twenty-first century, we suggest reading the books "Rupture: The Crisis of Liberal Democracy" by Manuel Castells (2020), "How Democracies Die" by Steven Levitsky and Daniel Ziblatt (2019), "The Hatred of Democracy" by Jacque Rancière (2009), "The Engineers of Chaos" by Giuliano Da Empoli (2019) and the Journal of Democracy, in Portuguese – Volume 8, Number 2, November 2019, edited by Bernardo Sorj and Sergio Fausto.

9 The Great, Possible and Necessary Democratic Convergence

So far we have visited the theoretical and ideological universe of the main currents of thought that inspired political projects and experiences of power in the modern world. We critically reviewed, in a somewhat summary and summarized way, the main theoretical matrices that dominated the universe of human thought and inspired political action since the seventeenth century: conservatism, liberalism, communism, social democracy, fascism, democratic socialism and authoritarian populism.

The world today is becoming different from Modernity. Great transformations have taken place, such as the end of the Cold War, dissolution of the Sovietic states, economic globalization, radical and rapid scientific and technological revolutions, emergence of digital platforms and social networks, emergence of new global players such as China, crisis of liberal democracy, radical changes in the working world, radicalization of Islamic fundamentalism, the challenge of mass immigration of poor populations, especially to Europe, and the consequent strengthening of far-right xenophobic nationalism, the global crisis in 2009 starting from the American financial market, the *Coronavirus* pandemic, the war in Ukraine, together with unprecedented cultural and social integration on a global scale boosted by communication technologies.

The classic ideas of left and right have long been problematized, but we believe that from the line opened by the Italian social-liberal thinker Norberto Bobbio in his "Right and Left: Reasons and meaning of a political dispute", from 1994, it is possible to reframe the concepts and realize that they still make sense in the contemporary world.

Nowadays, in Brazil and in the world, it is possible to detect the existence of three great mobilizing poles of thought and political action:

the right, the democratic center and the left. What are the decisive questions that guide the construction of divergences and convergences between currents of the most diverse shades?

First, the democratic question. There are broad sectors that defend democracy as a universal, definitive, unshakable and permanent value. Not everyone adheres to this conviction. Second, there is the challenge of environmental sustainability, which some support, others do not. Third, the fight against poverty, hunger and social exclusion, which is central in some formulations, and relegated to the background in others. After all, in a globalized world, why only capital and goods can freely circulate, not workers? Fourth, the national question, how to combine national identity and interests with the globalization of the world? Are we going to resurrect old retrograde nationalisms full of xenophobia? Responses range from the extreme right to the extreme left. The role of the State in the economy and society appears as a fifth parameter. And finally, the challenge of social organization around the technological revolution, which produces innovation, but also structural unemployment and leads to deep polemics of a moral, philosophical, economic and political nature. How to build a sustainable, fair, innovative, inclusive development model that does not harm the human person?

From there, after asking the essential questions, it is possible to advance in the discussion about the differences between right, left and democratic center in the contemporary world. The militants of the current insane and sectarian polarization do not want to know about questions, fundamentals and debates; they already have half a dozen ready-made phrases and thin convictions to stigmatize their "enemies" and build their narratives. Roughly speaking, with due regard for natural historical and national peculiarities, we can see the contemporary right in the experiences of Trump in the USA, Bolsonaro in Brazil, Erdogan in Turkey, Le Pen in France, Berlusconi, Salvini and Giorgia Meloni in Italy, Orbán in Hungary, Putin in Russia, among others. Commitment to democracy is low or non-existent, and it is up to institutions and society to resist. Concern for protecting the environment is minimal and is only present in the agenda thanks to international agreements and defense of sustainability, the planet and the climate. The concern with social equity is marginal for the extreme liberalism, but electoral concerns push them towards some social policy. In relation to globalization, they have a refractory, reactionary and radical nationalist position. As for the role of the state, contradictorily, they defend strong and authoritarian governments, which organize the economy and society from above, if possible transitioning to totalitarian and dictatorial experiences of power. As for new technologies, they haphazardly use modern digital forms of communication, but have a denialist attitude towards science and reservations about their achievements.

In the left field, the differentiating specificities are greater in their own terrain. It is difficult to identify in the same theoretical and political orbit State Capitalism in China, Maduro in Venezuela, the exotic experience of North Korea, Gabriel Boric in Chile, the democratic current of Bernie Sanders, Lula and the Brazilian Workers' Party (still not assumed as truely Social Democratic), Daniel Ortega in Nicaragua, the decadent Cuban regime, the Spanish *Podemos*, the 'Left Bloc' and the Communist Party in Portugal, Jean-Luc Mélenchon in France, the Italian 'Sinistra' and the 'Die Linke' in Germany.

There are many shades and different nuances. Some segments of the world Left radically defend democracy, but there are sectors that still carry the authoritarian traits of Marxism-Leninism-Maoism, even exercising totalitarian power in China, South Korea, Vietnam, Nicaragua, Cuba and Venezuela. Most militate and firmly defend the protection of the environment, but China, for example, was one of the most resistant poles to international treaties on greenhouse gas emissions. The search for social equity is one of the central points in the left's thought matrix, which is a seminal reason for its own existence, regardless of the effectiveness or otherwise of its experiences in power at this point. The national question has always been central to Marxist thought. The legacy of the theses on imperialism leads the left to take refuge in a left nationalism that is resistant to globalization, although China concentrates its energies on the movement of its integration into the great global productive chains. On the subject of the role of the State, they always tend towards the configuration of a Maximum State strategic planning, based on permanent distrust in relation to the signaling character of capitalist markets and derive, in varying degrees, to advocate strong state control of the financial system. In relation to the scientific-technological revolution, they tend to have a positive role in stimulating innovation, but a natural step backwards with its consequences on employment and on the poorest populations.

The democratic center remains. This diffuse field is where the great possible and necessary convergence can be born, bringing together social liberals, democrats, conscious catholics and evangelicals who are not extreme fundamentalists, social democrats or democratic socialists, a field that runs through the political and ideological spectrum of the center-right to the center left. This field can converge in the radical defense of democracy as a universal and permanent value, without ambiguities, as a regime that ensures checks and balances, experimentation, learning and self-correction of directions, and balance between powers and the guarantee of the fundamental rights of people and minorities; in defense of an inclusive and sustainable development model, with the preservation of the environment and protection of the climate; in the frontal fight against social inequalities based on strong and efficient social public policies, especially in the field of education and health, and income transfer programs; in defense of the socially necessary and democratically regulated State; in supporting negotiated national integration into the globalized world; and, in stimulating and democratically regulating technological innovations.

The German experience of the traffic light coalition – red for social democracy, green for the PV and yellow for the liberals – points the way. Emmanuel Macron, in France, has also struggled to consolidate a modernizing project, harassed on the right by Le Pen and on the left by Mélenchon. The Portuguese Socialist Party undid the famous Portuguese "Geringonça" (the alliance with the Left Block and the Communist Party) in the 2022 elections, due to differences over the budget and fiscal issue, and now governs with an absolute majority. The efforts of the Italian PD also converge in this direction, facing the Italian extreme right and extreme left. American Democrats, in their two-party system, continue to resist the onslaught of authoritarian populism by Donald Trump and are always a reference for this type of bet. In countries like Brazil and the United Kingdom, the democratic center used to have great expression, but it has been deflating in recent years.

The world presents complex and new challenges. We believe that neither the traditional right nor the traditional left today have adequate answers to face the agenda of the twenty-first century. The great convergence around a democratic and progressive pole at the center of the political system is what, in our opinion, has the best condition to generate answers. The contemporary scenario is so disruptive and lives in rapid permanent transformation that there is no ready and finished theory to generate definitive and undeniable alternatives. It is in the exercise of democracy – that universal and indispensable value – that we can, with an open mind, far from shallow sectarianism and fundamentalist bubbles, find these answers.

References

Aggio, A. (2019). *Gramsci no seu tempo* (Organização de Luiz Sérgio Henriques e Giuseppe Vacca, 2ª edição, 474p). Fundação Astrojildo Pereira.

Albright, M. K. (2019). *Fascism: A warning* (320p). Perennial.

Bernstein, E. (1988). *Evolutionary socialism* (72p). Pantianos Classics.

Bobbio, N. (1991). *The age of rights* (1st ed., 192p). Polity.

Bobbio, N. (1996). *Left and right: The significance of a political distinction* (1st ed., 160p). Polity.

Bosworth, R. J. B. (2014). *Mussolini* (546p). Bloomsbury Academic.

Cardoso, F. H., & de Oliveira, M. D. (2011). *A Soma e o Resto: um olhar sobre a vida aos 80 anos* (9ª edição, 192p). Civilização Brasileira.

Castells, M. (2020). *Ruptura: La crisis de la democracia liberal* (3.ª edición, 144p). Alianza Editorial.

da Empoli, G. (2019). *Les ingénieurs du chaos* (200p). LATTES.

de Tocqueville, A. (2002). *Democracy in America* (H. C. Mansfield, & D. Winthrop, Trans., 1st ed., 722p). University of Chicago Press.

Doria, P. (2020). *Fascismo à brasileira: Como o integralismo, maior movimento de extrema-direita da história do país, se formou e o que ele ilumina sobre o bolsonarismo* (1a ed., 280p). Planeta.

Engels, F. (2005). *Do socialismo Hostórico ao socialismo científico*. Centauro.

Fernandes, M. (2002). *Millôr definitivo – A Bíblia do Caos* (Edição de Bolso, 624p). L&PM.

Friedman, M. (2020). *Capitalism and freedom* (1st ed., 287p). The University of Chicago Press.

Giddens, A. (1999). *The third way: The renewal of social democracy* (166p). Polity Press.

Henriques, L. S. (2018). *Reformismo de Esquerda e Democracia Política* (303p). Verbena.

Jaguaribe, H., & Instituto Teotônio Vilela. (1998). *A proposta social democrata: a social democracia na atualidade europeia, hispano-americana e brasileira* (2.ed. atualizada, 297p). Jose Olympio.

Journal of Democracy (Portuguese). (2019). Bernardo Sorj e Sérgio Fausto, Vol. 8, No. 2, 111p.

Kant, I. (2012). *Kant's idea for a universal history with a cosmopolitan aim* (270p). Cambridge University Press.

Kautsky, K. (2021). *The social revolution* (A.M. and May Wood Simons, Trans., 200p). Legare Street Press.

Kershaw, I. (1999–2000). *Hitler* (2 Vol.). W.W. Norton.

Keynes, J. M. (2009). *The end of Laissez-Faire: The economic consequences of the peace* (300p). BN Publishing.

Kirk, R. (2009). *Edmund Burke: A genius reconsidered* (2nd ed., 300p). ISI Books.

Lenin, V. I. (2011a). *Imperialism the highest stage of capitalism* (124p). Martino Fine Books.

Lenin, V. I. (2011b). *State and revolution* (106p). Martino Fine Books.

Levitsky, S., & Ziblatt, D. (2019). *How democracies die* (320p). Crown.

Llosa, M. V. (2018). *La llamada de la tribu* (320p). Alfaguara.

Marx, K. (2014). *The civil war in France* (148p). Martino Fine Books.

Marx, K., & Engels, F. (2014). *The communist manifesto* (48p). International Publishers Co.

Mondaini, M. (2011). *Do Stalinismo à Democracia: Palmiro Togliatti e a construção da via italiana ao socialismo* (308p). Contraponto.

Rancière, J. (2009). *Hatred of democracy* (106p). Verso.

Reich, W. (1974). *Listen, little man!* (144p). Farrar, Straus and Giroux.

Rousseau, J.-J. (2012). *The social contract and other political writings* (Q. Hoare, Trans., 192p). Penguin Classics.

Scruton, R. (2018). *Conservatism: An invitation to the great tradition* (176p). Hardcover. All Points Books.

Scruton, R. (2019). *How to be a conservative* (New edition, 208p). Bloomsbury Continuum.

Smith, A. (2010). *The theory of moral sentiments* (1st ed., 494p). Penguin Books.

Smith, A. (2018). *The wealth of nations* (524p). CreateSpace Independent Publishing Platform.

Togliatti, P. (2017). *Lectures on fascism* (172p). International Publishers.

Vacca, G. (2001). *Riformismo vecchio e nuovo* (213p). Einaudi.

Vianna, L. W. (2011). *A Modernização sem o moderno: análises de conjuntura na era Lula* (192p). Contraponto.

von Hayek, F. A. (2005). *The road to serfdom* (140p). Institute of Economic Affairs.

Marcus Vinicius Pestana was Coordinator of the Academic Board of Economics and President of the Central Directory of Students at the Federal University of Juiz de Fora (1979/1982); Director of the Brazilian Committee for Amnesty – CBA/Juiz de Fora (1979); Economist and Professor of Economics at UFJF (1989–2019); Member of the City Council of Juiz de Fora (1983/1988), Coordinator of the *Diretas Já* Campaign in Juiz de Fora and Zona da Mata (1984); Coordinator of the Committee for a Free and Sovereign Constituent Assembly (1986); Secretary of Government of the Municipality of Juiz de Fora (1993/1994); Secretary of Planning and General Coordination of Minas Gerais (1995/1998); Chief of Staff of the Ministry of Communications (1999/2001); Executive Secretary of the Ministry of the Environment (2002); State Secretary of Health of Minas Gerais (2003/2010); President of the National Council of Health Secretaries (2005/2006); State Deputy for Minas Gerais (2007/2010); Federal Deputy for Minas Gerais (2011/2018); Special Advisor to RAPS – Political Action Network for Sustainability (2019); Special Advisor to BIOMM SA (2020/2022); Special Advisor to ABRAMGE – Brazilian Association of Health Plans (2020/2022) and Special Consultant at the Teotônio Vilela Institute (2019/2022).

Alfredo Pereira Jr is a Philosopher of Science, Professor at the Universidade Estadual Paulista Julio de Mesquita Filho (UNESP), Brazil, retired, with degrees in Philosophy and Administration. Currently working in the UNESP Graduate Program in Philosophy. Alfredo has around 300 publications in several areas of knowledge, mostly in the Philosophy of Neuroscience and Theory of Consciousness. Orcid: https://orcid.org/0000-0002-5960-041XAD, Scientific Index: 56th place in Top 100 Latin America Philosophers/Scientists.

Historical, Bioethical and Technical Aspects for the Creation of the Popular Online Forum

3

José Otávio Pompeu, José Luis Azpiazu, and Alfredo Pereira Jr

Abstract

The implementation of information technology systems for democratic government in several countries opens new possibilities for the interaction between the people and the public administration. The "fourth republican power" (the press and, today, other types of *mass media*) has been progressively replaced by privately managed social networks, in the task of forming public opinion. However, the interaction between the people and the state should not be based on private companies, because of commercial interests, vulnerability to private control, and manipulation of public opinion, while the democratic state should have different strategic goals, according to the directions of sustainable human development. We propose the creation of a digital public network, the "Popular Online Forum" (already named "FOPO" in previous publications), which can be integrated, with other useful functions for the citizens, in a smartphone application we call "All In a Nutshell" (AIN). This initiative has the purpose of inserting a participatory system to improve representative democracy, in the sense of *providing a two-way communication channel* between the government and the people. In Brazil, the FOPO can indeed be carried out using the capacity of the government to promote technological development, combining the people's "collective wisdom" – collected by means of a pool survey computational software, implemented in a public network of computers – with an artificial intelligence software – used to access public databases, checking the veracity and consistency of informational contents. For the end users, this public service can be offered in the AIN application, which would unite the will and needs of the population with the mechanisms for promoting and generating wealth. We're talking about a collective effort as big as building the Great Wall of China or the Apollo Program that took man to the moon. This type of initiative can inspire and push several generations to take advantage of all the cognitive, working, and dreaming potential to catalyze the development of a country like Brazil.

J. O. Pompeu
Federal University of Rio de Janeiro, Rio de Janeiro, Brazil
e-mail: pompeuesilva@nce.ufrj.br

J. L. Azpiazu
Fercomaz Ferragens e Complementos, Guaratinguetá, Brazil
e-mail: jose@fercomaz.com

A. Pereira Jr (✉)
State University of São Paulo, Botucatu and Marília, SP, Brazil
e-mail: alfredo.pereira@unesp.br

A. Pereira Jr, F. Sousa (eds.), *Principles for Governance*, Sustainable Development Goals Series,
https://doi.org/10.1007/978-3-031-40025-4_3

1 Introduction

The need for participation in social development is a constant in human history, and the freedom of the people in any given society can be related to the degree of participation in the development of the social fabric. In this mode of interaction, the individual enriches the group, and vice-versa, composing the basics of a democratic system. These conditions change and grow with the development of new tools, being present in the 'welfare state' that most modern societies have achieved. Going further, participation in public decision processes requires an update of the tools currently available for social networking, to achieve the ultimate goal of improving the people's life conditions. This is not simple, since the decision holders usually do not take fairly the transfer of their power to the people.

With the advent of the Big Data analytics, there is a good tool for understanding more the will of the population than before. Historically, democracies worked as a representation system, where the elected had to make decisions in the name of the people, because of the lack of direct and continuous communication of the people's will; however, the technologies of today permit to condense in various ways the discourse of a population, and this can be used as a new tool for democracy (Pereira Jr. et al., 2014).

The action of governing intrinsically brings with it acts of valuation, which can be for or against the preservation and promotion of life – both the life of the people and the other species that inhabit the ecosystem in which human society is inserted. Within the scope of autocratic governments, and even in limited representative democracy, we note the absence of dynamic and continuous forms of monitoring of rulers by the people. In this context, destructive actions against human life and the biodiversity of the ecosystem can be carried out indefinitely, which potentiates harmful consequences for life. This problem has become acute in the current health and environmental crisis, making it urgent to propose institutional measures that can reduce the distance between the people and their political representatives.

We propose in this book the establishment of Bioethical principles of Governance (BG), and a mechanism for mediation between the people and representatives, enabling, via information technology and Internet communication, the control of the actions of the rulers according to the established principles. We take medical bioethics as an example, which established parameters for evaluating professional conduct, in the face of vital questions, that would avoid harmful practices such as those practiced during the Nazi regime, in Europe, and in the Tuskegee case, in the USA (Hossne, 2006). The professional class council, based on these principles, can establish sanctions for health professionals who violate the principles.

In addition to outlining and discussing the philosophical foundations of Bioethics of Governance, we also propose to investigate the principles and good practices that best characterize this proposal in the macropolitical dimension, focusing on the formation of an interactive instance between rulers and people that enables the implementation of the BG as a state policy, to be obeyed by the rulers in the Three Republican Powers. This instance was previously titled Official Popular Online Forum (in Portuguese, abbreviated FOPO; Pereira Jr. et al., 2014), an instance that becomes possible from the use of information technology and communication via social networks on the Internet.

The FOPO we propose here, an updated version of the original FOPO proposal (Pereira Jr. et al., 2014) that includes recent developments in Artificial Intelligence (AI), is an instrument of participatory democracy in the digital age, complementary to the political mechanisms of representative democracy, without replacing the latter. More precisely, the FOPO, using trustable information gathered by large public databases and the various Observatories implemented by scientific researchers in the last decade, would be a mechanism for monitoring information, in both directions (from the rulers to the people, and from the people to the rulers), but *not* a decision-making instance in which AI *substitutes* human natural intelligence. The ultimate goal of the FOPO would instead be *using AI to make human collective intelligence more effective* at the political, economic and cultural dimensions of social life, avoiding the negative effects of the dissemination of "Fake News" in social networks.

The decision-making power, at least initially, would continue to be the prerogative of the Three Republican Powers, Executive, Legislative and Judiciary, each one within the respective Constitutional precepts. In other words, we aim to extend representative democracy into a more participative modality of democracy, by means of the use of information technology guided by human collective intelligence. This goal requires an institutional apparatus, to be created from existing plataforms; e.g., in Brazil, the *Gov.br* system, in Italy, the *IO* app (https://io.italia.it/), and the pioneer *E-Gov* in Estonia (see Astok, 2017).

2 The Opportunity of the FOPO

In the contemporary socio-political-ideological context, in which destructive conflicts that accentuate the crisis of the civilizing process predominate, alongside the exacerbation of forms of exploitation and oppression intrinsic to capitalism (such as neo-colonialism), it becomes important to formulate "positive sum games", in which the main social forces with the potential to generate human development and restabilization of ecosystems can establish synergies in theory and practice, composing a collective wisdom that is "smarter" than individualistic approaches (Surowiecki, 2004). This direction inspires the present proposal, in the sense that the computerized society should use technological resources in the public sphere in order to disseminate truthful information and collect feedback from the people regarding the actions of the rulers, in the three branches of government, and at the three national levels (federal, state, and municipal), so that political-economic interactions can compose a "positive sum game".

The "public sphere", a concept presented by Jurgen Habermas, refers to "the sphere of private people who come together as a public to regulate against the public authorities by engaging in a reasoned and critical debate (…) about politics and the state" (Singh, 2012). In the last decade, the public sphere was expanded with the dissemination of new ways of communication, such as the 'Smartphone', broadband Internet, YouTube and social networking applications (WhatsApp, TicToc, Tweet, Instagram, etc.).

As a consequence of this transformation, moving from the traditional unidirectional *mass media* to a free interaction of everybody with everybody, and vulnerable to the actions of automated mechanisms ("bots") used for the implementation of political and ideological massification, the (digital) public sphere has become an ideological battleground, in which true information is disseminated alongside with the so-called "Fake News". This situation constitutes a "negative sum game", in which information and communication technology resources are wasted in destructive conflicts. To overcome this situation and promote positive-sum games, there is a need for measures that favor not only the dissemination of truthful information about government officials and their actions, but also that allow for expanded participation in politics, through mechanisms that enable the dissemination of verified information and evaluative popular feedback on the actions of governments.

In order to constructively build an instance of interaction between government officials and the people, it is initially necessary to establish consensual or majority principles on the intended bioethical conduct of government officials, enabling their evaluation by the people. Besides this first task, we should take into consideration that participation by the people cannot be face-to-face, due to the size of the populations; the use of information and communication technology is essential, as long as it is reliable for governments and people, having the seal of approval of the federal state for its effective implementation, and using official databases. User authentication, in order to be reliable, requires the use of official data, shared with existing databases (in Brazil, the citizen registers, database of the taxation and revenue system, database of social programs as the *Bolsa Familia*, etc.). The site *Gov.br*, in

Brazil, has already been developed to allow communication of the government and federal systems with the people, *but was not formatted to allow efficient and efficacious communication of people with the government.*

The central function of the FOPO we propose here is based on the potential of crowd or collective intelligence that can facilitate very complex tasks such as writing an encyclopedia – that Wikipedia managed to do successfully, and the involvement of thousands of people, e.g., the Amazon Mechanical Turk (Mturk.com) that distributes microtasks to a crowd to solve and return to the companies that place the tasks. A government can use the collective intelligence of the citizens to receive micronews, or requests such as fighting a fire in an area of the Amazon Rainforest, or the need for agricultural machines to help family farmers in the Brazilian hinterland.

The interaction between people and the computer is an area of Human-Computer Interaction (HCI) that involves crowd computing technologies to obtain feedback from crowds. In a broader term we use the term *crowdsourcing* to leverage the perceptual, cognitive, or enactive skills of many people (crowd) to achieve a well-defined outcome such as solving a problem, classifying a dataset or producing a decision (Schneider, 2015).

The creation of the FOPO also involves a central issue of distributing state-issued money and public opportunities for social, economic and human development to the sectors of the population that need more support from the federal government in Brazil. One of the existing initiatives is *Bolsa Família*, a federal program of cash transfer that integrates benefits of social assistance, health, education, and employment, aimed at families in poverty. The program also offers tools for the socioeconomic emancipation of the family in a situation of social vulnerability. However, the integration of this transfer with educational, social and industrial entrepreneurship initiatives is still far from happening in Brazil.

The FOPO could also be a way of integrating the university education programs such as *Prouni* (a program that offers scholarships, full and partial, in private higher education institutions), and the democratization of access to universities promoted by the *ENEM* (a national exam that evaluates the performance of students who have completed high school). Other subsystems could be created to promote innovation and creation of new companies by layers of the population that did not accumulate capital in previous family generations and thus new companies could be born that would create sustainable businesses and linked to the needs of this broad layer of the Brazilian population. The economic issues, including the democratization of the distribution of currency for wealth generation by social layers with a history of social exclusion, are the biggest challenge of the FOPO, which can be a channel for receiving demands from the population and promoting the reduction of economic inequality. It has to be a two-way street, in which the population that needs economic support would have access to the outcomes of the state's money generation mechanisms, creating sustainable growth for decades, by democratizing wealth creation now for the part of the population that has always been excluded.

To create the FOPO, it will be important to use new information technologies such as AI and make connections between the most varied systems that already exist in the Brazilian State. The FOPO Platform can connect with courses offered by SEBRAE (Brazilian Service of Support for Micro and Small Enterprises), an autonomous entity of the Brazilian *Sistema S* that has existed with this name since 1990, previously created with the name of Brazilian Center for Support to Small and Medium Enterprises (*CEBRAE*) in 1972.

The FOPO can become a portal that connects the citizens to services that support entrepreneurship and innovation, composing a smartphone App that we call the "All in a Nutshell" (AIN) App, to be made available to the whole population. Using the same logic of *Bolsa Família*, which is operated by public banks such as Caixa Econômica Federal, the AIN App could be connected with this form of income distribution to increase the wealth and complexity of the Brazilian economy, providing the citizen, who wants to undertake and innovate, *the necessary access to money, training and knowledge* with the help of Brazilian Public Universities, strengthening new industries and enterprises in general that can sustainably and bioethically expand Brazil's development. The Information Systems of the Brazilian Government already charge the duties such as income tax from the citizens, now the expansion of access to rights can be the next challenge. The FOPO, made available with the AIN App, can unite the most basic needs of the citizen such as the right to food, guaranteed by the *Bolsa Família*, with the right to create new companies that increase the complexity and economic wealth of Brazil.

The AI behind this system must be used to satisfy the citizen's needs, opening space to register the needs of the most diverse sectors of the population, promoting the democratic participation of all Brazilians, especially those who need more support from the state to get out of the poverty zone and to be able to have capital for enterprises that increase national wealth, creating new companies based on the real needs of the Brazilian population. An economic algorithm that increases the liquidity of economic transactions all over the country, on the basis of state-issued and state-controlled digital currencies (such as the Central Bank Digital Currency, discussed in another chapter of this book), can favor the economic inclusion of the people in general and expand economic democracy throughout Brazil. This change, in turn, can be the solution to the current fiscal austerity, showing that among the popular layers the circulation of money can have multiplicative effects, producing for instance more than 300% of wealth from the nominal value of the issued digital currency. Working in areas such as sustainable and healthy food production, a virtuous chain of production, distribution, and food processing can be created that improves the quality of life and even the life expectancy of Brazilians. Other areas of biotechnology, electric mobility, and clean energy production can create countless national champions that would pull several other companies to have an economy with sustainable growth for many years.

Using the AIN App, citizens could get all types of relevant information from the public administration, and, conversely, the people would have a channel to report relevant facts, to make complaints and submit projects for public financial support. Once integrated in this App, the FOPO would attract the participation of a large number of people, as it would be linked to economic and social programs of central importance to these people, such as Bolsa Família, Aids, Housing and Business Financing, Federal Revenue Service, Transit Authorities, etc. With the help of social assistance public services, and the economic support already given by the *Bolsa Família* program, this App can be made available to the whole population. It is important to stress that such an information network should be two-way, enabling both the rulers to induce proper individual actions and the people to control the improper actions of the rulers, both sides obeying the legal constraints of the Constitution and BG principles that take into consideration the rights of the other.

3 Effects of Big Tech Monopolies on Democracy

Social interaction in a participative democratic, political, and economical system cannot be based only on privately owned social networks, because the goals of these companies are different from the goals of the democratic process. Contemporary society built democracy on four branches: Executive, Legislative, Judiciary, and the Media (initially, the Press; later, the Mass Media, and, more recently, Social Networks). The Press has been regulated and controlled, through public concessions and laws. States have state-owned Mass Media for transparency. The financing mechanisms of Mass Media were always vast, variating from different agents, including propaganda ads and printed editions. However, traditional media sources have been losing funding, due to the competition of the digital media. The imbalance of power has caused the erosion of the democratic fabric, in modern societies, exemplified by recent political actions using "Fake News", in some of the world's largest democracies.

Social media, driven by digital information technologies, have dominated communication in the twenty-first century. Initially, it had a positive effect in societies repressed by dictatorships (as in the Arab Spring in 2010; see Associated Foreign Press, 2011), but with its monetization it became a tool for market interests, and with the normal profit-driven accumulation of capital the companies responsible for social media have monopolized and controlled their new markets, replacing traditional media markets. Monetization has become a tool used to generate profits and accumulate capital, increasing economic inequality in the whole planet (Pilkington, 2016, p. 54).

The companies responsible for social media have monopolized and controlled their new markets, replacing traditional media markets. The financial resources used to manipulate social media users moved to targeted advertising, in which marketing effects are enhanced by knowing individual user values, obtained with the browsing of their histories. The use of 'bots' or non-human devices enhances in the human users the (false) perception of community and acceptance, creating an artificial bias of confirmation, amplifying fake narratives and perceptions among the population. The forced use of multiple versions of the same fact in modern asymmetric warfare creates societal conformity in the absence of truth in historical discourse.

Although this revolution in communication is recent, humans have shown resilience in channeling new technologies for the common good. The state must regulate free competition for free competition to exist (Galbraith, 1967). The diversity present in free competition increases the meta-equilibrium of the economic market ecosystem. Thus, monopolies in which Big Tech companies influence the thinking and behavior of users are questioned. There is already a movement in Europe to regulate the Internet, which arrived in Brazil in 2023. To achieve efficient regulation, the state must take the first step in breaking these monopolies, providing a quality, accessible, and free service for the common good, integrating the state with the citizens for governance and strengthening democracy.

Currently, all types of media focus exclusively on financial benefits achieved through the sale of advertising space and browsing history earned by their engagement. This is currently done by stimulating basic and primitive instincts, the most irresistible among the core needs of humans. This mechanism generates conflict among users, benefiting social media owners through engagement and increasing the consumption of their product. Concerningly, this mechanism also spreads destructive behaviors, which are blindly stimulated by algorithms seeking more interactions to sell advertised products. The spreading of behaviors not allowed by the penal code is also stimulated by this mechanism, where the algorithms, when seeking these inevitable interactions, provoke and foment illegal attitudes. This well-known problem makes, for example, the structural racism present in society to materialize even more strongly on social networks. The fact that social networks are protected by copyright owned by private companies makes the regulation of these responses difficult, and the reaction by the users may not be always efficacious to banish the illegal content. In current conditions even an open-source system would have difficulties in being regulated, due to the fact that the algorithms used in the system nowadays can no longer really be understood by humans, being considered 'black boxes' for practical purposes.

The only real way to control this response is understanding the results of the algorithms by mapping patterns, and policing the generators of the illegal content. The argument contrary to the public control of social networks is that this is ultimately a restriction of democratic free speech. This may be a valid argument, since the narrow AI algorithms of today, as said, are not comprehensible by humans; we can only police the result of these algorithms. Public regulation and control can be helpful, but would be too extensive to be applied in advance, and to avoid injustice it would be necessary to interact in controlled situations as a way to prove the induction of harmful behaviors. Pretending that humans would be able to completely adjust the contemporary algorithms of narrow AI is unrealistic. Thus this algorithm will need to be open source and even constantly modified, in order to suit and protect the evolving and dynamic societies.

The practical end of the lack of control of social networks is a clear segmentation of society into urban tribes, which are not well stratified by questions of social and financial classes as it was in the past, but rather a segmentation originated in the very conflict promoted by these networks, where, in fact, tribes almost define themselves more by their enemies and their oppositions than by their true interests. We can exemplify this situation with current polarizations in Brazil, in which the left-right wing political battle is not defined by proper Fascist and Communist political ideologies, but by using these terms to define others. Being "communist" is not a true political standpoint, but a nominal pole in opposition to being "fascist", and conversely being "fascist" applies to everyone that is likely not to be a "communist".

The format of social media moves towards an integration of profiles and direct communication, like chats and group chats, in apps like Telegram, WhatsApp, and WeChat. The profiles in this media are allowed to operate with cash flows and payments, but as a media product, the useful financial functions go together with narrative interactions of small stories called "Reels", which, in addition to being important for the subjectivity of the persons behind the profiles, is fundamental for the perception of how to integrate the different virtual communities to form a common goal. This can be established, in interactions within the media itself, with an influence unconsciously driven by the algorithms that select contents and connections in the computational systems used for this type of public communication. The social network sells leisure, and the product of consumption is digital content, which can appear in various formats, but always represents a story that must have a place and a context of representation, and aspires to provoke engagement to become a 'meme' and reproduce itself digitally.

In this situation, the democratic system requires the possibility of the users editing some parameters consciously, to increase the people's control of their social experience, and facilitating the communication with the government for the society to achieve strategic goals. The emotional situation of a person is very variable, and these states of mind influence the desires and corresponding motivated behavior of the individual, so we must be able to edit our profile in the social networks according to this mood or disposition, if the objective is to faithfully achieve our intended goals. Departing from the current state of affairs in social communication, the proposal of the FOPO would be to create a portal where a citizen can create a user profile where he/she considers his/her preferences to be represented, and actually publish content, initially in the format of text entries, and possibly evolving to the Reels format. On the basis of the entries, the system uses AI to generate public reports, and feedback to the persons behind the profiles. For instance, a musician who expresses a message that resonates with the collective desires of a community could be suggested by the system to be hired to an event of the same community. Thus, increasing the efficiency of social communication is not only a question of public self-organization or autonomy, but it also affords the effectiveness of execution of social actions, both by a "crowd-wisdom effect" (Surowiecki, 2004).

4 The Institutional Space of the FOPO

Brazil already has a national registry of individuals, included in the *Gov.br* portal, where each citizen can have a profile. In this portal, the government already provides services to citizens, including access to different types of documents and information for the registered profiles. Allowing citizens to publish text messages and Reels, and interact politically, would be a computationally easy task, which requires establishing an open-source interaction algorithm, in order to select information from society and afford social integration activities.

As there are legal instances of discussion and decision in a complex society, the proposition is that only strategic features for the people's quality of life and human development should be taken into consideration in the FOPO. There are issues of vital importance to be technically discussed, regarding the safety of the algorithm, the secrecy of the individuals' input on the system, and individual freedom; there must also be some kind of rule to constrain possible unconstitutional proposals. Still, for practical purposes, it could also be a powerful security system using standard procedures.

As a basis for achieving the degree of reliability necessary for the proper functioning of this intermediary instance between rulers and the people, the performance of Observatories built on the basis of scientific research is necessary (see, for example, the *Fiocruz* Observatory at: http://observatorio.fiocruz.br/). However, currently the results posted in these Observatories are restricted to a few users. In the mechanism to be implemented, it is necessary to connect the Observatories with the FOPO, making reliable information accessible to the majority of the people.

Electronic Government is the term used to designate the set of services offered by the government and government agencies to citizens, intergovernment relations, and commercial operations provided by the Internet. Brazil has a history of using information technology and the Internet that began in the last century. In 1995, the Revenue Service in Brazil created its website and, in 1997, it began to receive Brazilian income tax returns via the Internet. It was followed by countless other sectors of the federal government, and the creation of sites and services on the Internet by state and municipal governments also began. In 2000, the Executive and Policy Committee for *E-Gov* was created. *E-Gov* is the name of the set of services offered by the government and government agencies to citizens, intergovernment relations, and commercial operations provided by means of the Internet, and also the electronic address of the site (see *E-Gov*, 2023) that aggregates the services.

The Electronic Government Department was created in 2004, in charge of coordinating and articulating the implementation of unified and integrated actions of electronic government, the provision of public services by electronic means, in addition to standardizing and disseminating the development of government actions and information. Accessibility for people with disabilities was foreseen and started to be implemented in 2005. In 2011, the Electronic Government Portal was implemented with the aim of promoting society's participation and making Electronic Government actions more transparent. The great challenge for the future will be to expand this participation.

In 2018, the federal government publishes the *Brazilian Strategy for Digital Transformation* (E-Digital) document, which outlines future actions in the *E-Gov* area in Brazil, to improve a digital platform for social participation as a privileged space for dialogue between the federal public administration and civil society. In 2019, the government launched the *E-Gov* portal, which unified most of the Brazilian federal government's online services. The Brazilian federal government was successful in creating a platform where government services can be carried out online and easily accessible via the Internet for all Brazilian citizens, but it has not been able to advance digitally in popular participation in the political and economic decision-making processes, and access to public resources for better quality of life and human development.

Since the implementation of the Federal Constitution of 1988, Brazil has had a tradition of holding national conferences that start at the municipal level, move on to the state level and even bring together delegates in Brasília to create guidelines for public policies in the most different areas of the federal government. These conferences have a high cost and only happen every 4 years – if governments encourage them to happen. There are several isolated initiatives such as the *Participa + Brasil Platform*, which acts in the dissemination of consultations and public hearings, research, and the promotion of good practices. However, there is no functional analogue of these Conferences in the digital world of the Internet, as part of the *E-Gov* Portal or in another information technology system. As a consequence, during the pandemics with social isolation, and/or under an autocratic government, popular participation in the democratic process is compromised.

5 The 'Modus Operandi' of the FOPO

Here we present some provisory ideas on the functioning of the FOPO and the steps necessary for the implementation of the computational system for the Brazilian democracy. The basic concept of the FOPO is the following:

1. The entries in the system would be simple text messages. Each person with a Federal State Citizen Register Number (in Brazil, the *Cadastro de Pessoas Físicas*, abbreviated *CPF*, composed of a 11 digits number) would be entitled to post once a day, using a cell phone or computer, by entering a message that fits with a keyword that corresponds to a principle of the Bioethics of Governance (see Chap. 1 of this book);
2. A purpose-built pool survey software would then categorize the responses and respective predicates people use, to build a succinct weekly report;
3. In the process of building the report, the system converts the entries from natural to symbolic language, and accesses the federal databases and the scientific observatories, operating on reliable information to make a synthesis of the popular manifestations;
4. The weekly report is to be sent to all representatives in the three republican powers, as a feedback from the people to guide the actions of the representatives. The reports can be useful for state managers and politicians as a feedback from the people, and for the people in the case of submitting projects for public funding.

The AI mixture that uses symbolic computation can be complemented by generative AI Engines such as *ChatGPT* or another. Engines that use symbolic databases can make the system's response to the request at the data entry prompt very precise and allow a mix of natural and computational language (Wolfram, 2023). In the beginning, the system would be a prompt for request entries in which all Brazilians above a certain age (to be defined) can participate. A simple example would be when a citizen reports that there is a hole in the asphalt in front of his house; the system would return that his request would be at the municipal level; that it was forwarded to his municipality and could be followed at the link sent to her email or cell phone number.

Brazil is a federative country and has tripartite responsibilities between municipalities, states, and the federation. In the above example, information about the hole would go into a database managed by symbolic computing that would be extremely accurate. It would generate a request for the street camera or an employee to check the hole and confirming it could generate a service order to repair the asphalt layer on the street. Another notification from a citizen could be about their difficulty in finding jobs in their city of domicile; this information would be distributed to the municipal, state, and federal levels and if there were many requests in this sense the federal symbolic database and other sources of information would confirm that this is a problem in that city or region. The confirmation can trigger incentive programs for companies that generate jobs for that profile of people, who have their right to employment and income at risk. The system can unite economic complexity indexes to generate a virtuous cycle that creates wealth and well-being in the country. Decision-making by politicians elected as representatives of the people would be much more accurate and closer to Brazilian citizens with the use of the FOPO.

The FOPO system receives information inputs posted by citizens who would have the right to a daily entry reporting some abnormality or suggestion to the federal government. The posting would have a line of text with a character limit and geolocation to know where this information happens and to be able to have a real-time Forum Map of the citizens' perception of the country. These posts can be classified as linked to themes such as the Bioethics of Governance - that works with large public issues such as environment, hunger, income distribution, and others. We can choose another theoretical matrix such as the UNESCO Declaration of Bioethics and Human

Rights (DBHR) that lists 12 fundamental principles that can be the keywords linked to the requests posted by citizens. These fundamental principles (explained in Chap. 1 of this book) are:

1. Human Dignity and Quality of Life
2. Peace and Personal Safety, including emergency warning for catastrophes
3. Availability of Employment and Income
4. Non-Discrimination
5. Non-Pollution
6. Access to Water
7. Right to Education
8. Right to Health
9. Availability of Energy for Domestic Purposes
10. Mobility for People and Commerce
11. Preservation of Biodiversity of Biomes and Stability of Ecosystems,
12. Respect and Support for Practices (Social Technologies) and Knowledge of Communities

Several other classifications could facilitate the indexing of these data into themes and sub-themes that intertwine to construct a living Map of the needs and aspirations of the Brazilian population. Thus, one could in real time see issues such as the issue of clean and sustainable energy and in which regions this theme appears in Brazil. The government could launch real-time surveys on important topics and receive feedback from a significant portion of the Brazilian population, expanding the power of census systems such as the *Instituto Brasileiro de Geografia e Estatística* (IBGE), which could be connected to FOPO to guide the questions to be asked. Participation in FOPO would be reinforced with distinctions for the participatory people who spend their time helping to build FOPO in Brazil, such as support for their submitted projects. This would be a way in which the citizen would inform their demands to the Government, expanding participatory democracy and collective intelligence to solve Brazilian problems and promote better choices that benefit everyone. The other way would be the Government organizing its social programs and income distribution to reach the population that needs it most.

The FOPO proposal is yet another functionality for the Brazilian *E-Gov* that would organize the existing Conferences in person, and expand to new possibilities for democratic popular participation, introducing functionalities in which citizens can report their needs and also good government practices, and use new technologies of Large Language Models (LLMs), the most popular of which is the *ChatGPT* Open AI, to make a synthesis and validation of the information entered by the citizens.

6 Implementation of the FOPO

You readers must be wondering how the FOPO can be implemented, what would be the structure of this system, and who would build it. In order to have these answers, we need to think about collective intelligence and the great undertakings already carried out on our planet. The AI Revolution is even more powerful than the Industrial Revolution, which with its steam and later electric machines changed the economy and the way of life on our planet. We can go further in history and find the collective effort to build the Chin Dynasty more than 200 years before Christ and which lasted almost 2000 years and was essential for China's security and brings reflections to the present day. A collective effort using all the capacity of the Brazilian people to develop complex solutions that use the nascent AI industry can be the driving force of development for the next 2000 years in Brazil.

In order to build the FOPO, we first need to aggregate in Brazilian universities the young people who are willing to unite their natural intelligence in this new venture, creating many new companies based on AI technologies, which will be the basis of the FOPO. The FOPO has a great challenge of uniting the Symbolic Computing that coordinates existing databases and information and that will be created about Brazil with the Broad Language Models and natural language processing that use neural networks and statistical parameters to process in parallel these gigantic databases and bring answers, reports, and maps in real time of the needs of the country and citizens.

This democratic and humanitarian undertaking can indeed be carried out using the capacity of the Brazilian government to promote technological development, which here would have the clear intention of creating this device that would unite the will and needs of the population with the mechanisms for promoting and generating wealth that the Brazilian State has. We're talking about a challenge as big as building the Great Wall of China or the Apollo program that took man to the moon. It is known that this type of initiative can inspire and push generations to take advantage of all the cognitive, work, and dream potential to catalyze the development of a country like Brazil.

We should detail the implementation of the FOPO in Executive steps. We ask the reader to follow us and understand that each step will only be possible using different areas of human and economic development in our country.

The first step is educational; we will need to understand the technology of AI and introduce in our schools and universities, from an early age, the necessary training that enables the human development of the potential for work in AI for the next generations. Our minds use neural networks that have been developed over billions of years of the development of life on our planet and AI technologies mimic artificial neural networks that connect with statistical parameters that simulate how our brain solves problems. Brazil has many problems to be solved, including hunger, the increasing complexity of the economy, the right to education, and health, among hundreds of other Brazilian problems. The AI revolution will change the field of jobs and the future of new generations of Brazilians, so why not bet on an industry that uses AI as a driving force for national development? For this, we will need to unite universities and basic education, taking the knowledge of university students who study AI to young people in schools across the country. With the help of this cognitive technology, we can develop the skills and competences of future professionals who will be the group of workers to develop the FOPO.

Implementing the first educational step is not that difficult, because using modern communication technologies we can connect basic schools with universities, and we don't need more than one weekly class to bring skills and competences of basic AI usage (such as *ChatGPT* queries) to children and young people throughout Brazil. Then we would move on to the second step, which is to diagnose among our young university students a group of girls and boys who can create companies that would deliver the necessary solutions for the development of the FOPO. It is very important to know that the nascent AI industry is interdisciplinary and diversified in terms of employment opportunities: we will need the most diverse university areas such as Computer Science, Mathematical and Natural Sciences, but also the areas of Humanities and Arts as well as all diversity of human culture to create the parameters that structure the basis of AI behind the FOPO.

ChatGPT, which is currently the state-of-the-art example in the development of Broad Language Models that processes natural language and gives responses in real time, has more than 100 billion statistical parameters based on the most diverse areas of human knowledge in its architecture version 3.5 and walks to trillions of parameters and even bigger numbers in its future versions. The development of this intelligence began in 2016 and employed many specialists from the most diverse areas. To implement in Brazil, we can start with the open-source model of version 3.5 of ChatGPT and implement parameters linked to the reality and knowledge of Brazil and for that we would need both young university students and specialists who know Brazilian problems.

The third step will be to create a mechanism to promote the AI industry in Brazil using the BNDES (National Bank for Economic and Social Development), FINEP (Financier of Studies and Projects), and Public Banks to promote this nascent sector of the Brazilian economy who would be paid to deliver the solutions for creating the FOPO. See that in addition to the development of the FOPO, this promotion can create wealth in the most diverse areas of Brazil. Imagine that one of the companies that will develop the integration between symbolic computation and LLM in the AIs of the FOPO is young university students from the Brazilian hinterland and they use the same solution to solve how to increase the productivity of the Brazilian *Caatinga* and preserve the biome of *Caatinga* at the same time. The development of the AI industry can be a virtuous cycle for the entire Brazilian economy.

With these three steps, we will have the basis for the development of the FOPO, and will need to overcome some bottlenecks in the development of technologies and conceptual issues related to the AI industry. The most important of these bottlenecks is the union between symbolic computation and LLM, both using parallel processing in artificial neural networks. Only in the year 2023 did we see the emergence of commercial applications that deal with this problem, as was the case with the plugin that united the symbolic computation used in the *Wolfram Alpha System* and *Wolfram Language*. They deal with the mathematical language and relational databases consolidated with the architecture of the ChatGPT, processing natural language based on statistical parameters fed by neural networks that deliver the best response; in the early development of these new AIs, the responses are imprecise. The union of symbolic computation will make these AIs much more accurate and precise. When a citizen places the location of the hole in front of his house, the AI of the FOPO will consult the symbolic computation and store this information in a database that will be able to use ways to verify the information later or in real time and will communicate with the citizen in natural language, making the interface very easy to use.

Brazil has a history that dates back to the 1990s with the development of information technologies linked to the Internet and open-source software, and this path was to support national information technology companies that delivered orders placed by the government. We can follow the same path for creating the FOPO, by creating an interministerial working group that would manage the creation and implementation of the FOPO at the federal level and suggest to states and municipalities the integration of this technology at the national level.

7 An Example of Possible FOPO Support to Popular Cooperatives

The FOPO is a two-way communication channel, that is, basically, it transmits information about public administration to the people and the people provide feedback on how they evaluate the government's action. This would be done in the system that uses information technology for the benefit of democracy, let's say, not to manipulate people, but to make the collective wisdom effective in politics and economy. It is possible to use more sophisticated AI to organize the flow of information, and generate reports that can be used for the democratic process. The final objective would be, in the economic-financial plan, to subsidize the government in its primary function of the state, which is to finance human development, not necessarily creating state-owned companies, but affording the people to create new enterprises with social and environmental responsibility, financing health, education, and a series of activities that promote human development.

One of the main aspects of our planet's environmental crisis, which manifests itself in climate change, is the change in the stability of ecosystems, caused by the reduction of biodiversity. To face this process, it is necessary to attribute economic value to the native species of ecosystems, so that people start to interact constructively with the species that make up the biodiversity of the biomes that they inhabit and have fed them for many decades. With the initiatives that we propose here, instead of despising these species because they are not recognized, certified, and catalogued, people may come to recognize their value for human consumption, replanting them and thus guaranteeing the continuity of the region's biodiversity in which they live. This is an ecological technology model that can be extended to all regions of our country, or other regions of the planet that offer natural resources to be recognized by their population.

Ecosystem technology is, therefore, an innovative technology for the simple fact that it uses resources obtained from the biodiversity of biomes, unknown or neglected by the country's production system and consumer market. Let us take as an example the fruits Umbu, Cajá-Umbu, and Maracajá from the Caatinga, from which the pulp can be processed, making flour from the seeds, or even extracting essential oil with multiple applications of commercial interest. Therefore, the focus of ecosystem technology is on the ecosystem resource to be transformed, which demands new technological processes that are adequate to its intrinsic characteristics and also to the product to be developed, which, in turn, must be in accordance with the society's demand for healthy food, produced with environmentally friendly and socially inclusive methods: "Local communities have a culture that already has valuable knowledge and practices for environmentally sustainable enterprises. For this reason, it is vitally important to finance transformative projects, not only to bring scientists and technologists closer to local populations, but also so that, after this contact, there are resources to implement production and consumption systems that would generate the dynamism of local economies. Without such dynamism, the concept of sustainability becomes an empty abstraction" (Pereira Jr., 2019).

The *Caatinga* is an ecological region that lies in the semi-arid hinterland of northeastern Brazil and covers 912,529 km^2. It is a seasonally dry tropical forest that presents heterogeneous subregions, with a total population of around 30 million people. A significant shift in the semi-arid *Caatinga* region towards sustainable development requires not only persistence and creativity, but mostly the consistent financial and political support from public administration to establish a robust connection between the improvement of human livelihoods and the conservation of natural landscapes.

One of the pioneering initiatives in this direction has been the cooperative *Ser do Sertão* which, among other initiatives, has developed technological adaptations to generate food for school-age children, using native fruits (Umbu and Passion Fruit), producing not only pulp but also taking advantage of pulp production residues to generate other types of food, which can contribute to the eradication of hunger and provide the necessary nutritional ingredients for human health.

To face the challenges of access to education, health, food security, and access to water, public policies and joint solutions were needed. It is in this sense that in 1996 the *Pintadas Network* (*Rede Pintadas*) was born. Currently formed by a collective of 15 organizations, the network emerged as a forum for civil society entities and collectives and was formalized in 2003, with its registration in the government office. It acts in an articulated manner at the local and territorial level, in the elaboration and execution of programs and projects to empower communities, especially women. *Rede Pintadas* has associated with other regional, national, and international networks to enhance their plans and projects focused on sustainable local development and also on the issue of climate change.

Along this trajectory, the *Rede Pintadas* organizations have already implemented a set of actions in partnership with national and international entities from different countries, highlighting the achievement of 100% of cisterns in the rural area, the first municipality in Brazil to reach the mark; management of the *Public Center for Solidarity Economy*, which has already served more than 200 productive groups in the region; the *Sicoob Sertão* credit union; the *Delicias do Sertão* Restaurant and Bakery managed by the Women's Association; the fruit pulp agroindustry of the cooperative *Ser do Sertão*; the Cooperative of Goat and Sheep Producers – *Frigbahia*; and the Solid Waste Recycling Cooperative *Reação*. These are work and income generation initiatives that create and promote economic opportunities, with community engagement and participation.

These initiatives should receive financial support from the federal state to improve the production methods and distribution of products to consumers, but it does not fit in the format required by public banks (Banco do Brasil – BB, Banco do Desenvolvimento Econômico e Social – BNDES) to receive special credits. They may be able to use the FOPO to submit projects and receive funding in a non-bureaucratic way. As in this case, thousands of other popular organizations in the country could benefit from the channel of communication between rulers and people provided by the FOPO.

8 Towards an "All in a Nutshell" Custom Interface App

In popular language, the expression "All in a nutshell" means a summary of an extensive matter into a synthetic statement. The 1949 Disney movie "All in a Nutshell" presents a plot in which Donald Duck operates a machine that extracts nuts from their shell and crushes them to produce butter. Could a public government create a machine (a smartphone app, in this case) to condense all the relations with the citizens in a single interface? This would be the final solution for the challenge of putting state-issued money in the people's pockets, democratically triggering the process of human development in the country; the empowered people would then control the conduct of the politicians, to ensure the respect for all life forms in the planet.

In the digital era, the most practical way of implementing a new social paradigm is using information technology. We propose a cell phone application that contains the necessary features for the citizens to have access to the state-issued money for survival and productive occupations, in an integrated governance system, which would be the way for us to promote better economic conditions for the people, and also a popular control of public finances to prevent or minimize corruption, and increase the efficiency of public management at the federal, state, and municipal levels.

It is a new idea that we are registering here, for an application that people would download on their cell phones and that would then enable them to participate in the co-management of the state at the three levels (city, state, and federation) and without overlapping with the three powers (Executive, Legislative, and Judiciary), by creating a new pathway for the people to be able to interact with the government and receive resources to survive and have the opportunity to be productive, submitting proposals for special credits to be offered by the federal state. It can be helpful for the people to solve the main problems of our country, because we have a lot of natural wealth and human wealth, but there is really a lack of effective public management and financing of social enterprises. A big problem limiting economic and human development of Brazil is the (in)action of the public administration. For this reason, the solution to our underdevelopment cannot come from the 'top-down' direction only; it also has to be a solution that enables 'bottom-up' popular self-organization, that enables each person to participate in their own way. This is only possible through information technology. So, that's why the concrete form that our proposal acquires is that of a cell phone application in which all the important functions of this new process would be present there, at the tip of every citizen's finger.

This would imply a public financial system parallel to the existing private banking system, to make the country's wealth available to all people, because the currently existing system, which operates under the capitalist mode of production and concentration of property, is a system of accumulating wealth in the hands of a few. It gets to the absurd point where we have big companies fighting to get the spare change that people still have; so we need a parallel system to make an opposite movement, to *reduce inequality*, since the current system leads to *increased inequality.* We need a system to reduce inequality and this could only be done by changing the way in which money is issued and distributed in the society in which case we would have in Brazil another currency – the Digital Real – parallel to the conventional Real, combined with social currencies and special credits, as discussed in another chapter of the book. So, the base of the social pyramid would be the greatest beneficiary of the effort to provide survival conditions and financing to productive enterprises of the cooperative type, in which people are productive, having the opportunity to work to generate the goods they need to survive, such as food, housing, clothing, transport and health infrastructure, education, and energy. All these necessary conditions for social life could benefit from public investments in popular enterprises, through this application.

So the AIN app is multifunctional, centered on an interactive page containing the four main functions:

1. *Basic Income Balance of the Citizen*: according to approved laws, every living Brazilian person would receive approximately a monthly minimum wage, from the day of birth until the day of death.
2. *Operating a Personal Account in a Public Bank.*
3. *The FOPO*, with Two Sections: *Receiving Information* and *Sending Information.* The App affords access to external data on transparency, where all municipal, state, and federal public expenditures would be registered. So, if the user clicks the Transparency button in the FOPO *Receive Information* section, she has access to her city, state, and federal budgets; she would have all the links in the FOPO. So we would know how currency is being issued, where it is going, and how it is being spent. This would make secret budgets and the corruption that exists linked to public funds unfeasible. Because here, then, any citizen could know what currency is being issued at the federal level and also what is being collected through taxes and fees at the municipal, state, and federal levels. Besides receiving information, citizens can, once a week, make a complaint, a proposal, or a suggestion, within allowed categories, by clicking on *Send Information.* These propositions would be inside a list of items, which would be in the folder; clicking there would take you to a page, with a list of items of subjects that can be commented on, and which propositions can be made, as exposed in a previous section of this chapter. So, for example, if a person experiences hunger, this would need to be recorded, because it is the responsibility of the governors that no Brazilian goes hungry. If there is, for example, open sewage, if there is deforestation, or illegal burning, then all these pertinent events should be reported, as well as propositions, comments, and complaints. Weekly, the FOPO system builds a report, which also appears in the *Receive Information* section.
4. *Submitting Proposals of Special Credits for Cooperative/Social Enterprises.* The release of funds for productive undertakings is conditioned to an approval by the Development Agency (DA). So here there would be a DA icon in the AIN App, and by clicking it, the citizen could submit a proposal to receive funding. So, for example, she wants to create a company to produce beans to meet a certain demand. The demand should be identified on the basis of the FOPO's reports, so the user forwards the proposal importing FOPO data to justify the financing. Weekly we would have reports of propositions that people make, compiled by AI, because it is a very large amount of data, it cannot be done manually. And then, based on these reports, proposals can be made here for the DA. Of course, it is difficult to do something alone, so a group of cooperative people would be formed and forwarded, through this application, to have a budget from the DA. If, for example, there is a neighborhood that does not have basic sanitation, the inhabitants can self-organize, and set up a company to carry out basic sanitation in that neighborhood. They can make a proposal by clicking on the DA icon, forwarding it directly to the DA; if it is approved, they will receive enough money to carry out that project, carry out that sanitation work in that neighborhood, by means of the same AIN App, in the section of *Operating a Personal Account in Public Bank.* It does not need to go through a municipal, state, or federal bureaucracy. This is not the provision approved by the Congress in the annual budget. This is a *parallel* budget, and not a *secret* budget; it is *a totally transparent budget*, because each citizen can monitor it, with the tips of her fingers, clicking on the AIN App on her cell phone.

9 Computational Technicalities of FOPO and AIN

The FOPO, departing from existing government informatics systems, to be made available to the people in the AIN App, is designed to be a platform that seeks to unify all government databases and services in Brazil, providing citizens with a simple interface to request services or provide information in natural language. The FOPO can be built on a new programming language called *SuperPhyton* (*SP*), which combines high-level symbolic computation with LLMs for natural language processing.

SP is designed to be easy to use and to provide high-level abstractions for data manipulation and symbolic computation. It includes a large library of pre-built functions for data manipulation, statistical analysis, and machine learning, as well as integration with popular scientific computing libraries such as *MXNet* for neural network training, *GMP* for high-precision numeric computation, and *LAPACK* for numeric linear algebra.

One of the key features of SP is its integration with LLM such as ChatGPT, which allows for natural language processing and understanding. This integration enables FOPO to interpret citizens' requests and generate appropriate responses in natural language, making the platform more accessible and user-friendly. To ensure the security and privacy of sensitive information, FOPO should utilize cutting-edge cryptographic techniques such as homomorphic encryption, secure multi-party computation, and zero-knowledge proofs. Homomorphic encryption enables computations to be performed directly on encrypted data, preserving the privacy and security of sensitive information. Secure multi-party computation enables multiple parties to compute on the same set of data without revealing their individual inputs, protecting against data breaches and privacy violations. Zero-knowledge proofs allow for the verification of information without revealing any underlying data, thus protecting against unauthorized access and data leaks.

In addition to these security features, FOPO requires all data storage to be located within Brazil, in compliance with local data protection regulations. The FOPO data storage system utilizes blockchain technology for secure and decentralized data storage, ensuring the integrity and immutability of all stored data.

Overall, the FOPO platform and its underlying SP language represent a significant advancement in government service delivery and citizen engagement. By combining advanced symbolic computation and natural language processing with cutting-edge security measures, FOPO is able to provide citizens with a seamless and secure interface for accessing government services and information.

The FOPO system will also make use of high-level symbolic computation to provide citizens with trustworthy and accurate information. By utilizing reliable and secure databases, the system will significantly reduce the risk of fake news and misinformation, which are prevalent on other social media platforms. This will ensure that the information provided to citizens is based on facts and backed by trustworthy sources. Additionally, the use of symbolic computation will enable policymakers to analyze and interpret vast amounts of data, allowing for more informed and effective decision-making. The integration of these advanced computing technologies makes FOPO an essential tool for ensuring transparency, citizen engagement, and effective governance in the digital age.

Why *SP* was chosen as the best candidate to support FOPO? The story of *SP* is one of innovation and forward-thinking. It all began with a course for gifted and neurodiverse children at the Federal University of Rio de Janeiro (UFRJ), led by Professor Carlo Tolla. The course used *SP* as a language for creating games in Python, giving the students a chance to explore their creativity while developing their coding skills, but its impact didn't stop in the classroom. It was also brought into the community of Costa Barros in Rio de Janeiro, where the world of gaming and programming offered a new reality beyond the violence and drug trafficking that plagued the children's daily lives. Now, *SP* has evolved beyond being just a language for creating games. It is the backbone of the FOPO platform, a powerful tool that enables the integration of various cutting-edge technologies. *SP*'s unique design allows it to manipulate vast amounts of data efficiently, connect with *LLM* systems, and understand natural language inputs. Moreover, SP boasts features that make it ideal for creating mathematical and scientific applications. Its programming paradigm includes declarative constructs that eliminate thousands of lines of code, making it easier to write and maintain complex programs. And with its ability to interface with reliable databases and incorporate mathematical formulas and models, amplify the capacities of the computer language.

The following features indicate the adequacy of *SP* to support the FOPO and the related AIN application:

1. SP is a groundbreaking language that represents the future of computational linguistics. Its versatility and power make it a valuable tool for a wide range of applications, from game development to scientific research. With *SP* programmers research and similar efforts we will be able to achieve breakthroughs that were once thought impossible.
2. SP is a high-level programming language designed to be taught to children in Brazilian schools and comprehended in a simple and intuitive way by new generations worldwide. It integrates cutting-edge concepts from the fields of neuro-linguistics, bioethics, and indigenous mythology to provide a unique programming experience.
3. SP's syntax is designed to be human-readable, with functions and variables named in Portuguese and drawn from the country's rich cultural heritage. For example, the "rainforest" function can be used to connect multiple databases intelligently, while the "samauma" function teaches other trees of knowledge. The language also incorporates concepts from indigenous languages, which emphasize flexibility and adaptability. For example, the "caipora" function can be used to create a new data type that evolves over time, like a living entity.
4. SP's flexibility and adaptability make it an ideal language for integration with LLM and AI systems. Its functions can be used to manipulate large amounts of data and analyze complex algorithms, making it an essential tool for data scientists and AI researchers. In addition, SP's syntax is designed to be easily translated into other programming languages, facilitating integration with existing systems and tools.
5. SP also incorporates principles of bioethics and environmental responsibility. The language includes functions such as "biodiversity" and "sustainability", which can be used to analyze data related to these important topics.
6. SP is a language of computing that can adapt to the native language or choice of the programmer or AI that interacts with it, just like modern LLMs. Its neural translation function allows for seamless communication with other programming languages and platforms, making it an ideal language for global use.

In conclusion, SP is a programming language uniquely designed for the Brazilian context, incorporating elements of indigenous mythology, neuro-linguistics, and bioethics. Its intuitive syntax and flexibility make it an ideal tool for data manipulation, AI research, and environmental analysis. As the language continues to evolve and adapt, it will remain an essential tool for building a better future for Brazil and the world. To build a successful SP community, it's essential to learn from successful examples, such as the communities behind Linux and Wikipedia. These communities have shown that openness, collaboration, and transparency can lead to powerful, innovative outcomes. To achieve these goals, SP's development should follow a transparent, community-driven approach, with a central committee responsible for overseeing the core language design. The community should be encouraged to provide feedback and contribute code, with the help of AI algorithms that can identify errors and suggest improvements. The committee can then review the contributions and decide which ones to integrate into the language. Through this process, SP can become a powerful, flexible tool that can help solve some of the world's most pressing problems, from climate change to social inequality. By combining the power of AI with the wisdom of crowds, we can create a new paradigm for language development that's inclusive, innovative, and impactful.

The integration of SP and FOPO with the Brazilian government's systems is a crucial aspect of the project's success. FOPO is built using a microservices architecture, which allows for greater scalability and flexibility, as well as easier maintenance and deployment. Each microservice is responsible for a specific function, such as authentication, data processing, or AI integration, and can be updated and deployed independently. Moreover, FOPO's architecture is designed to comply with the Brazilian General Data Protection Law (LGPD), which ensures the privacy and security of citizens' data. With FOPO's microservices architecture, each service is responsible for its own data and security, reducing the risk of data breaches and ensuring compliance with the *LGPD*.

The integration of SP and FOPO with the Brazilian government's systems is a game-changer for the country. With SP's capabilities for processing large amounts of data, connecting with AI and LLM, and its high-level declarative programming features, the Brazilian government will have access to a powerful tool for developing innovative solutions to complex problems. One of the main challenges for FOPO is to enable its installation in large parallel computing structures stored in Brazilian clouds while complying with LGPD and cybersecurity requirements. The use of microservices architecture and compliance with LGPD ensures the security and privacy of citizens' data, while also providing greater scalability and flexibility in the development of new services. With the integration of SP and FOPO into the Brazilian government's systems, we can expect to see a significant improvement in the efficiency and effectiveness of government services, leading to a brighter future for Brazil. In conclusion, the combination of FOPO and SP, leading to the AIN App for the end user, represents a significant step forward for government service delivery, citizen engagement, and effective governance in the digital age. With its advanced computing technologies, FOPO is a powerful tool that can help build a more transparent, secure, and trustworthy society.

10 Concluding Remarks

We can use systems with the modern technology of LLMs to communicate directly with the citizen through their single access to the Brazilian government's digital services. This system can take a large state investment and thereby create new companies in the area of AI and information technology that could create even more wealth with this technological development. Supporting students from popular layers who have recently graduated from state, federal, and private Brazilian universities is one of the ways these services could be developed, creating opportunities for new companies created by these students to have government orders for the development of technology.

Many priority areas of national development could take advantage of this new ecosystem to be created in Brazil to follow the great technological revolution of AI that will change the world and the economy in the coming decades of the twenty-first century. We conclude this chapter by exemplifying that the FOPO could lead a national Brazilian effort to reforest forest areas, change the energy matrix to clean energy, and create new companies based on socio-environmental bioethics that will combat global warming. These efforts for reforestation are based on the ideas of pioneers such as Antonio Nobre and Marina Silva, who proposed that a sustainable society can fight global warming by valuing the Amazon rainforest, the other Brazilian biomes, and the native peoples.

An important remark to conclude this proposal is that the base principles and algorithms of the Intelligent System of the FOPO need to contain the precepts of the Brazilian Constitution, the Humanitarian and Ethical Treatises of which Brazil is a signatory, and the socio-environmental precepts of local and global governance we suggest in this book, to create a better future for Brazilians and humanity, in addition to containing cybersecurity mechanisms and protection against unconventional wars to maintain the integrity and security of the FOPO System.

Acknowledgments We are grateful to the co-authors of the original FOPO paper for the groundbreaking proposal (Valdir Gonzalez Paixão, José Monserrat Neto, Luis Alfredo Chinali, and Enidio Ilario), to Antonio Nobre for several suggestions that enriched the proposal, to Nereide Segala for useful information about popular cooperatives in the Brazilian *Caatinga*, and for her productive and organizational work in this region.

References

Associated Foreign Press. (2011, February 27). Social media, cellphone video fuel Arab protests. *The Independent*.

Astok, H. (2017). *Introdução ao Governo Digital: Governo digital no apoio às metas de desenvolvimento sustentável*. Published by E-Governance Academy. Tallinn, Estônia. Available at https://estoniahub.com.br/wp-content/uploads/2020/12/manual-egov-portugues.pdf. 27 Apr 2023.

E-Gov. (2023). https://www.gov.br/governodigital/pt-br/estrategia-de-governanca-digital/do-eletronico-ao-digital. Accessed 30 Apr 2023.

Galbraith, J. (1967). *The new industrial state* (576p). Princeton University Press. ISBN 978-0691131412.

Hossne, W. S. (2006). Bioética Princípios ou Referenciais? *Mundo da Saúde, 30*(4), 673–676.

Pereira Jr, A., Ilario, E., Paixão, V. G., Chinali, L. A., & Monserrat, N. J. (2014). Tecnologia, Democracia e Socialismo: o Encontro do Século? In *Perspectivas em Ciências Tecnológicas* (Vol. 3, pp. 88–108). Available at http://www.fatece.edu.br/revista/perspectivas/volume-3.html. 27 Apr 2023.

Pereira, A., Jr. (2019). Perspectivas para o Investimento em Economia Ecológica no Brasil. *Cadernos de Campo/Revista de Ciências Sociais da UNESP-Campus de Araraquara, 1*(27), 195–209. E-ISSN 2359-2419.

Pilkington, P. (2016). *The reformation in economics: A deconstruction and reconstruction of economic theory* (368p). Palgrave MacMillan. ISBN 978-3319407562.

Schneider, D. S. (2015). *Uma Abordagem de Computação Social para a Construção de Histórias e Tramas Noticiosas por Meio da Curadoria Social*. Tese de Doutorado em Engenharia de Sistemas e Computação. Universidade Federal do Rio de Janeiro, COPPE, Rio de Janeiro.

Singh, M. N. (2012). Jurgen Habermas' notion of the public sphere. *The Indian Journal of Political Science, 73*(4), 633–642.

Surowiecki, J. (2004). *The wisdom of crowds: Why the many are smarter than the few and how collective wisdom shapes business, economies, societies and nations*. Little, Brown and Company. ISBN 0-316-86173-1.

Wolfram, S. (2023). *Wolfram | Alpha as the way to bring computational knowledge superpowers to Chat/GPT*. Available at https://writings.stephenwolfram.com/2023/01/wolframalpha-as-the-way-to-bring-computational-knowledge-superpowers-to-chatgpt/. Accessed 30 Apr 2023.

José Otávio Pompeu, an interdisciplinary professor at the Federal University of Rio de Janeiro, is the creator of the discipline "Maps", impacting a thousand students per semester by revealing the individual potential of each one. Representative of Brazil at the United Nations Conference on the Rights of Persons with Disabilities in 2017 and 2018, he also carried out a postdoctoral internship in Finland, exploring E-Health and Artificial Intelligence. His experience and dedication make him stand out as an influential leader in education and technology.

José Luis Azpiazu is an engineer on industrial technology trained in the Universidade de Burgos (UBU) with a background in the automotive industry, he has made a name for himself as a hardware dealer and entrepreneur, his interests extend to the study of cognition to its nuances on modern society. https://orcid.org/0000-0003-4113-063X

Alfredo Pereira Jr is a Philosopher of Science, Professor at the Universidade Estadual Paulista Julio de Mesquita Filho (UNESP), Brazil, retired, with degrees in Philosophy and Administration. Currently working in the UNESP Graduate Program in Philosophy. Alfredo has around 300 publications in several areas of knowledge, mostly in the Philosophy of Neuroscience and Theory of Consciousness. Orcid: https://orcid.org/0000-0002-5960-041XAD, Scientific Index: 56th place in Top 100 Latin America Philosophers/Scientists.

Part II

How to Achieve Economic Democracy?

Money to the People: The Challenge of Financing Human Development

4

Alfredo Pereira Jr and Francisco Sousa

Abstract

The promotion of human development in sovereign countries, under an elected democratic government, can greatly benefit from information technology tools to direct monetary flows from the State's financial system (Treasury, Central Bank, Sovereign and Green Funds, Public Banks) to the social infrastructure. Besides the regulatory framework and policy goals, a Central Bank Digital Currency should also serve as reference and support of special credit offered to the people by public and community banks (following the original proposal of Brown, Banking on the people: democratizing money in the digital age. The Democracy Collaborative, Washington, DC, 2019), to promote the creation of cooperative and sustainable productive enterprises. In this approach, the state induces effective demand by creating means of benefits for the poor, as suggested in working papers published by the International Monetary Fund on (Universal) Basic Income, which are endorsed by many authors, per Standing and Orton (Development and basic income: an emerging economic model. In: United Nations (Org) The ins and outs of inclusive finance: some lessons from microfinance and basic income. UNCTAD (United Nations Conference on Trade and Development), 2018) on the basis of the success of the Brazilian *Bolsa Família* program. In this chapter we propose a general and more current mechanism for the expansion of the existing monetary system with new features to enhance access to financial resources, leveraging the state-issued digital money to the people, putatively reducing economic inequality and boosting productive enterprises more compatible with current socio-ecological standards as well as inclusion. The selection of qualified projects, performance metrics and audit procedures should be made by a committee of experts recognized by peers in their professional communities, as to prevent issues related to fraud, abuse and waste. The goal of the social budget should be exclusively the promotion of human development, such as generating jobs, producing income, ecological stability and improved quality of life.

A. Pereira Jr (✉)
State University of São Paulo, Botucatu and Marília, SP, Brazil
e-mail: alfredo.pereira@unesp.br

F. Sousa
MFA Business Office, George Washington University, Washington, DC, USA

SAIS John Hopkins University Washington DC – Senior Staff, Johns Hopkins University, Baltimore, MD, USA

A. Pereira Jr, F. Sousa (eds.), *Principles for Governance*, Sustainable Development Goals Series,
https://doi.org/10.1007/978-3-031-40025-4_4

1 Introduction

The induction of human development should be a key policy goal for democratically constituted governments in sovereign countries. Thus, how can the federal state provide incentives and access to credit to support production of consumer goods and services in an open economy, while maintaining fiscal and budgetary discipline pre-negotiated with major stakeholders?

Although the sovereign state has, in principle, the capacity of generating 'Fiat' and digital money 'out of nothing', in some countries – including Brazil – existing laws and political constraints prevent authorities from issuing stimulus payments directly to the people. These constraints may involve increasing public debt, risk of corruption, as well as inflation pressures resulting from high cost of credit intermediation by commercial banks, which are major obstacles to sustainable production of goods and services supporting human development.

The pandemics era and the political pressure towards a "New Green Deal" revealed the necessity of public spending above tax revenue, to avoid recessionary and humanitarian crises. The central question became: How to make public spending generate economic and human development, instead of increasing the health of those who are already privileged, or opening a new opportunity for corruption, diversion and fraud?

According to Modern Monetary Theory (MMT) defenders, the solution may seem to be simple: "When the government spends money, that money goes somewhere: into the broader economy. As (Stephanie) Kelton puts it, 'their red ink is our black ink'" (Carrillo, 2020). However, as pointed by Brown (2019), on the basis of the analysis of the split-circuit identified by Huber (2017), the existing financial system often prevents public spending from reaching the 'real economy' and inducing human development. There is a "Web of Debt" implemented by the intermediary layer of commercial banks and obscure financial entities (called the "Shadow Market" by Brown, 2019) operating on their own interest, preventing state-issued money to reach productive enterprises and the poor.

As a counterexample, China has a financial system that allows public spending to finance techno-industrial enterprises, placing this country at the top of economic growth and allowing a large part of the population to overcome poverty. Is it possible to reproduce useful aspects of this system in a democracy? *How* can a democratic state guided by a popular government issue money to finance development, either through productive investments in social, cooperative and sustainable companies, and/or through the implementation of a substantive Basic Income (BI) program to reinforce popular consumption power, combined with public spending in strategic areas of infrastructure, health, education, security and social assistance? An answer to these important questions demands a series of reasonings, presented in the next sections of the chapter.

2 Theoretical Background

This chapter presents an interdisciplinary work that makes a synthesis of several scientific lines of reasoning and interpretation of the meaning of empirical data, addressing the field of Philosophy of Public Administration. Macroeconomics is not the main focus, but the "state concept of money" (the *Cartalism* of Knapp, 1905) and MMT, a branch of Macroeconomics (see Mitchell et al., 2019) is central to the proposed strategy. We focus on the conceptual framework underlying the choice of strategies in public administration, building on suggestions advanced by compatible theoretical perspectives (Lietaer, 2001; Huber, 2017; Barrowclough, 2018; Standing & Orton, 2018; Kelly & Howard, 2019) and technical suggestions (Lonergan, 2016; Lagarde, 2018; Brown, 2019; Brainard, 2020), which can be encompassed under the umbrella of *economic complexity* (Hidalgo & Hausmann, 2009; Hausmann et al., 2014) and *self-organizing systems* approaches (Debrun, 2018).

In orthodox models of economic development, macroeconomic stability and job creation – which are within the goals of the state financial institutions (such as Central Banks and the

Federal Reserve Bank – FED – in USA) – are limited by constraints put on fiscal balance. There is a false assumption that money to finance human development *must come from the society*, e.g. from the collection of taxes and fees, and/or social security savings. In fact, social funds and other savings can be legally used in public management to promote human development, as argued by Drucker (2003), but when they are scarce, or when it is not convenient to use them for non-profit causes, nothing prevents the sovereign federal state to issue the quantity of money that is necessary for national strategic programs. This conception overcomes the risk of, e.g. using pension funds to promote development, at the cost of the rentability of the funds for their owners, who depend on these values to finance their living during retirement.

As argued by MMT theorists, the sovereign state can issue money for public spending and investment on development projects, without depending on tax and fee collection. If the money circulating in the society is issued by the state, why should the state take the money back from society to promote human development? "New" money can be issued for legitimate purposes, as currently done in several developed countries for cases of environmental emergencies and financial crises. According to MMT, the main form of money issuance is (or should be) public spending, and the power to issue currency is not limited by revenues, but (as advanced by Cartalism) by *legal* and *political* constraints.

The orthodox 'neoliberal' dogma (that is, following the economical policies and advices of Margaret Thatcher in UK and Ronald Reagan in USA) is that a nonlinear increase of 'new' money issued and spent by the federal state, as in "New Deal" or "quantitative easing" strategies (see e.g. Brown, 2019), *automatically* causes undesired effects, as the uncontrollable increase of public debt, leading to the insolvency of the state, and/or the increase of inflation of prices in the market. However, these are not necessarily the cases, as shown by MMT theorists and Brown (2019), with abundance of empirical data.

The effects of public spending depend on *how* it is made, or, in other words, it depends both on the *quality* of the distribution of money (as in the concept of *qualitative easing* advanced by Brown, 2019) and proportionality with the increase of the Gross Domestic Product (GDP). For instance, if public investment using 'new money' is targeted to produce food for human consumption, then, according to the intrinsic laws of the market (the informal "Law of Supply and Demand"), the price of these goods is likely to remain stable or decrease. If the state also finances a BI program to reduce poverty and hunger, people can use this money to buy the food, and consequently there will be effective demand, the goods will be sold to the popular consumers, increasing their quality of life and at the same time triggering a cycle of development in the economy.

3 Should Governments Use Central Bank Digital Currencies?

A Central Bank Digital Currency (CBDC) is basically a digital banknote. As a retail note, it can be used by individuals in direct payment transactions, while as a wholesale note a CBDC would be generally used in business-to-business or institutional payment transfers. Governance in a solidary and democratic society can greatly benefit from information technology applied to public banking. Aiming at a financial system serving the people's interests, Brown (2019) endorses the International Monetary Fund (IMF) movement towards adoption of Central Bank Digital Currencies (see Lagarde, 2018). In this effort, the Bank of England has carried public consultations, and the U.S. Federal Reserve Bank has partnered with the Massachusetts Institute of Technology (MIT) for developing the U.S. digital dollar. However, this project is facing prolonged debates in the Congress over the role of bank intermediation in the digital transactions. The U. S. is leaning towards a wholesale, institutional model, not allowing to individuals the direct access to CBDC. The access must be interfaced through the individual's digital wallet in a financial institution, which then would settle all transactions with the Federal Reserve system. Several other countries and research institutions are developing and implementing this new technology both for institutional functions and for retail usage.

As a digital platform, the CBDC applications may eventually be expanded, e.g. encompassing online payments of federal, state and municipal taxes and fee, thus adding greater velocity to the entire financial system. The intended design is that the CBDC cannot serve for the accumulation of private wealth, as it pays no interest and may also carry an expiration date. The Bioethical Principles presented in this book lead to the suggestion that the emission of CBDC should be limited to specific purposes, such as funding projects related to the social infrastructure, ecological sustainability and supporting BI or other types of benefits.

In USA, the FED currently maintains 'direct credit' programs to mitigate economic risks of crises and emergencies authorized by the Executive, but still did not approve the usage of CBDC in this type of public spending. When the law allows, as in the USA, "direct deposits of cash by the Fed into individual accounts is becoming increasingly probable...the central bank...will use digital money apps...to transfer money directly to US consumers" (Derden, 2020). The 'direct circuit' advocated by Huber (2017) has been implemented in the USA to fight the sanitary and economic crisis, but not using the CBDC. In China, it has been used to implement their CBDC (the *Digital Yuan*), by means of sweepstakes and awards.

In Brazil, there are Sustainability Funds available, such as the Amazon Fund, so far provided mainly by Germany, and BRICS contributions for sustainable development. The CBDCs can greatly facilitate these investment transfers and currency conversions, as well as establish reliable reporting on these transactions. Considering its advantages, the CBDC (in Brazil, the Real Digital, recently named DREX) can be a link in the financial chain to make the money available to the people. As the implementation of DREX in 2024 has been planned for institutional use only, the desired impact on popular economy should be mediated by public banks.

In a recent paper, Ozili (2023) argues that "the emergence of central bank digital currency...provides an opportunity for central banks to make an important contribution to the transition to a circular economy...in two ways: first, by making central bank digital currency accessible to circular businesses and other players in the circular economy sector; and second, by looking into how the design features of CBDC can support circular economy goals" Among the advantages of this strategy, he lists:

(a) CBDC offers "a better payment option for circular economy financial transactions".
(b) CBDC "can lead to greater financial inclusion for 'unbanked' informal workers in the circular economy".
(c) CBDC "can create a gateway that allows a central bank to offer financial assistance to distressed circular businesses".
(d) CBDC "can reduce illicit activities in the circular economy".
(e) CBDC "can be used to provide stimulus funding to support circular businesses during crises".
(f) CBDC "can offer low transaction cost for circular economy financial transactions" (Ozili, 2022).

CBDC do not need private banks as intermediaries, because financial profit-directed institutions increase the cost of the transaction for the end user; however, according to the laws of many countries, including Brazil, the CBDC cannot be directly transferred to citizens, producers and/or consumers. *Alternatively, it can be combined with financial entities, such as public banks, to reach, at the end of the chain, the people.* More to the point, it might operate *as a link for public banking programs offering special credits, or to guarantee existing social currencies* that serve the people. This thesis, sketched by Brown (2019), is developed in this chapter. In spite of the current legal controversies and temporary unregulation of CBDC, this innovation has the potential to become a revolution that will enable the state to control the flow of money it emits, so that under a democratic government it can be channeled to productive activities and to the reinforcement of popular income.

This chapter focuses on the possibilities of the implementation of CBDC, combined with public banking, social currencies, social enterprises, BI programs and 'New Green Deal' strategies. These resources open for public administration an opportunity to reduce economic inequality and promote human development. In Brazil, departing from the perspective of Lara Resende (2020), the main state budget in sovereign money (the Real) can be under a fiscal policy, while a parallel budget using CBDC (the Real Digital, or DREX, and eventually a Latin-American integration currency and/or a curency issued by the BRICS bank; see a critical perspective in Rodrigues, 2023) can be created and managed by development and social public banks (in Brazil, mostly the BNDES – the Bank for Economic and Social Development and the *Caixa Econômica Federal*), to face the demands of the people. Considering the limitations of the BNDES to provide money to the people, we suggest the mediation by cooperative, public, federal banks, to offer special credit to small businesses and assistential programs.

The strategy to provide money to the people, reduce economic inequality and promote human development, is more complex than it may seem to the non-specialist. To bridge the existing gaps, at least in the theoretical domain, we present a general model, exemplified by the roles attributed to Brazilian institutions, under the new government that began in 2023, to extend the financial system for the satisfaction of popular needs. We elaborate on an extension of the existing monetary system – based on information technology – that is potentially able to overcome the obstacles, presenting a legal solution to the bottlenecks met by efforts to "democratize money".

4 The Need of a Social Human Development Function of Public Banks

Public banks have a social responsibility of promoting economic and human development, which implies fighting financial inequality. Currently, the Central Bank of Switzerland, for instance, is one of the big players on the US stock and commodity exchanges; shareholders of Apple, Amazon, Tesla, etc. However, this function does not exactly fight economic inequality. We argue for the need of a social function of public banks to democratize money, and for the creation of new institutions with this proper function, when the older are limited in this regard. For this purpose, CBDC (and Sovereign Funds after converted to CBDC) can operate on three bases:

(a) When connected to a social currency, CDBC can exponentially increase 'money velocity' (https://fred.stlouisfed.org/categories/32242) and, therefore, exponentially increase the *availability of money for the people*, as in the partially successful case of PIX (an interbanking discuount instant transfer mechanism, using digital technology; see https://www.bcb.gov.br/estabilidadefinanceira/pix) in Brazil;
(b) It helps to fight the 'scarcity of money' (Lietaer, 2001) in the real economy, which ultimately leads to recession. Although Lietaer did not assume an explicit Cartalist view, his prediction can be made possible by means of the strategy we develop here:

> I have come to the conclusion that one of the most promising approaches to shape our future … is to bring conscious choice into the arena of our money system. The reason is that money could be described as the universal social incentive … (The) change in our money systems is in fact already well under way, irresistibly driven by the social and technological forces of the Information Age. The real issue is not whether widespread changes will happen or not, but how much awareness there will be about where these changes are leading us. The real question is whether we are even conscious that we have a choice in the matter. (Lietaer, 2001)

(c) To bridge the gap created by the 'split-circuit' (Huber, 2017) financial system, which leads to extremely speculative financial practices that increase economic inequality in the society. Hudson and Brown (2020) trace the possible scenario of Western capitalism if the current financial system is not improved:

> The Federal Reserve is the official Ponzi scheme that keeps finance capitalism operating in the United States. Obviously, at some point every exponential growth scheme has to stop, because otherwise you'd have an infinite amount of debt. So at a certain point, the Fed will sit down with the main Wall Street firms and the main billionaires that are behind these firms, and say, "Well, you know, the game is over. We've got to let it go". These investors will say, "OK, we'll take the money and run". They're just going to drop everything, sell out and there will be a crash with the pension funds and the small savers who aren't in on the game, losing whatever they have. (Hudson & Brown, 2020)

MMT tells us that a potentially infinite amount of debt is always payable by the sovereign federal state, by means of the (also potentially infinite) emission of new reserves, but in the split-circuit financial system these reserves need to be converted to bank money to reach the 'real economy' (producers and consumers). If commercial banks and the financial institutions of the "investors" referred by Hudson and Brown (2020) 'take their money and run', there will be a financial crisis and possibly the crash of the popular savings and retirement systems, because of the lack of liquidity in the bank system. These are, of course, speculative scenarios, projected from actual tendencies of the financial system.

As an alternative, CBDC operations open new possibilities to improve social functions of the monetary system. The primary movement of CBDC is internal to the computers of the state and may be kept under control. This system, combined with other financial and social institutions, makes it possible to extend the capabilities of the existing financial system to transfer money to the people (producers and consumers), as proposed in the 'direct circuit' concept advanced by Huber (2017) and similar approaches. The advantage of this strategy is the possibility of directing and monitoring the pathways of the money in the society, avoiding misuse of the credits, and blocking unwanted transactions; guaranteeing that money is going to accomplish the designed social functions, being directed exclusively to production and consumption of goods, which induces the generation of wealth in the society. Therefore, CDBC is adequate for programs aimed to increase economic complexity, resulting in human development. It can be used, in combination with other financial mechanisms and institutions, to provide income for poor citizens, creation of new enterprises that add value to national products for export, opening new jobs and financing educational opportunities for the new generations.

This line of reasoning leads to a new understanding of the difference between *quantitative* (state money for banks and capitalists) and *qualitative* (Brown, 2019) easing (selective spending, aimed at goals of the elected government: in our case, public investments in the real economy, aimed at popular development). "Qualitative easing" in the sense of Ellen Brown comes from a proposal advanced by Jeremy Corbyn, candidate of the British Workers' Party in 2015. In his program, he referred to as "'quantitative easing' for the people", meaning the public investment made in the domain of interaction between a democratic state and the real economy (producers and consumers), not in the original sense of the term, in which it would be an internal procedure within the Central Bank-Financial Institutions circuit (see https://files.stlouisfed.org/files/htdocs/conferences/annual/Farmer-QuaE-R2.pdf, access in February 2023).

5 The Combination of CBDC with Social Currencies

Here we propose a solution to overcome the difficulties in democratizing money, based on a combination of CBDC with already existing public banking programs and social currencies, as suggested by Brown (2019). The idea is to increase the leverage and "social velocity" of sovereign money – the CBDC – by means of a combination with special credits offered by public banks and social currencies, in benefit of efforts towards human development.

The creation of CBDC in several democratic countries follows the guidelines of two institutions, the BIS (Bank of International Settlements) and the IMF (International Monetary Fund). Beyond national issues, there seems to be a difference between the IMF and BIS guidelines about the CBCD; the first is more focused on the impact of the currency on the real economy, while the second is more focused on issues internal to the financial system. The main agreed functions of the CBDC are: to be a reference of value, to facilitate payments between financial institutions and to contribute to "macroeconomic stability". The last one depends on what is meant by "macroeconomic stability"; in the Keynesian sense, the use of state sovereign money to balance the production and consumption of goods, and promote full employment, must be a major target. CBDC projects in many countries mentions "macroeconomic stability", not only because of what is feared (e.g. it might compromise the Central Bank and private financial institutions, causing a flight to other currency a related exchange rate crisis, or migration from traditional deposits to CBDC causing commercial banking crisis etc.), but also because macroeconomic stability is related to the social function of money, or, in other words, it is not only a matter of financial accounting balances.

In Brazil, under the influence of Keynesianism, one of the legal goals of the Brazilian Central Bank (BCB) monetary policy is "to promote full employment". However, the Brazilian Constitution (Article 164, first paragraph) prohibits the BCB to deliver the money that is created (with reserves from the Treasury or Sovereign Funds) directly to the citizens or legal persons; only "financial institutions" can redeem the reserves with "bank money". This constraint implies that the CBDC cannot be directly used by the people, in retail transactions.

For more than a decade, an important way that socially responsible enterprises and popular cooperatives found to fight money scarcity is the creation of *social currencies*, as the African M-Pesa (Eijkman et al., 2010; Mas & Radcliffe, 2011). There are several already existing digital and non-digital social currencies in many states and cities of Brasil (Segundo & Magalhães, 2009), as well as in many other countries. These social currencies make money avaliable to the people involved with the production and consumption of goods and related services, promoting human development. To avoid excessive risk, they should be backed upon sovereign money, but there are two mechanisms that may allow a multiplication of currrencies circulating in the real economy:

1. Special laws appliable to small businesses allow the proportion between sovereign money and social currencies being not 1 to 1, at the act or creation and/or during the operation of popular financial institutions. These laws may allow financial institutions (such as community banks or a financial department of cooperatives) connected to popular activities to issue social currencies backed up on sovereign money in the proportion of one to many, e.g. 1 to 5, that is, one unit of sovereign money backs up 5 units of the social currency. Although the main initiative of community banks in Brazil, as the *Banco Palmas*, began with a 1 to 1 proportion, for the issue of the social currency, in practice – during the operation of the bank and its subsidiaries – the proportion has changed (for a review, see Garcia, 2009);
2. There is a systemic "multiplication of money", analogous to the "multiplication of bread" in the Bible (in the slang of jazz musicians, "bread" is money). *For the individual* who exchanges sovereign money with a social currency, e.g. Africa's *M-Pesa* (provided by Vodafone; see https://www.vodafone.com/about-vodafone/what-we-do/consumer-products-and-services/m-pesa), there is 1 to 1 exchange. However, *for the system, there is a multiplication* of the currencies. A single unit of sovereign currency generates an 'offspring' unit of the social currency each time it is used to acquire one unit of the social currency. The circulating social units multiply in the system, derived from a single unit of sovereign money!

> Using this artifice, *the people can do the same "miracle" that the banks do when they generate credit*. So the title of this chapter, "Money to the People", actually means giving the people the same power as banks to multiply state-issued money, according to a set of agreed rules, which address activities that afford a proportional increase in the society wealth, including the "Human Factor" (Drucker, 2003).

The combination of CBDC with social currencies, using both above mechanisms, is the crux of the solution we propose to multiply money in society, similar to Jesus Christ's multiplication of bread in the Bible! Our solution to democratize money in favor of human development is to use the CBDC (in Brazil, the Real Digital, called DREX) as a guarantee to back up social currencies. The usage of these currencies by the people automatically increases the quantity of money available in the real economy, boosting the production and consumption of goods. As the CBDC cannot circulate in people's accounts, it would be operative between financial institutions (the Central Bank and the institutions that issue the social currency), being converted to sovereign money (in Brazil, the Real) *if and only if* there is a risk of bankruptcy of the institution issuing the social currency. For instance, if the institution that issues the social currency finances agricultural production, but the harvest is lost due to natural factors, the institution can redeem the warranty by converting the CBDC reserve into usable money to pay the creditors.

There is room for innovation in the digital public space, a new frontier of risks and opportunities. CBDC legislation is still nascent and evolving in many countries. Brazilian jurisprudence in this area includes domestic and foreign considerations. We need to be aware of both sides of this new coin, that is, financial and social! For example, the IMF admits that some public accounting issues related to CBDCs remain under scrutiny. In USA, October 2022, the Fed remained undecided on how to address the legality of CBDC before the US Congress (https://www.rollcall.com/2022/10/18/can-congress-buy-in-to-digital-dollar-without-legislation/, access in February 2023). We are facing a 'brave new world' and the difficulties related to this new monetary platform may be considered as challenges or constraints, depending on the chosen strategy: to promote human development by democratic popular states, or to maintain the scarcity of money and economic inequality, in the case of oligarchical states.

Focused on the international finance market itself, the CBDC project in Switzerland (the *Helvetia*) is the most advanced. For cross-border foreign trade, there are initiatives from China, Thailand, Hong Kong and United Arab Emirates. The CBDC program in China mixes nationalization of instant retail payments (in line with what Pix does in Brazil) with commercial diplomacy and environment protection initiatives. An example that is relevant for our discussion is the Nigerian case. Nigeria was the first African country issuing a CBDC, the *eNaira*. The Nigerian model has precedence and relevance for the specific case of Brazil. A recent paper argues for the extension of the currency, redesigning "features which the eNaira should possess for it to become very effective in offering payment solution and for macroeconomic stability" (Ozili, 2023). In addition to financial inclusion (which is very important), it may also finance development initiatives, such as cooperatives and social enterprises.

Brazil still hasn't officialized its CBDC standard; a pilot project was published in March, 2023 (https://www.bcb.gov.br/detalhenoticia/667/noticia, access in April 6, 2023). Operational decisions about the Real Digital are not exclusive prerogatives of Brazialian Central Bank (BCB). In the USA, the Federal Reserve meets regularly with the Treasury and the National Security Council to decide about the development of the 'Digital Dollar'; after all, these decisions have ramifications in all sectors of the country, besides issues related to the monetary system.

In the financial and economical-political landscape of Brazil in 2023, there is also a parallel proposal of the Latin American currency advanced by Economy Minister Fernando

Haddad (for a critical perspective, see Rodrigues, 2023). The first step to relate both trends is to understand the definition of a 'social currency', which is different from the CBDC (except in China, where the Digital Yuan is given to and used by the people). The case of Nigeria, a country with more than 100 social currencies circulating and which need to be consolidated into a social digital model, is relevant to Brazil, because our country also found social currency solutions for the scarcity of money in several local economies. The idea is to use the Real Digital as a reference of value and guarantee for the social currencies issued by several financial institutions, including the Latin American currency proposed by Minister Fernando Haddad, which is still not assigned to a financial institution to be responsible for the emissions. There is a possibility of financial innovations being backed up by the BRICS Bank using the Digital Yuan.

While the federal government maintains a neoliberal agenda of money scarcity for the people, the 'monetary balkanization' in Brazil continues to increase; see, for instance, the banking system in the "favela" (https://america.cgtn.com/2021/05/17/brazil-favelas-create-their-own-banking-system; access in March 17, 2023) and several currencies issued by small cities, as the Cocais (https://economia.uol.com.br/noticias/redacao/2014/03/20/sem-nenhum-banco-cidade-do-piaui-cria-moeda-propria-para-girar-economia.htm, access in March 17, 2023). The country's monetary future is, definitly, partially decentralized. We need a 'currency exchange' to convert these values between municipalities and locations. One solution would be to 'validate' these existing social currencies linked to the 'Real Digital', allowing local issuance; after all, these currencies are mostly authorized by BCB. People will no longer wait for government solutions… thus, more and more create their local monetary solutions!

In practice, the connection of the CBDC with social currencies implies *the establishment of a parallel budget for the Digital Real.* The CBDC circuit may eventually be expended, e.g. payment of federal taxes and fees may be accepted, thus keeping the system running. The intended design is that the CBDC cannot be privately accumulated, pays no interest, and may have an expiration date. The emission of the CBDC should be limited to exclusive purposes, internal to the financial system, or related to public programs, such as backing up BI programs and/or public investment in human development.

In sum, there are technical, economic, political, legal and social issues in implementing a digital currency, and three types of formats: institutional, retail (for producers and consumers) or hybrid with some restrictions. This has not yet been resolved in Brazil; the new government wants greater social inclusion, as Brazil has about 1.200 municipalities out of the 5.575 national ones that do not have bank branches. The South American currency for Mercosur may facilitate trade within the continent, becoming a platform that can accommodate the interchange of currencies from several countries. In Brazil it would be 'Real Digital'; in Argentina, it would be 'Peso Digital', but the standard would be the same… There is a political issue behind the digital standard!

The functional view of monetary policy, one of the conceptual contributions of MMT, finds the CDBC as the ideal information technology tool for public administration to induce human development, including both the reformatting of the production system (e.g. by means of "Green Deals") and social assistential BI programs. The possibility of such public investments, at the end of the day, implies not putting straitjackets, like the recent limitation to public spending in the annual federal budget, in Brazil, or applying the rules of the Basel Accord for secure private banking on public banks that promote development.

6 A Possible Institutional Structure for Public Banking with CBDC in Brazil

There are several implications to be addressed for a human development program based on the democratization of money, in the Brazilian institutional and legal conditions (about this structure, see Arellano, 2020; Conti, 2020a, b). The effort to democratize money for popular production and consumption, promoting human development and fighting hunger, would require the usage and reformatting of several public and nonprofit private institutions.

Although all of us are aware that sovereign Fiat Money can be created out of nothing, the federal government has the ethical obligation of justifying the emission of 'new' money (not previewed in the annual public budget voted in the Congress) for the democratically elected representatives of the people. For this task, depending on the legal structure of the public financial system, the federal government can eventually use existing reserves and public assets to justify the issue of CBDC; however, if the CBDC is to be issued by an independent entity, as the FED in USA, public resources should not be used to back up it.

The issued CBDC can be used to guarantee special credits to the people, offered by public banks, for instance in cases of sanitary and environmental emergency, as well as to fight crisis in strategic areas, as in the current cases of deindustrialization and excessive interest charged by commercial banks (in part due to a large margin of risk in their operations, which creates a snowball of increasing interest). Ideally, special credits can be directed to promote human development by boosting socially and environmentally responsible production and consumption of goods, or, in the Keynesian terminology used by our lawmakers, "macroeconomic stability" and "job creation". The reason for opting for this strategy is quite obvious. Without sufficient production of food for popular consumption, the price of food in the market will increase, causing macroeconomic *instability*. Without public investment in the productive system, there is no job for the population, and the popular income decreases, causing a crisis of demand (not all produced goods will be sold, and then the capitalists reduce the amount of goods that is produced, causing unemployment and economic recession).

The central bank has the authority of converting reserves, which can be created by the issue of bonds by the Treasury, or conversion of assets from any existing public funds. In both cases, the BCB exerts the authority delegated by specific laws passed by the Congress. The money is transferred to financial institutions, in exchange for 'bank money' – the money created by commercial banks when they make loans to the people, putatively using their federal reserves (this is the money that circulates in the market; see https://www.britannica.com/topic/bank/Bank-money). The bank money (not the reserves or the CDBC) is also the money used in public spending, by public banks; in Brazil, the main public banks are the Caixa Econômica Federal (CEF), Banco do Brasil (BB) and Banco Nacional do Desenvolvimento Econômico e Social (BNDES).

Investment in economic development and social assistance is the goal of the BNDES. It lends bank money to the population, both physical and legal persons, for economic and social enterprises. The congress authorizes, by law, the transfer of public reserves to the BNDES. The interface with the people can be done by public banks, but not by BC itself, because the "direct circuit" (Huber, 2017) is illegal in Brazil (see the Federal Constitution, Article 164, first paragraph). There are constitutional barriers that prevent the Digital Real from being a social currency. According to the constitution, only "financial institutions" can buy the BCB's reserves, that is, the Digital Real cannot circulate in the market (individual and company accounts) as the Digital Yuan does in China. Even with lower interest rates, the money issued by the BCB would not automatically reach the productive system and people's pockets, because the structure of the current financial system separates the bulk of money issued by and/or owned by the state (bonds and reserves) from the real economy, placing banks merchants and brokers as intermediaries (working as barriers, not bridges).

The issue of bonds by the Treasury *cannot* be done to pay an increase of the state's own expenses (e.g. with the public employees), but it can be done for the proper goals of the BNDES and/or public (in some cases only) and commercial banks directed to promote human development. For a Brazilian 'New Deal', what can be done is to design loans in such a way that they can be used in *qualitative* easing, or, more specifically, for *special credits,* which are facilitated *exclusively for strategies of human development,* with low interest, no interest or even 'negative interest'.

A monetary policy for development *independent of fiscal policies* of obtaining superavit can be implemented by means of the redemption of Treasury-issued reserves authorized by prior law by the Central Bank, issuing the Real Digital and then, through the BNDES, supporting special credits used by people to finance productive enterprises, the consumption of the produced goods and related services. The crux of this strategy is that – for the purposes of human development (referred by means of Keynesian terms as "macroeconomic stability" and "creation of jobs" in the laws) – it is possible to guarantee both special credits and social currencies issued by popular financial institutions in the proportion of many units of social currencies for each unit of *Real Digital*, as explained in our previous section.

The institutional digital currency is not for use by producers and consumers, but can serve as a 'promissory note' backing special credits. In USA, the FED offers a "discount window facility" for some state programs, as in the case of emergency and crisis. Municipes are elligible to subsidized credits for urgencies, as in cases of food unsafety. In Brazil, the BNDES has lines of credit for agricultural cooperatives, which could be used as a leverage to promote a "New Green Deal" towards food safety, financing productive cooperatives in the interior of the country and the distribution of their products to urban consumers. In the past federal government (from the beginning of 2019 to the end of 2022), the BNDES did not use even half of its reserves to support cooperatives. The Bank of Brazil (BB), on the other hand, has offered facilitated credit for agrobusiness, but the production of these large companies is directed to exportation, being not affordable to the hungry people in the periphery of the big Brazilian cities.

Using the proposed strategy, it is possible to promote a "New Deal", in which the federal state finances popular cooperative productive enterprises, mostly for internal food security and exportation, and the distribution of the products all over the country; for example, the production of Umbu (a native Brazilian fruit) juice by a popular cooperative in the Bahian countryside, and the incentive to the creation of a public supermarket network to deliver the products to urban populations, to fight hunger. In this case, a structure should be organized, to allow the popular cooperatives, together with their financial institutions issuing the social currencies, to guarantee their issues with the Real Digital, by means of projects submitted to the BNDES and evaluated by a scientific committee, or by means of special credits offered directly to citizens, by another public institution to be created.

There is an important role for public banks in the development strategy. In many countries, Central Banks can generate reserves from the bonds issued by the Treasury, but cannot give this money directly to the people. In other words, the reserves should be exchanged with bank money (the money used in the market) issued by financial institutions. In other to direct the money to the people, the state has to use public banks. The implementation of this strategy requires another component for the structure: a technical support organization responsible for making the directions accessible to the public, associated with a new commercial financial institution able to make the conversion of *Real Digital* reserves into *Retail Reais*.

The first function may be assumed by the Brazilian Micro and Small Business Support Service (SEBRAE) that is present in practically all places in the country (https://sebrae.com.br/sites/PortalSebrae). This second function would require *the creation of a new Cooperative Bank, possibly* associated to the SEBRAE and financial branches of cooperatives, to provide special credits to small enterprises and assistential projects by non-profit organizations ("Third Sector"), possibly using public banks for direct interface with users (the people).

To finance human development through the promotion of ecological agriculture, guaranteeing internal food security and exports, together with necessary improvements in the health system, housing, transportation, social assistance etc., it will be necessary to authorize food production and other cooperative institutions, which operate within the principles of sustainability, to obtain special credits (with low interest, zero interest or even 'negative interest'), mediated by the new Cooperative Banks, based on strict meritocratic criteria (approval of scientific-technological projects, in accordance with a national human development project, by a qualified committee, in partnership with public universities).

It may eventually be the case of changing the first paragraph of Art. 164 of the Federal Constitution (which requires a PEC approved by 2/3 of both houses of the National Congress), which states that the issue of money by means of Treasury bonds can only be redeemed by "Financial Institutions", to allow individuals and legal entities to have access to the money issued by the BCB. With this change, it is possible redeeming Treasury bonds destined for the purposes of human development, and using public banks as vehicles for depositing money in the holders' accounts.

The problem of giving money to the poor people, non-profit assistential organizations and small businesses could also be solved without changes in the Constitution, through a financial engineering – the abovementioned "Many to One Strategy" – that combines CBDC with social currencies and special credits operated by public financial institutions, e.g. the abovementioned by creation of a SEBRAE-associated Cooperative Bank, delivering the money directly to citizens, non-profit assistential organizations and small businesses. The financial engineering requires an institutional structure involving public banks, sovereign funds, reserves, social fintechs (from cooperatives), social currencies, and state-people interface apparatuses (as the SEBRAE), once these services are democratically made available to the public, in all regions of the country.

Public banks in Brazil, although having a central position in the financial system, were not designed to support economic and human development programs with massive investments. Other countries, such as Sweden (with the *MedMera* Bank, founded in 2007) and Germany (with a network of *Sparkassen* and *Landesbanken*), have a financial system capable of financing cooperatives and small businessses. International cooperation for productive investments in Europe has been financed by the *European Investment Bank*.

One way to overcome the historical limitation of public banks to support popular enterprises in Brazil was the creation of cooperative banks, as the *Sicredi* (https://sicredi.gupy.io/), today the seventh bank in the Brazilian ranking, after 120 years of operation (see https://www.sicredi.com.br/trajetoria). However, this bank is not connected to the public administration system, but is an autonomous profit-seeking institution that operates with the resources of its associates. The Cooperative Bank to be supported by the federal state issue of CBDC, in our proposal, should be designed as a non-profit financial institution directed to public goals of economic and human development, in accordance with the principles of the Bioethics of Governance.

The public-private interface also requires a coordination center to evaluate and monitor projects executed with directions from the SEBRAE and financing by the related Cooperative Bank, with federal support from CBDC reserves. Lara Resende (2020), has suggested a *Development Agency* composed of scientists, evaluating projects on their merits within the adopted development strategy. This would be the central coordination of a distributed system, with the SEBRAE making the operative interface with the public. Therefore, we argue for the necessity of creation of a Development Agency, in sovereign states, with the power to redeem extra-budgetary bonds for public spending with selected projects of human development, to implement an ecologically sustainable and socially solidary economic system model.

7 Basic Income as an Igualitarian Share of Common Wealth

There is a colossal terminological problem in the discussion of the "public debt" that would ultimately result from the concepts embedded in public accounting, which, in Western capitalist countries, derive from private accounting; more precisely the 'Double Entry' system (in China, for example, the financial system is different, which greatly facilitates investment in strategic goals). If every debit corresponds to a credit, and vice versa, when currency is issued out of thin air (through encryption in state computers), then who is a debtor and who is a creditor? By convention, it is understood that the state would be the debtor, and that the amount that is issued adds to the public debt. As for the creditor, since the issuance is not made based on someone's wealth (ballast, guarantee), we have *an unknown* in financial equations.

We all know very well that in the financial system of countries like Brazil, creditors are those who receive interest in exchange for acquiring securities issued by the Treasury: banks, brokerages and other private financial institutions, which have privileged access to the money issued; and also, of course, the beneficiaries of public expenditures defined in the annual budget voted by the National Congress, in operations mediated by public banks. It does not go unnoticed, for the most critical analysts, such as the *Positive Money* movement in England and Europe, the *Public Banking Institute*, by Ellen Brown, in the USA, for the American adviser to the Chinese rulers, Michael Hudson, or for the Brazilian economist Ladislau Dowbor, that the money thus issued hardly ever reaches the pockets of the people, contrary to what happens in China, where the existing financial system enables strategic selective allocations for the money issued, including the Digital Yuan, which has been used to encourage environmentally friendly actions.

Private companies cannot create money. For them, what we can call the "Law of Conservation of Money" prevails, in which the financial amounts only change headings; the ultimate creditor is either the owner of the business, or some natural or legal person to whom the business owes. In the case of the federal state, when Fiat money is issued, who is the creditor? The creditor, in a democracy, cannot be the state itself or any or its rulers (this is typical of absolutism, in which those in power thought that the money would be theirs), nor can it be a power external to the state (this would be a case of neo-colonialism, incompatible with national sovereignty). There are two interpretative possibilities in democracy:

1. For the capitalist ideology, the state's creditors are the people who hold (they are the owners or have the possession) the wealth of society, and, therefore, the capacity of buying bonds and having the corresponding rights over the public debt; therefore, it is justified that the sovereign state, in order to be able to use such money, pays interest to these people;
2. For the social state ideology, as a matter of social justice, all citizens would be equal beneficiaries of "Fiat" money issues. This approach justifies the proposal by Guy Standing, from the University of London, in his books, that the payment of the Universal Basic Income would be a way for the state to remunerate its final creditors, promoting intergenerational justice, in which all citizens benefit equally from the wealth accumulated by their ancestors.

The main argument of capitalist ideologues against such an equitable distribution of state-issued Fiat money is that it would necessarily generate inflation. We will not here review this fallacious argument, and the valid counter-arguments presented in Brazil, with extreme competence, by André Lara Resende, as the subject has been exhaustively debated. However, we cannot fail to comment on an unfounded belief, which runs through the minds not only of economists who are against economic democracy, but also of a part of the population that experienced the problem of inflation during the period of military government and after the restoration of democracy in the 1990s. Many people in Brazil think that the 'Economic Miracle', which lasted until 1973, caused the superinflation that happened after the increase in the dollar and oil in 1977. Until now, no one has explained how, in the 4 years between, Brazilian economy violated the Law of Supply and Demand (increased production causing prices to rise rather than fall). Based on this false belief, some authors seem to think that large and selective productive investment now would cause inflation to rise. They olympicly ignore that the most economically prosperous countries, like the USA and China, issue huge amounts of Fiat currency and invest in the economy, in strategic technological research and in the well-being of their citizens. When challenged, they say that we cannot compare Brazil with these countries, when in fact what has been compared is the *type* of economic policy.

The naive Left, in turn, believes that the Executive power's ministerial team will manage to eradicate hunger, recover the productive system and infrastructure destroyed by the previous government, and improve the lives of the people, through taxes for the rich and cuts in government spending. They do not think it is necessary for the state to issue billions of Reais, and invest that amount selectively in the productive system, education, health, security, transportation, housing, basic sanitation, social security, social assistance etc., in short, in Human Development. Those who bet that a progressive tax reform will solve the problem of financing development do not take into account that many productive capitalists are financially tight, because the people do not have money to buy what they sell. In Brazil the debt for working capital pays the highest interest rates on the planet and, moreover, those who already evade current taxes will find ways to evade the additional. In turn, civil servants, who are the workers who can leverage human development in the country, will not run a program funded by reducing their wages (which for the most part are already behind schedule).

Maintaining the regime of scarcity of money for the people, while financial institutions make high profits by charging interest, is nothing more than maintaining slavery, by other means. The people are obliged to serve those who have money, in unfavorable conditions, which impede Human Development. In addition to recognizing all this, we also need to understand that the currency issued by the state cannot be just an asset of the "private sector", but of all citizens, who arguably have the right of receiving their share of the common wealth through the Basic Income, as argued by Guy Standing. In Brazil, the emission of money, by the federal state, in the format of the Real Digital, will most probably be needed to pay the Bolsa Família (a type of conditional Basic Income) and finance other public investments in human development, such as job guarantee, the public health system, basic sanitation, public transportation, social assistance, food safety, public security, affordable retirement programs etc.

According to Guy Standing, the main problems of contemporary society can be overcome by means of Basic Income (BI) programs (Standing, 2017). This democratic strategy has been criticized by some defenders of the Modern Monetary Theory (MMT), for the reason of (supposedly) being a *liberal* economical political program. We argue that the BI strategy to reduce inequality is not only compatible with MMT, but also necessary in any application of MMT principles to the induction of human development in public administration.

A study conducted by UNCTAD (see Barrawclough, 2018) arrived to the conclusion that the most effective way of boosting economic growth is the transfer of money to the poor, instead of lending the money in the form of microcredit (Standing & Orton, 2018). The transfer may be unconditional, as the Universal Basic Income, or in the form of benefits conditioned to the satisfaction of some requirements to enroll in the program, as in the cases of the Brazilian successful program Bolsa Família. A parallel public budget, controlled by a committee of experts indicated by the scientific and technological community, can be applied to finance these monty transfer programs, together with educational efforts towards the formation of a new generation of entrepreneurs to create new enterprises adequate to the available natural and human resources, to produce the goods to be bought by the enrolled people receiving the money transfer.

By implementing the Basic Income in dignified values, the mechanism of raising inequality is broken, triggering a transition process that could lead to a solidary society, which would be a system in which economic inequality would not reach the point of forcing a class of miserable people to submit to the exploitation of another class, in order to survive. In a financially democratic society, based on the autonomy of each person, recipient of the Basic Income, there would be a group of institutions that would make possible a participatory democracy, in which the people could monitor and control the action of the rulers, so that they behave according to bioethical principles collectively defined by the society.

In the Universal modality, all living people would receive a certain monthly amount, regardless of any conditions. In the Conditional modality, criteria are established for receiving benefits, as currently happens with the Bolsa Família and in other countries, such as benefits in England. There would no longer be the need for people to be exploited in order to survive. Money changes its function, which is currently a means of accumulation, and becomes a means of livelihood. The market continues, controlled by the democratic state, which prevents the private appropriation of collective work from generating an increase in inequality. This is done by identifying the profiteers, those who divert the fruit of collective work for their own and exclusive benefit. Entrepreneurs who generate wealth for the community should instead be rewarded for their contributions.

8 Towards a New Green Deal

The *New Deal* was a US development program that was implemented by President Franklin Delano Roosevelt to recover the economy from a period of recession after the war. So, when we talk about the New Deal, it means a development plan that is financed by the Federal State, with ample resources that the Federal State makes available for this new model to be implemented. Because the idea is that a new model will not emerge spontaneously, it is necessary to have this funding, the money that, so to speak, motivates people, with facilitated funding for certain types of venture, that which motivates people to undertake. Because nobody works miracles of being able to create a new type of enterprise on their own, it is necessary to have the support of the state to have a change of model, of paradigm. The most viable option for funding a New 'Green Deal' in Brazil would be *Green Bonds* issuing by the National Treasury, for international investors. Such bonds could be denominated in International Monetary Fund's Special Drawing Rights (SDR), for reducing cost of financing ecological programs. High interest rates make ecological bond financing via commercial markets too expensive for public policies; therefore, the multilateral institutional option should be further pursued. In addition, it is essential for ecological programs to be managed by qualified professionals, experienced in financial operations, risk management and auditing procedures.

In terms of production incentives, if we have a New Deal, it cannot simply be giving money to large companies, which are already well established, or to businesses that already exist and are not environmentally friendly. There has to be criteria. In a New Deal, funding has to be for those who meet certain requirements. And then, there is a need for the elaboration of projects and qualified personnel to evaluate these projects.

In the Brazilian context, projects to create new social enterprises should meet requirements such as the usage of natural resources in a sustainable way (for example, cultivation, industrialization and commercialization of products from our biomes); high employability; organization in cooperative form; use of adequate scientific and technological knowledge; originality and/or innovation of the products and services; respect to local communities and their knowledge; international sale of products. So what would be done?

From all over the country, mostly from the educational, scientific and technological institutions, people would make projects to be funded with low or no interest by the federal government. So, for example, from a university classroom, the professor could get together with interested students and other colleagues and elaborate on a project for the region they live, creating jobs for the people in that region. The project is to be evaluated by a committee of qualified scientists and, if approved, the money would be given to the people to create, to set up a company, to generate employment, to generate a product having value for the people.

A company of the type that we are talking about should have the criteria we mentioned, of generating employment, of using local resources, of being sustainable, as this will not generate much profit at the beginning. So, there has to be special lines of funding, similar to the scholarships we already have in Brazil. The Federal state is in charge of creating an agency; the agency must have state-issued money to finance approved projects. People in capable institutions, all over the country, can elaborate projects, e.g. a course conclusion work that will meet the criteria, send this project to be evaluated by a competent committee, and if it is approved, receives a grant or special funding to create a company that will, for example, extract, process and sell Pequi (a native Brazilian fruit) oil. This would be done by people who know the subject from their region, in partnership with research centers (in Brazil, mostly the public university). The university needs to get closer to the people, the communities, to see what is viable. We can't, for example, do a project to grow rice there in the Northeast. It's good there for the Brazilian South. So, we'll only know what's viable in a region if we have contact there with the countryside people, with the workers who would become partners in the enterprise.

9 Final Comments

The adjustment of production and consumption can be achieved by means of control of the flux of state-issued digital money in social networks. This control, to be done by the people together with the state managers, requires the apparatus described in the previous chapters: the Bioethics of Governance to establish principles of conduct; the Popular Forum to gather information and manifestation of the popular opinion; information technology for the tracking of the financial operations. At each region of the country, human work using artificial intelligence would track, for instance, which products the beneficiaries of Basic Income cash transfers are spending the money on, and then, through a Development Agency, the state would make efforts to finance the producers of those consumer goods, expanding the scale and improving production methods, so that their supply is increased in the respective region where there is demand. Thus, the increase in demand would not press for an increase in the price of the products.

These concerns lead to the task of rethinking the state's means of investing in popular production and consumption, human development, and the relation between the concept of human development and theories of complexity and self-organization. This type of program requires the definition of how to combine conservation and production, how public universities can give support to local communities, and ultimately about what is the role of science and the public university in a new economic model that has an ecosystem perspective. In the social area, we should also re-conceptualize how to achieve well-being and well-living, coping with information technology issues, quality of life and satisfaction indicators.

An important characteristic is employability; for instance, the mechanization within the agribusiness model generates unemployment. In the other model, in which we are trying to solve environmental and social issues, there should be activities that generate employment and income, including more people in the economic system. So, employability is an important issue, and for this reason, we should value the knowledge that people already have, not treating the native people as if they were ignorant. Within a model of ecological economics, the knowledge that they have of plants in the region where they live, much greater than ours, has value for the ecomonic system. So, scientists and technology experts working on the altrnative model need to value that knowledge, as for instance it already happens in the field of Etnobotanics (not without ethical concerns of refunding the native people for their knowledge, used in industrial products). Giving fair value to popular knowledge and social technologies is an important feature of the new model.

Another characteristic is that they must be projects aimed at generating popular consumer goods. It makes no sense, in a social perspective, to generate a very expensive product that only 0.001% of the population can buy, that is, a very restricted market This will not have a production scale, that is, we will not enable the generation of many jobs, nor a great transformation in the relationship with nature. So, the target of these enterprises should be popular consumption, and this implies having a distribution network for these products, to have the products placed all over the world, because there is a demand, there is interest in Brazilian tropical tasty foods, exclusive to our biome. What is missing is public investment to produce human healthy food using sustainable methods in a large scale, and a commercialization network for the flow of these products to a wide range of consumers.

Acknowledgments We are grateful to Rafael Bianchini Abreu Paiva, Frederico Carvalho, Cesar Tholedo, Luis Otávio Bau Macedo and Bruno Mader Lins for informative discussions, and to André Lara Resende for his original ideas and approval of an earlier version of this proposal.

References

Arellano, L. F. (2020). *Teoria Jurídica do Crédito Público e Operações Estruturadas* (Open Access E-Book). Ed. Blucher. Available at: https://www.blucher.com.br/livro/detalhes/teoria-juridica-do-credito-publico-e-operacoes-estruturadas-emprestimos-publicos-securitizacoes-ppps-garantias-e-outras-operacoes-estruturadas-no-direito-financeiro-1629

Barrowclough, D. (2018). Starting with the poor. In United Nations (Org.), *The ins and outs of inclusive finance: Some lessons from microfinance and basic income*. UNCTAD.

Brainard, L. (2020). *An update on digital currencies*. Board of Governors of the FED. Available at: https://www.federalreserve.gov/newsevents/speech/brainard20200813a.htm

Brown, H. (2019). *Banking on the people: Democratizing money in the digital age*. The Democracy Collaborative.

Carrillo, R. (2020). *Our money where our mouth is*. Current Affairs. Available at: https://www.currentaffairs.org/2020/08/our-money-where-our-mouth-is

Conti, J. M. (2020a). *O Planejamento Orçamentário da Administração Pública no Brasil* (Open Access E-Book). Ed. Blucher. Available at: https://www.blucher.com.br/livro/detalhes/o-planejamento-orcamentario-da-administracao-publica-no-brasil-1640

Conti, J. M. (2020b). *Dívida Pública* (Open Access E-Book). Ed. Blucher. Available at: https://www.blucher.com.br/livro/detalhes/divida-publica-1509

Debrun, M. (2018). The concept of self-organization. In Pereira Jr., et al. (Org.), *Systems, information and self-organization: A interdisciplinary perspective*. Routledge.

Derden, T. (2020). *Fed's "direct money transfers" are coming: Brainard says fed collaborating with MIT on "hypothetical" digital currency*. Zero Hedge. Available at: https://www.zerohedge.com/markets/preview-feds-coming-direct-money-transfers-brainard-says-fed-collaborating-mit-hypothetical

Drucker, P. (2003). *On the profession of management*. Harvard Business Review Press. ISBN 978-1-59139-322-1.

Eijkman, F., Kendall, J., & Mas, I. (2010). Bridges to cash: The retail end of m-pesa. *Savings and Development, 2*(XXXIV), 219–252.

Fernandez-Villaverde, J., Sanches, D., Schilling, L., & Uhlig, H. (2020). *Central bank digital currency: Central banking for all?* (Working Paper No. 26753). National Bureau of Economic Research.

Garcia, D. B. (2009). *A contextualização teórica de Bancos Comunitários de Desenvolvimento*. Instituto Banco Palmas. Disponível em: http://www.institutobancopalmas.org/wp-content/uploads/A-contextualiza%C3%A7%C3%A3o-te%C3%B3rica-de-Bancos-Comunit%C3%A1rios-de-Desenvolvimento.pdf

Hausmann, R., Hidalgo, C., & Bustos, S. (2014). *The atlas of economic complexity: Mapping paths to prosperity*. MIT Press/Harvard University.

Hidalgo, C., & Hausmann, R. (2009). The building blocks of economic complexity. *Proceedings of the National Academy of Sciences, 106*(26), 10570–10575.

Huber, J. (2017). Split-circuit reserve banking: Functioning, dysfunctions and future perspectives. *Real-World Economics Review, 80*. Available at: http://www.paecon.net/PAEReview/issue80/Huber80.pdf

Hudson, M., & Brown, E. (2020). *Let the banks go under and put money into the real economy*. CounterPunch. Available at: https://www.counterpunch.org/2020/07/23/let-the-banks-go-under-and-put-money-into-the-real-economy/

Kelly, M., & Howard, T. (2019). *The making of a democratic economy: How to build prosperity for the many, not the few*. The Democracy Collaborative.

Knapp, G. F. (1905). *The state theory of money*. Simon Publications.

Lagarde, C. (2018). *Winds of change: The case for new digital currency*. International Monetary Fund. Available at: https://www.imf.org/en/News/Articles/2018/11/13/sp111418-winds-of-change-the-case-for-new-digital-currency

Lara Resende, A. (2020). *Consenso e Contrassenso: Por uma economia não dogmática*. Portfolio Editora.

Lietaer, B. (2001). *The future of money: Beyond greed and scarcity*. Random House.

Lonergan, E. (2016, February 16). Debt-free money: A brief reply to Randall Wray. In *Philosophy of money* (Website). Avaliable at: https://www.philosophyofmoney.net/debt-free-money-a-brief-reply-to-randall-wray/

Mas, I., & Radcliffe, D. (2011). Mobile payments go viral: M-PESA in Kenya. *Journal The Capco Institute Journal of Financial Transformation – Applied Finance, 32*, 169–182.

Mitchell, W., Randall Wray, L., & Watts, M. (2019). *Macroeconomics*. MacMillan/Red Globe.

Neto Segundo, J. J. M., & Magalhães, S. (2009). Bancos Comunitários. *Em: IPEA Mercado de Trabalho, 41*, 21–26. Available in: https://repositorio.ipea.gov.br/bitstream/11058/4059/1/bmt41_10_Eco_Bancos_41.pdf

Ozili, P. K. (2022). Circular economy and central bank digital currency. *The Circular Economy and Sustainability, 2*, 1501–1516. https://doi.org/10.1007/s43615-022-00170-0

Ozili, P. K. (2023). *Redesigning the eNaira central bank digital currency (CBDC) for payments and macroeconomic effectiveness*. https://www.researchgate.net/publication/364358844_Redesigning_the_eNaira_central_bank_digital_currency_CBDC_for_payments_and_macroeconomic_effectiveness

Rodrigues, B. C. (2023, February 9). *La Cara Sur de la Moneda*. Coyunturas. Available at: https://coyunturas.com.ar/la-cara-sur-de-la-moneda/

Standing, G., & Orton, I. (2018). Development and basic income: An emerging economic model. In United Nations (Org.), *The ins and outs of inclusive finance: Some lessons from microfinance and basic income*. UNCTAD (United Nations Conference on Trade and Development).

Alfredo Pereira Jr is a Philosopher of Science, Professor at the Universidade Estadual Paulista Julio de Mesquita Filho (UNESP), Brazil, retired, with degrees in Philosophy and Administration. Currently working in the UNESP Graduate Program in Philosophy. Alfredo has around 300 publications in several areas of knowledge, mostly in the Philosophy of Neuroscience and Theory of Consciousness. Orcid: https://orcid.org/0000-0002-5960-041XAD, Scientific Index: 56th place in Top 100 Latin America Philosophers/Scientists.

Francisco Sousa is an Economist for social enterprises: structures, operations management and market development. Mr. Sousa's work experience includes risk management, trading markets and consulting services with client assignments in the United States and overseas. Mr. Sousa has worked with IBM, Canada's Manufacturers Life Insurance, CACI International, George Washington and Johns Hopkins Universities. In recent years, Mr. Sousa served on the 'Community Housing Finance Commission' in Arlington, VA appointed by the County Board.

5 Ways to Reshape the Economy Towards Solidarity and Ecology: Impact Investment and Social Impact Bonds

Mathilde Pellizzari and Pauline Boulanger

Abstract

This chapter suggests that a more sustainable version of the world economy is possible. Productive activities need to align with sustainability requirements in terms of caring about the planet and the people. Since in modern economies financial investment is the main channel through which production is shaped, this chapter argues for a financial system that finances and favors activities that seek to deliver social good. It addresses the growing trend of impact investing. As defined by the Global Impact Investing Network (GIIN), impact investments are investments made with the intention to generate positive, measurable social and environmental impact alongside a (positive or zero) financial return. This agenda calls for further collaboration between all economic actors, namely, financiers, governments, businesses, social organizations, and citizens. This chapter introduces and discusses social impact bonds (SIBs) as one mode of collaboration in which the public sector, the social sector, and the financial sector align to experiment social programs while assessing their social impact. Through the study of impact investing and SIBs in the French context, this chapter additionally reflects on the transformations impact measurement and outcome funding imply on policymaking.

M. Pellizzari (✉)
Centre de Sociologie de l'Innovation, Mines Paris-PSL, Paris, France

P. Boulanger
PROPARCO, Agence Française de Développement Group (AFD Group), Paris, France

1 Impact Investing and the State: What Is Impact Investing?

In his 2020 book, Sir Ronald Cohen, one of the founding fathers of impact investing, defines impact as "the measure of an action's benefit to people and the planet. It goes beyond minimizing harmful outcomes to actively creating good ones by creating positive impact. It has social and environmental dimensions" (Cohen, 2020). Impact investing is a kind of finance that actively supports such actions. Impact investing seeks the creation of positive impact. Thus, it translates a stronger commitment to society than other financial practices such as Socially Responsible Investment (SRI) and approaches looking at ESG criteria (i.e., Environmental, Social, and Governance) (see Arjaliès et al., 2022). As this chapter discusses, impact investing is attached to a political project that aims to transform the world economy.

According to the Global Impact Investing Network (GIIN), the global impact investing

A. Pereira Jr, F. Sousa (eds.), *Principles for Governance*, Sustainable Development Goals Series,
https://doi.org/10.1007/978-3-031-40025-4_5

market would amount to USD 1.164 trillion assets under management (GIIN, 2022). This number has been growing at a fast pace over the last 15 years, but the amounts remain limited in comparison to the global financial sector: global assets under management account for more than USD 112 trillion (BCG, 2022). Impact investment remains a drop in the ocean of finance. However, it has shown incredible resilience, including in the aftermath of the COVID-19 crisis. It has consistently supported the development of a different, more inclusive, more just, and more environment-friendly economy.

Impact investing can contribute to pave the way for a new economy while complementing the public financing of social activities. What is often called the Welfare State has taken various forms across Europe. In this kind of political system, the state provides social services for the benefit of all citizens, including the most deprived. Over the last two decades, impact investing has emerged as an additional and complementary way of serving social purposes. Since 2007, when the practice of "impact investing" was first coined during a meeting hosted by the Rockefeller Foundation in Italy, this movement has gained momentum (E.T. Jackson & Associates Ltd., 2012). It originally aimed at channeling institutional investors' money towards impactful social initiatives. The GIIN set a definition for the impact investing practice, which is now recognized and adopted worldwide: "Impact investments are investments made with the intention to generate positive, measurable social and environmental impact alongside a financial return." Building on the GIIN's definition, a group of French impact investing practitioners provided an extended definition of impact investing for the French context. This definition rests on three pillars, namely, intentionality, which requires the formalization of an impact strategy by investors; additionality, that is, proactive asset management towards the maximization of the impact produced by the investees; and effective and transparent measurement (FIR & France Invest, 2021).

In France, impact investing was established long before the term "impact investing" itself appeared. The so-called "solidarity-based finance" (i.e., "finance solidaire" in French) came to life in the 1970s and is highly intertwined with the French Welfare State. The first 90/10 fund, a flagship tool in solidarity-based finance presented in further detail below, was born in 1994. One peculiarity and strength of this solidarity-based finance resides in the sourcing of the funds used to finance social projects: the savings of private individuals. Amongst many savings options, and thanks to enabling and protective state regulations, any French citizen can decide to put his or her savings into products that transparently invest in social and/or green businesses.

There are many different ways to articulate government and private actions, from some level of independence between both, to close collaboration or facilitation. Although this chapter does not take for granted the global narrative of public funding scarcity calling for systematic support from private finance,[1] it acknowledges that impact investors are *de facto* supporters of social-purpose organizations and social services. We recognize that a diversity of setups is useful to address a diversity of challenges.

1.1 Inherent Diversity

Impact investors share essential features such as being funding providers – in the form of invest-

[1]This narrative is exhibited in, for instance, Social Impact Investment Taskforce. (2014). Impact Investment: The invisible heart of markets. Harnessing the power of entrepreneurship, innovation, and capital for public good (p. 51). https://impactinvestingaustralia.com/wp-content/uploads/Social-Impact-Investment-Taskforce-Report--FINAL.pdf [visited on 03/04/2023]. See also The OECD. "Blended Finance." https://www.oecd.org/development/financing-sustainable-development/blended-finance-principles/ [visited on 03/04/2023]. For critical discussions about this narrative, see Chiapello, E., & Knoll, L. (2020). Social Finance and Impact Investing. Governing Welfare in the Era of Financialization. Historical Social Research, 45(3), 7–30. https://doi.org/10.12759/HSR.45.2020.3.7-30; Golka, P. (2019). Financialization as Welfare. Social Impact Investing and British Social Policy, 1997–2016. Springer; and the introduction of Pellizzari, M. (2022). *Feasibility first: Trials and tactics of feasibility in the implementation of social impact bonds* [PhD dissertation]. Mines Paris – PSL.

ments – and having a thorough understanding of the business models and overall activities of social businesses, non-profits, or impact entrepreneurs. They explicitly seek social impact and promote the growth of impact investing.

Beyond these points of commonality, impact investors vary in terms of identity, strategies, and the products they use. First, there are "pure players" such as private equity fund managers (e.g., INCO, among many others, in France) or innovative ad hoc organizations (e.g., France Active, a network of territorial agencies offering solidarity-based financing and advisory support) that are fully dedicated to impact investing. Second, there are banks with strong ties with the French social and solidarity economy (e.g., Crédit Coopératif, a French cooperative bank) as well as more conventional financial services providers (e.g., the commercial bank BNP Paribas, insurance firms such as Aviva). These institutions allocate part of their resources to impact investments, often with a dedicated impact investing team in-house. Third, state-owned banks (e.g., the public bank Caisse des Dépôts et Consignations, the development agency Agence Française de Développement) also play a meaningful role in the impact investor community.

Investment strategies vary across impact investors. Some focus on well-established organizations in their scale-up process, while others rather support early-stage ventures. Some choose to invest in one specific sector (e.g., decent housing, sustainable agriculture), or one kind of beneficiary organization (e.g., cooperatives, entrepreneurs from deprived neighborhoods), while others opt for a diversified approach. Importantly, some investors expect financial returns in line with mainstream finance, while others prioritize positive impact over financial return and may thus expect lower financial performance.

Impact investors also resort to a variety of financial tools, from debt to equity and quasi-equity or guarantees. They are usually referred to as "patient" investors (who invest "patient capital"), which means that they expect returns after a longer period of time than mainstream investors. Hence they use products with longer maturities. Moreover, specific impact investing tools have been developed. For instance, in France, "titre associatif" (i.e., literally, associative security) and "titre participatif" (i.e., literally, participative security) were created by law to enhance the financial strength of, respectively, associations and cooperatives. Associations and cooperatives are key legal forms within the social and solidarity economy. Associations are nonprofit organizations which, by nature, do not have equity. Cooperatives do have equity, which is exclusively owned by the members of the cooperative. Through associative and cooperative securities, associations and cooperatives can raise external funding. These products aim to strengthen the balance sheets of such organizations, in order to build up resilience and facilitate growth.

As this chapter extensively discusses below, social impact bonds (SIBs) are a financing product for impactful social programs that has been on the rise since 2010. It involves not only investors and social organizations, but also public authorities. In SIBs, impact investors provide a social organization with upfront funding for the implementation of a social program with precise social goals. After a period of implementation (from 1 to 7 years), the state (central government or local authorities) pays according to the effective social results. If the results reach the predefined goals, then the investors are fully reimbursed with a financial return. If the results are below the objectives, the investors may lose part or all of their investment.

To conclude, the impact investing sector is multi-faceted. As such, it intends to reflect and support the complexity of social and environmental issues and the resulting diversity of market players in the social and solidarity economy.

1.2 The State, an Active Player in the Impact Investing Ecosystem

By establishing an enabling policy framework, the state can foster the growth of impact investing. Legislation and regulation are important levers for developing an impact economy in which impact investors are able to adequately support all types of organizations dedicated to creating impact (social businesses, non-profits, impact entrepreneurs, social, and solidarity enterprises).

In a 2019 publication on social impact investment, the OECD proposes a policy toolkit to promote impact investing (OECD, 2019). It identifies different ways for states to do so: from defining national strategies, to setting up fiscal incentives or enabling legislation or regulation, to allocating public funds to impact investment initiatives, to communicating and disseminating accurate financial and impact-related information.

In France, policy has facilitated the growth of the impact investing market. Firstly, a unit dedicated to the social and solidarity economy and impact investment has been in place for several years within the French administration. This unit drives interesting initiatives for the sector. For instance, the administrative unit administrates and delivers an accreditation to social businesses that are considered to be part of the social and solidarity economy, as defined by the French law. This accreditation facilitates the identification of social businesses and the channeling of dedicated funding to them.

Moreover, the regulation of company savings plan has been a big step towards the development of impact investing in France. Indeed, all big companies (i.e., employing more than 250 people) are required to offer its employees a company savings plan. Each company may propose several funds. Since the adoption of a dedicated law in 2008, companies must include a 90/10 fund into the savings options, that is, a savings product in which up to 10% of the funds are directed towards social businesses.

More recently, in 2020, the French government decided to launch a "second wave" of SIBs, after a dozen of SIBs were launched between 2017 and 2020. Three dedicated funding envelopes of about EUR 10 million were pledged by, respectively, the Agency for Ecological Transition, the Ministry of State for the Social and Solidarity Economy, and the Ministry of Labor. Concretely, this funding is allocated to innovative social programs that aim to deliver effective social outcomes in the sectors of, respectively, circular economy, equality of economic opportunities, and inclusive work.

1.3 The Value of Assessing Impact

Measuring impact is a way for a social organization to report on the outcomes of its actions to its financiers, to government, or to any other stakeholder. It is also, and this is at least equally important, a way to drive and continually improve its actions towards enhanced impact. As such, it follows what can be called a "prove and improve" logic.

The United Nations' Sustainable Development Goals (SDGs)[2] provide a widely used framework to categorize the nature of the impact delivered. There is not one established methodology for social organizations to report on their impact. Common methodologies include the following:

- Establishing a *theory of change*, that is, a framework in which an organization defines how it will achieve its goals and explains the causality between the inputs, activities, and expected outcomes.
- Defining a set of *indicators*, possibly from the IRIS+ database developed by the GIIN,[3] to be monitored regularly by the social organization, but also by impact investors.
- *Monetizing* its social impact, for instance with the SROI methodology (Social Return On Investment), which calculates a ratio between the cost of a project and the monetized social benefits of this project.
- Using *counterfactual* methods, in which the outcomes of a group of program beneficiaries are compared with those of a control group that does not benefit from the program.

As impact investing expands and the number of stakeholders involved in this market grows,

[2] Source: United Nations' Sustainable Development Goals website: https://sdgs.un.org/ [visited on 03/04/2023].

[3] Source: IRIS+ website: https://iris.thegiin.org/ [visited on 03/04/2023].

conversations on impact measurement focus on two main points: first, the quality of measurement in terms of ensuring impact integrity, that is, ensuring that impact investments are indeed directed towards impactful projects; second, the level of measurement complexity that is possible and desirable to implement.

Many initiatives worldwide have been focused on standardizing impact criteria. This is a slow and contested enterprise given the complexity and granularity of the social and environmental effects produced by the actions of impact businesses. On the one hand, systematic impact measurement can effectively help showcase the benefits brought by impact businesses. The financial industry would benefit from further standardization in order to scale impact investment by comparing different organizations more easily and transparently. But on the other hand, the downside of standardization resides in the risks of rigid categorization, financial exclusion of certain organizations that may not seem to produce enough or the right type of impact, and a biased normalization of what “good” impact ought to be. To add to this debate, we underline here one interesting idea behind the systematization and standardization of impact measurement, which is to shift investors’ interests towards impact at the same time as to incentivize businesses to manage their impact. Several initiatives are taking steps in this direction, such as ecological accounting, led by the Chaire comptabilité écologique of AgroPariTech in France,[4] or impact-weighted financial accounts led by Harvard Business School.[5]

To conclude, impact assessment has raised much attention over the last decade. The path forward is not only to have it adopted widely by social organizations without compromising the roll-out of their activities, but also to have other stakeholders aware and involved. Further standardization is necessary for the coherence of the sector, but this should not mean standardizing social action, in order to make sure to preserve the diversity of social services and support.

2 Social Impact Bonds: A Partnership for Social Outcomes

At the 2020 Global Impact Summit organized by the Global Steering Group for Impact Investment (GSG), the head of the Division for Social and Solidarity Economy and Impact Investing from the French government joined an online workshop about outcomes-based financing. He expressed in a meaningful way the position of the French administration regarding social impact bonds (SIBs). He said that SIBs have made possible multi-stakeholder partnerships over the medium term to experiment innovative social services, which put together several administration departments in an unprecedented manner. As an example, he mentioned a France-based SIB program that develops an alternative to jail for homeless people with severe mental health issues who are brought to court. This program, he explained, put together different levels of court, the Ministry of Justice, and the Health Administration. To him, such a level of coordination would have never happened without an engaging contractual framework such as the *contrat à impact* (i.e., “impact contract,” the French translation of SIB).

This section introduces SIBs globally and then focuses on SIBs in France. It builds on a feedback report on SIB implementation from 2016 to 2019 based on 21 interviews, drafted by one of the co-authors of this chapter and published by Impact Invest Lab (iiLab) in December 2019 (Pellizzari & Sebag, 2019). It additionally analyzes SIB resilience after the COVID-19 crisis (see Pellizzari, 2020).

SIBs are multi-stakeholder partnerships between policymakers, investors, and social service providers to implement social programs that are subject to impact evaluation. SIBs aim to shift public expenditures towards preventative services and effective social outcomes. Indeed, in

[4] More information available from Chaire Comptabilité Ecologique website https://www.chaire-comptabilite-ecologique.fr/?lang=en [visited on 03/04/2023].

[5] More information available from Impact-Weighted Accounts Project website https://www.hbs.edu/impact-weighted-accounts/Pages/default.aspx [visited on 03/04/2023].

this scheme, an "outcomes payer" from the public sector pays for social outcomes. Private investors provide upfront funding to a social service provider for the implementation of a supportive program for a targeted population. Programs can relate to employment, but also to child welfare, criminality, or homelessness, for instance. In case of good social performance, investors are reimbursed and receive a bonus. Outcome objectives are defined upfront and the results are certified by a third-party evaluator.

The first SIB program was implemented in Peterborough, UK, in 2010 for reducing reoffending rates among short-sentenced prisoners in the local prison. Since then, 283 SIBs have been implemented in 38 countries around the world.[6] SIB development has been supported by narratives arguing that, as a consequence of the 2008 crisis, public funding is too scarce to address the funding need in the social service sector (Social Impact Investment Taskforce, 2014). Some argue for better efficiency in public funding and better performance of social services (Carter et al., 2018; Dear et al., 2016; Gustafsson-Wright, 2018). In this context, collaboration between governments and impact investment practitioners is often praised (Carter et al., 2018; Dear et al., 2016; La Torre et al., 2019; Social Impact Investment Taskforce, 2014), in that SIBs, by channeling welfare spending to effective outcomes, could generate public savings from preventative programs. Indeed, the payment for outcomes mechanism is supposed to create financial incentives for investors and social providers towards the achievement of social outcomes while introducing flexibility in operational management, which can additionally fuel social innovation. In this view, multi-actor collaboration for funding and designing social programs could enhance outcomes.

However, by all accounts, SIBs have remained a niche market. A minimum (because data is missing) of USD 753 million have been invested. Comparatively, the global impact investing market size is estimated to be USD 1.164 trillion in 2022.[7] Social spending represents USD 12 471 billion in 2019 in the OECD only (USD 9 143 per capita).[8] Moreover, the 283 registered SIBs have reached a minimum of 2.1 million people while for instance, in 2019, 8.5% of the global population suffers extreme poverty at USD 2.15 per day (about 658 million people), according to the World Bank.[9] Some scholars identify some reasons why the SIB market may be developing at a slow pace. Arena et al. (2016) identify four key issues that have slowed down the diffusion of SIBs: the legislative framework, public procurement procedures, the public accounting framework, and the measurement infrastructure. SIBs put together social policymaking and private investment, which indeed in many respects has not been a straightforward process. To Williams (2020), SIBs are indicative of a collision between a financial logic using standardized tools and risk scales, and complex local realities, which fundamentally hinders SIB development. Harsher critiques have arisen in the literature. SIBs are seen as a new form of marketization of state policy (Berndt & Wirth, 2018; Joy & Shields, 2013; Mitropoulos & Bryan, 2015), that could deepen the financialization of the sector (Cooper et al., 2016; Tse & Warner, 2020; Warner, 2013). In this view, SIBs are riding

[6]This is a minimum amount since financial data are missing for a number of projects. To be more accurate, this number of projects include proper SIBs and development impact bonds (DIBs). The latter are implemented in developing countries with development banks and foundations paying for outcomes instead of governments. Source: INDIGO Impact Bond Dataset https://golab.bsg.ox.ac.uk/knowledge-bank/indigo/impact-bond-dataset-v2/ [visited on 25/08/2023].

[7]The GIIN. (2022, October 12) "GIINsight: Sizing the Impact Investing Market 2022" https://thegiin.org/research/publication/impact-investing-market-size-2022/ [visited on 25/08/2023].

[8]OECD Data. "Social spending." https://data.oecd.org/socialexp/social-spending.htm [visited on 25/08/2023].

[9]The World Bank Data. "Poverty headcount ratio at $2.15 a day (2017 PPP) (% of population)." https://data.worldbank.org/topic/poverty; and "Population, total." https://data.worldbank.org/indicator/SP.POP.TOTL [visited on 25/08/2023].

the wave of neoliberal reforms, which could undermine social policymaking.

In sum, there are "narratives of promise" and "narratives of caution" about SIBs, as Fraser et al. (2018) phrase it. Here, we intend to add on these debates by analyzing the concrete implementation of SIBs.

2.1 Where Do SIBs Come From?

SIBs were first mentioned in a 1988 paper by Ronnie Horesh, a New Zealand-based economist, pushing for what he called "Social Policy Bonds" (Agribusiness and Economics Research Unit, Lincoln College, Canterbury, 1988). Horesh described a financing model in which social policy outcomes can be rewarded. He proposed to implement tradable non-interest-bearing bonds, to be issued by public authorities, in order to incentivize the market to achieve social and environmental goals. Actual SIBs took a different shape. The first SIB was launched in Peterborough, UK, to finance a program to prevent ex-prisoners from the Peterborough jail from reoffending and returning to jail. It was a contractual partnership that involved a high number of stakeholders: seven social providers gathered as the "One Service" consortium, a group of 17 investors, the Ministry of Justice and the Big Lottery Fund as outcomes payers, an intermediary and program manager, three independent evaluators, a number of local professionals and 2000 beneficiaries. The stakeholders defined the objectives as reducing the reoffending rate whether by 10% in one cohort of 1000 beneficiaries or by 7.5% in the three cohorts over 7 years. These objectives conditioned the payment from outcomes payers to investors, including the principal and a variable return from 2% to 13%. Finally, the program was implemented for 5 years on two cohorts instead of three cohorts over 7 years because a national program was implemented for the same purpose. The final assessment accounted for a 9%-reduction in reoffending overall. The results were considered successful, so the investors received a payment of their initial capital and a return of 3% per annum. The Peterborough partnership established the SIB as a model to replicate. It differs from Horesh's Social Policy Bond in at least two features. First, a SIB is a partnership and not a tradeable asset, contrary to what the term "bond" suggests. Second, the return on investment in SIBs is variable depending on the social results whereas Policy Bonds would have a fixed interest depending on the achievement of a social objective.

SIBs crossed the Atlantic Ocean to the Ricker's Island jail in 2012 in New York City, USA, and then developed across the USA. The first Europe-based SIB was set up in 2014 in Belgium, led by Brussels' Employment Agency Actiris, with the objective of reducing the unemployment rate among young immigrants. In France, SIB development started in 2016, when the government issued a call for SIB proposals addressed to social organizations. In 2013, UK Prime Minister David Cameron set the Social Impact Investment Taskforce within the G8. France took part in this international taskforce and formed a National Advisory Board (NAB) to examine impact investment in the country. The French NAB was led by the Vice-President of the cooperative bank Crédit Coopératif, and was composed of 29 people from banks, venture capital, social entrepreneurship, independent expertise, administrations, academia, international organizations, etc. (Comité français sur l'investissement à impact social, 2014). The NAB submitted the final report in September 2014 to the government. The report provided an assessment of the impact investment sector and included 22 recommendations to foster its development. One of the recommendations was to "experiment Social Impact Bonds while adapting them to the national context" (Comité français sur l'investissement à impact social, 2014). In parallel, the government started investigating SIBs and how these could fit in with the principles and rules of the French welfare state.

2.2 The Adie CIS

The government decided to experiment SIBs through a call for proposals, which was launched in March 2016. The call was open to all organizations from the social and solidarity economy and to all policy areas within the legal frame of social innovation, that is, activities dedicated to populations that are not or are not adequately served by existing services. The government received about 60 applications. Thirteen proposals were selected as eligible programs for SIB financing. Ten of these programs were effectively launched as SIBs between 2017 and 2021.

The first French SIB, launched in 2017, was implemented by the French association Adie (i.e., the social provider). It aimed to develop a microcredit offer for entrepreneurs in targeted rural areas. Adie is a renowned French organization that supports professional inclusion through microcredit. Adie aimed to support entrepreneurs in rural areas who could not access on-site services due to mobility issues. Thanks to the SIB, Adie developed remote support for this population in the form of a 6-year experimentation. The Ministry of Economy and Finance is the outcomes payer. The objectives of this SIB are twofold: (1) providing microcredit and support to 269 to 500 people and (2) achieving sustainable professional inclusion for 172 to 320 people. A group of five social investors provided a total of €1.3 million upfront to cover the operational budget and SIB-related costs. Among the investors were the French public bank Caisse des Dépôts and the CSR department of the commercial bank BNP Paribas, which also played the role of the structurer, that is, the intermediary in charge of legal and financial structuring and stakeholder coordination. Out of these €1.3 million, €1.2 million were dedicated to the operations, and the remaining €100,000 was for legal and financial structuring and evaluation. The outcomes payer pledged an additional €200,000 as a bonus for investors but also for the social provider in case of overperformance.

2.3 Lessons Learned and Perspectives

2.3.1 What Were the Motivations of Social Providers in France to Develop SIBs?

First, SIBs allow for innovation outside of sometimes rigid ordinary law and siloed public budget. SIB financing helps providers overcome difficulties to find tailored funding for their programs. As one provider explained: “Even before looking for funding, we knew it would be complicated because we wanted to develop the project on three different territories” (Pellizzari & Sebag, 2019). This implied to target three or more different public budgets, which would have required intensive efforts, and maybe in vain. Besides, SIBs secure an adequate amount of money to carry out an experimentation over several years whereas grants are usually given on an annual basis. Second, SIBs allow social organizations to create a dialogue with public authorities in order to influence public policy. “[Using a SIB to finance our client’s program] allows the public stakeholder (here the department[10]) to test an innovation while paying only if it works,” a structurer explained (Pellizzari & Sebag, 2019). To date, there has been no clear commitment from outcomes payers to keep supporting the programs after SIB financing ends. Still, SIBs have brought knowledge on social issues and adapted solutions. Third, SIBs value social impact. On the one hand, impact measurement sheds light on outcomes. On the other hand, impact is granted an economic value through the financial model. Fourth, social providers are attracted by the novelty of SIBs, and in some cases, they take it as a stimulating challenge.

2.3.2 How Is Collaboration Implemented into the Governance Structure?

SIBs are one mode of collaboration in which public authorities, social organizations, and investors align in order to finance social programs and assess their social impact. The exam-

[10] Local authority in France.

ples of the Peterborough SIB in the UK and the Adie SIB in France show that SIBs involve a diversity of actors, but also cover different financial and legal arrangements, and different forms of impact measurement. Such collaboration mechanism is highly complex in terms of legal structuring and require heavy negotiation processes. Indeed, SIBs bring together stakeholders with very diverse backgrounds, languages, and tools, which must align to build a common contract for social purpose. As such, the SIB model is an unprecedented collaboration model. In France, the SIB mechanism is highly collaborative and aims to develop bottom-up innovation coming from social providers. Thus, SIB contracts and impact indicators are tailored to each program. The structuring phase involves all the stakeholders, who discuss the social objectives and the actual contracts. Needless to say, such level of co-construction entails to dedicate 1 to 2 years to the design of each program and involves high transaction costs.

During the phase of implementation, governance committee that involves all the stakeholders regularly gathers – at least once a year – to take stock of progress and resolve problems as needed. In fact, the governance committee is entitled to make decisions about all that the contract does not explicitly cover.

Both the evaluation and the dialogue engaged through SIBs last beyond the scope of the programs. SIBs can deeply transform outcomes payers' intervention models. An official from a public agency involved in two SIBs expressed this idea very clearly: "We ourselves build the Agency's modalities of intervention along the way. The SIB has an impact at the level of management, at the budgetary level, at the level of the articulation with our historical intervention devices, at the level of ownership by our network, including deconcentrated state services and local authorities. So we need to create a tailored intervention model [...]" (Pellizzari & Sebag, 2019).

SIBs articulate different actors' practices to co-construct a model focused on impact. They are complex arrangements in which building trust among the stakeholders is essential. "There was no blockage, simply because everyone was looking in the same direction. [...] In fact, everyone wants impact [...]", as one provider expressed (Pellizzari & Sebag, 2019).

2.3.3 Are SIBs a Sustainable Way of Doing Public Policy, Especially Since COVID-19?

When COVID-19 hit in late 2019, almost all productive activities stopped within a few months, besides massive home confinement. This situation led to not only a health crisis but also an economic crisis and a social crisis. The most vulnerable populations before COVID-19 have been and will be the most affected. Populations that are poorly housed and homeless have suffered from the impossibility to quarantine in proper conditions, and they have received less support from charities. According to International Labor Organizations in 2018, 60% of the worldwide labor force works informally (Bonnet et al., 2018). Many people thus lost their livelihood due to lockdown. The crisis has made inequalities more salient in terms of poverty, housing, education, but also loneliness and access to internet.

In this context, impact investing proponents have pushed for urgent and structural actions in response to the crisis. In this view, SIBs and more generally payment by result could accelerate research and innovation (Cohen, 2020), but it could also ensure public accountability, flexibility, and robust monitoring for future social programs (Lopez Taylor & Bode, 2021; Palandjian & Brest, 2020). However, SIBs as such may not be suited for emergency responses. First, they are complex contracts. Second, SIBs are based on the prediction of future impact, while in times of crisis, the future becomes uncertain.

Although SIBs may not be suitable to all contexts, two aspects proved useful during the crisis: data on the one hand, and governance on the other. First, sound data analysis can bring valuable information on the effect of the crisis on vulnerable populations. Second, collaborative governance ensures the resilience of SIB partnerships, since governance committees can make decisions on any contract alteration. In fact, SIBs have proved quite resilient during the crisis. The programs that could be adapted despite the sanitary measures kept going whereas others had to pause. In some SIBs, the stakeholders preferred to temporarily switch to direct subsidy instead of outcomes financing. In some cases, the objectives were modified by the governance committee. To allow for further flexibility, contracts could become more relational and less transactional, that is, they could include modalities of collaboration instead of fixed deliverables (Ball, 2020).

3 Conclusion

To conclude, this chapter introduced impact investing and SIBs as initiatives that put forward a timely political project: that of gathering cross-sector forces to intentionally serve the most vulnerable thanks to financial and economic levers. As economic inequality is rising in the context of an ecological, sanitary, social, and diplomatic crisis, we believe that the financial system should shift towards the systematic generation of positive social and environmental impact (possibly by reducing profit expectations). Still, an important challenge lies in the possibility to use the vocabulary and principles of finance (i.e., profitability, viability, investment tools, and future performance) for the benefit of social and environmental impact instead of transforming welfare and ecology for the sake of finance.

Impact investments are investments made with the intention to generate positive social and environmental impact alongside financial profits. Impact investing is a way to transform the financial system towards a fairer and transparent distribution of resources to solitary activities. More widely, the transformation of the financial system goes along with the transformation of businesses: the latter could not produce as much negative impact if shareholders were all committed to a positive impact strategy.

SIBs, as a partnership between public authorities, social services providers, and investors focused on the generation of social impact, ensure cross-sector collaboration, shared governance, and the exploration of social policy outcomes. Although SIBs may pose many other difficulties such as complex contractual framework, partial measurement, and financial pressures, they are an example of public-private collaboration for social good.

References

Agribusiness and Economics Research Unit, Lincoln College, Canterbury. (1988). *Papers presented at the New Zealand Branch Australian Agricultural Economics Society Conference* (Vol. 1, p. 174). New Zealand Branch Australian Agricultural Economics Society. https://researcharchive.lincoln.ac.nz/bitstream/handle/10182/848/aeru_dp_121_vol1.pdf?sequence=1&isAllowed=y

Arena, M., Bengo, I., Calderini, M., & Chiodo, V. (2016). Social impact bonds: Blockbuster or flash in a pan? *International Journal of Public Administration, 39*(12), 927–939. https://doi.org/10.1080/01900692.2015.1057852

Arjaliès, D.-L., Chollet, P., Crifo, P., & Mottis, N. (2022). The motivations and practices of impact assessment in socially responsible investing: The French case and its implications for the accounting and impact investing communities. *Social and Environmental Accountability Journal*. https://doi.org/10.1080/0969160X.2022.2032239

Ball, N. (2020, September 9). *Reflecting on the social outcomes conference 2020*. GOLab. https://golab.bsg.ox.ac.uk/community/blogs/reflecting-social-outcomes-conference-2020/

BCG. (2022). *Global asset management 2022 – 20th edition. From tailwinds to turbulence*. https://web-assets.bcg.com/c8/5a/2f2f5d784302b945ba1f3276abbc/global-asset-management-2022-from-tailwinds-to-turbulence-may-2022.pdf

Berndt, C., & Wirth, M. (2018). Market, metrics, morals: The social impact bond as an emerging social policy instrument. *Geoforum, 90*, 27–35. https://doi.org/10.1016/j.geoforum.2018.01.019

Bonnet, F., Leung, V., & Chacaltana, J. (2018). *Women and men in the informal economy: A statistical picture* (3rd ed.). International Labour Office. https://www.ilo.org/wcmsp5/groups/public/%2D%2Ddgreports/%2D%2D-dcomm/documents/publication/wcms_626831.pdf

Carter, E., FitzGerald, C., Dixon, R., Economy, C., Hameed, T., & Airoldi, M. (2018). *Building the tools for public services to secure better outcomes: Collaboration, prevention, innovation* (p. 37). Government Outcomes Lab. https://golab.bsg.ox.ac.uk/documents/BSG-GOLab-EvidenceReport-20190730.pdf

Cohen, S. R. (2020, April 18). *How to mobilize a global testing effort: Pay for success*. Barron's. https://www.barrons.com/articles/how-to-get-enough-coronavirus-tests-pay-only-for-success-51587221753?mc_cid=4719cef9ba&mc_eid=501c497e1d

Comité français sur l'investissement à impact social. (2014). *Comment et pourquoi favoriser des investissements à impact social? Innover financièrement pour innover socialement* (p. 150). Comité français sur l'investissement à impact social.

Cooper, C., Graham, C., & Himick, D. (2016). Social impact bonds: The securitization of the homeless.

Accounting, Organizations and Society, 55, 63–82. https://doi.org/10.1016/j.aos.2016.10.003

Dear, A., Helbitz, A., Khare, R., Lotan, R., Newman, J., Crosby Sims, G., & Zaroulis, A. (2016). *Social impact bonds: The early years*. Social Finance.

E.T. Jackson & Associates Ltd. (2012). *Final report of the strategic assessment of the Rockefeller Foundation's impact investing initiative*. Rockefeller Foundation.

FIR, & France Invest. (2021). *Investissement à impact. Une définition exigeante pour le coté et le non-coté*. (p. 17). FIR & France Invest.

Fraser, A., Tan, S., Lagarde, M., & Mays, N. (2018). Narratives of promise, narratives of caution: A review of the literature on social impact bonds. *Social Policy and Administration, 52*(1), 4–28. https://doi.org/10.1111/spol.12260

GIIN. (2022, October 12). *Sizing the impact investing market 2022*. https://thegiin.org/research/publication/impact-investing-market-size-2022/

Gustafsson-Wright, E. (2018, April 16). *Impact bonds could offer a paradigm shift towards more effective public services*. PIRU. https://blogs.lshtm.ac.uk/piru/2018/04/16/impact-bonds-could-offer-a-paradigm-shift-towards-more-effective-public-services/

Joy, M., & Shields, J. (2013). Social impact bonds: The next phase of third sector marketization? *Canadian Journal of Nonprofit and Social Economy Research, 4*(2), 39–55.

La Torre, M., Trotta, A., Chiappini, H., & Rizzello, A. (2019). Business models for sustainable finance: The case study of social impact bonds. *Sustainability, 11*(7), 1887. https://doi.org/10.3390/su11071887

Lopez Taylor, J., & Bode, M. (2021). *Service delivery: With key considerations for the recovery from COVID-19* (p. 31). Global Partnership for Results-Based Approaches, World Bank Group. https://golab.bsg.ox.ac.uk/documents/OBF_Covid19_04_040521_WEB.pdf

Mitropoulos, A., & Bryan, D. (2015). Social benefit bonds: Financial markets inside the state. In G. Meagher & S. Goodwin (Eds.), *Markets, rights and power in Australian social policy*. Syndey University Press. Open Science Framework. https://doi.org/10.31219/osf.io/pcujx

OECD. (2019). *Social impact investment 2019, the impact imperative for sustainable development*. https://www.oecd.org/dac/social-impact-investment-2019-9789264311299-en.htm

Palandjian, T., & Brest, P. (2020). *After the Pandemic: Addressing the Permanent Crisis With Pay for Success Programs. Stanford Social Innovation Review*.

Pellizzari, M. (2020, June). *Que révèle la crise du Covid-19 sur le paiement aux résultats?* https://www.finance-fair.org/sites/default/files/2022-05/FAIR-crise-covid-paiement-r%C3%A9sultats-2020.pdf

Pellizzari, M., & Sebag, R. (2019). *Retour d'expérience sur les contrats à impact social en France. Les acteurs impliqués font le bilan après 3 ans d'expérimentation multi-facette*. Impact Invest Lab.

Sir Ronald Cohen. (2020). *Impact, reshaping capitalism to drive real change*. Morgan James Publishing.

Social Impact Investment Taskforce. (2014). *Impact investment: The invisible heart of markets. Harnessing the power of entrepreneurship, innovation and capital for public good* (p. 51). https://impactinvestingaustralia.com/wp-content/uploads/Social-Impact-Investment-Taskforce-Report-FINAL.pdf

Tse, A. E., & Warner, M. E. (2020). The razor's edge: Social impact bonds and the financialization of early childhood services. *Journal of Urban Affairs, 42*(6), 816–832. https://doi.org/10.1080/07352166.2018.1465347

Warner, M. E. (2013). Private finance for public goods: Social impact bonds. *Journal of Economic Policy Reform, 16*(4), 303–319. https://doi.org/10.1080/17487870.2013.835727

Williams, J. W. (2020). Surveying the SIB economy: Social impact bonds, "local" challenges, and shifting markets in urban social problems. *Journal of Urban Affairs, 42*(6), 907–919. https://doi.org/10.1080/07352166.2018.1511796

Mathilde Pellizzari holds a PhD in Science and Technology Studies (STS) focused on social impact bond (SIB) setup in France, Colombia and Chile. She conducted her PhD research from 2018 to 2022 at FAIR (formerly iiLab) and the Centre de Sociologie de l'Innovation, Mines Paris-PSL, after a Master's Degree in General Science and Executive Engineering from Mines Paris.

Pauline Boulanger was the deputy managing director of iiLab from 2018 to 2020. She is now an investment officer at PROPARCO, a subsidiary of the Agence Française de Développement (AFD) devoted to private sector funding. Prior to that, Pauline had worked for 6 years at AFD, being mostly based in South Africa. She holds a MSc in management from ESSEC Business School, with majors in corporate finance and social entrepreneurship.

6 Popular Self-Organization and Cooperativism: Towards Economic Democracy and Human Development

Enidio Ilario, Alfredo Pereira Jr, and Heleno Rodrigues Corrêa Filho

Abstract

We depart from a brief review of the history of cooperativism worldwide, particularly in Brazil. We show the conceptual framing of genuine cooperation and solidarity, listing the Brazilian-related existing organizations with movements of popular self-organization and solidary economy. As Brazilian examples, we highlight the cases of three types of cooperativism: recyclable waste picker cooperatives, financial, and agricultural. Although different, they can be synergic, providing ways of popular participation and economic democracy. In the general framework of this book, the three types of cooperatives can be the catalysts for economic democracy, executing three essential social functions: to provide a form of organization for the poor and unemployed and to afford the capture of public money for productive investment, environmental sustainability, and other aspects necessary for human development.

1 Introduction

Dealing with a subject as revisited, such as cooperativism, is always a challenge in adding something new. Our option in this chapter is to deal with its political, ethical, and contemporary aspects. The year 2012 was declared by the United Nations, the International Year of Cooperatives (Source: www.ica.coop), with the epithet: "cooperative companies build a better world." It is worth asking: Ten years later, did cooperatives manage to construct this new world?

What can be observed, to answer this question, is that cooperativism is successful, despite more slowly than one would like. It has grown all over the world and always in a creative way, facing the challenges imposed by the economic and political-ideological scenario. Its political arm is the International Cooperative Alliance (ICA) founded in 1895, with approximately 100 countries and one billion members, employing around 100 million workers. Given the magnitude of the phenomenon, what is new in this process is the urgency of viable solutions to a practically terminal crisis of "unproductive capitalism" (Dowbor, 2017).

E. Ilario (✉)
University of Campinas – Center for Logic Epistemology and History of Sciences, Campinas, Brazil
e-mail: enidio@unicamp.br

A. Pereira Jr
State University of São Paulo, Botucatu and Marília, SP, Brazil
e-mail: alfredo.pereira@unesp.br

H. R. Corrêa Filho
ESCS-FEPECS – Escola Superior de Ciências da Saúde, Brasília, Brazil
e-mail: helenocf@unicamp.br

A. Pereira Jr, F. Sousa (eds.), *Principles for Governance*, Sustainable Development Goals Series,
https://doi.org/10.1007/978-3-031-40025-4_6

We are aware of the risks of systemic crises, in the condition of emptiness in which the old crumbles and the new is still in deliverance. But what, after all, would be new in such circumstances, in terms of the organization of work and production, in a world in which a true revolution has been taking place regarding Digital Information and Communication Technologies (DICTs)? Wouldn't such a revolution be precisely the expected and desired new thing already there? We don't think so, that what presents itself as a revolution in the general sense is nothing more than a late stage of the Scientific and Technological Revolution (STR) in the development of capitalism.

Although the profound transformations of the production process resulting from this stage of the STR, contribute to the outbreak of contradictions that characterize the crisis of contemporary capitalism, from the full update of the potential of financial capitalism and the so-called global value chains, what we have is a form of dystopian globalism, known generically as "finance capitalism" or "neoliberalism." It is a dystopian reality insofar as it suffocates the production potential in its human, social, ethical, artistic, political cooperation, and social organization aspects, reducing it to pure technique or technoscience.

Even at the end of the last century, these conditions were already clearly visible to the attentive eyes, for example, of the French journalist Viviane Forrester, who published in 1996 the emblematic book "The Economic Horror" (Forrester, V. São Paulo: Ed. da UNESP, 1997). The author describes in the small volume of great editorial success worldwide how neoliberalism, from the governments of Margaret Thatcher in the United Kingdom and Ronald Reagan in the United States, managed to impose a repugnant narrative. This narrative claims that the unemployed are to blame for the situation themselves, due to their inability to keep up to date and cope with high productivity demands.

The publication anticipates notions that would become classics from the following century onwards, such as the "precariat," a term coined by Guy Standing (2012). The term "precariat" defines the condition of workers with no ties or with very fragile employment ties, in the labor context. The end of the 20th century, after the disintegration of the Soviet Union and the end of the so-called Cold War, was prodigal in globalizing the impoverishment of workers. However, the novelty in this globalization was the inclusion of the middle class in the central countries, particularly in the European Union and the United States. In the case of the European Union, with the insidious abandonment of the principles of the 'Welfare State'. In the case of the United States, with the loss of jobs, both due to automation and the systematic export of jobs to countries with abundant and cheap labor, such as emerging countries AsiaTigers, and later China.

In this period, the seeds of fascism, irrationalism, and xenophobic nationalism, which put democracy at risk and – from the 2008 crisis – capitalism itself, intensified. Such a crisis, however, opened previously closed spaces for debating, economic divergent thinking from the orthodoxy, fueled by the interests of rentiers worldwide. In this scenario, a brilliant and young French economist, Thomas Piketty, collected data in gigantic research work, in 2013 (Piketty, 2014). He exposed in numbers and statistics the viscera of neoliberalism, clearly manifested in the unprecedented income concentration throughout the world. In 2019, the same author launched the also rich and well-stocked "Capital e Ideologia" (Piketty, 2020), in which he proposes "an economic, social, intellectual and political history of ideologies and unequal regimes."

In the latter work, the author mentions some historical facts, for example, the attempt to develop cooperatives, launch ambitious literacy programs, and impose some minimum respect for work in the State of Pernambuco (Northeast Brazil), by the governor democratically elected and who was overthrown in a coup:

> This is true especially for the Northeast, especially in the Pernambuco region, whose democratically elected governor—a man who had tried to develop cooperatives, launch ambitious literacy programs, and enforce some minimal respect for work rules—was violently overthrown by putschists after the coup. (Id. p. 249)

The French economist also dedicates an entire section to the treatment of the theme "Cooperatives and Self-Management: Capital, Power, and Voting Rights" (Ibid pp. 510–513), and further on in "On the Role of Private Property in a Decentralized Social Organization." He recalls the failed experience of Perestroika, which, among other objectives, intended to prioritize authentic cooperativism as an alternative to the privatization of the Soviet economy. Citing the famous interviews of Vladimir Putin, conducted by Oliver Stone in 2017, Piketty will draw attention to aspects that are often ignored in that historic moment, from which we highlight the following excerpt (Ibid.):

> Putin mocks Gorbachev's egalitarian illusions and his obsession with saving socialism in the 1980s, especially his liking for "French Socialists" (an approximate but significant reference, since French Socialists at the time represented what was most socialist in the Western political landscape). In substance, Putin concluded that only an unambiguous renunciation of egalitarianism and socialism in all their forms could restore Russia's greatness, which depended above all on hierarchy and verticality in both politics and economics (Ibid. pp. 603–604).

Ironically, in the process of transition from the centralized economy of real socialism to a capitalist economy, it is not improbable that Mikhail Gorbachev had in mind Lenin's belief (Lénine, 1975) in cooperativism as a path to develop socialist agriculture and thereby eliminate the last germs of capitalism. It is worth remembering that the Leninist theory saw cooperativism as a path towards the constitution of large units of agricultural holdings, highly technified and of high productivity, concerning peasant culture, allowing a dignified life for the former serfs of the era of tsarist feudalism.

Another prominent author is economist Yanis Varoufakis. He was the finance minister of Greece in 2015 and later coined the term techno feudalism (Varoufakis, 2020). It defines better than any other expression, the contemporary dystopia, given the capacity of control and molding of society by DICTs (Digital Information and Communication Technologies), the enormous concentration of income (and power) in a few private hands. In such a scenario, the so-called big techs constitute powerful corporations capable of manipulating the hearts and minds of huge contingents of the population, consequently, being able to manipulate the state itself for its benefit. Hence, it is from such a panorama that we will seek, in the present approach, to deal with cooperation and the solidary economy as an alternative for the self-organization of society in general and workers in particular, to face the challenge of overcoming the contemporary dystopia of the so-called post-industrial and post-employment society.

2 The Conceptual Framework

2.1 A Brief Look at the History of the Cooperative Movement in the World and Brazil

According to the Dictionary of the Other Economy (Viveret et al., 2013), semantically the term cooperation means working together, in a collectivity, to produce social life. Cooperation, in a strict sense, is the basis of the economic-social relationships that associated workers seek to establish in the work process, denoting ethical-political value, resulting from a vision of the world and human beings that attributes disposition, commitment, and solidarity to the collective subject (id. pp. 92–99).

Authentic cooperativism has a lapidary principle in self-management to search for dignity at work understood in the terms described by the International Labor Organization (ILO) and in education for autonomy and social emancipation. There is a certain consensus among historians of cooperativism that its origin is related to the Industrial Revolution. Noteworthy, forms of cooperation in work and social organization are as old as humanity. Besides, the horrors imposed by the Industrial Revolution in its infancy, with the rural exodus, misery, and violence such as child labor, inhuman working hours, deaths, and mutilations, without guarantees of any kind and starving wages, indeed updated the potential cooperative and solidary, dormant or even eliminated by steam engines and later by the Taylorist and Fordist organization of work.

Although the ideas that guide cooperative work already existed in a structured way, at least since 1817 with Owen's plan, presented to the British government for the construction of cooperative villages (Faria, 2005), only about half a century after the first registration of the patent for the steam loom, in England, the "Rochdale Society of Equitable Pioneers in Rochdale" appeared in 1844, which instituted the so-called ideals or principles of cooperativism. This laid the groundwork on which co-ops around the world continue to operate[1]: Free and voluntary membership; Democratic management; Economic participation of members; Autonomy and independence; Education, training, and information; Intercooperation and Interest in the community. It was in this context that two influential names in political and economic philosophy, which are often seen as antipodes, Karl Marx and John Stuart Mill, converged on the understanding that cooperativism would be a privileged form of organization of work and production in the fight against ills of the Industrial Revolution. We know that Marx devoted an entire chapter of "Capital" to the theme of cooperation. The concept of cooperation is defined in chapter XIII of the first book (published in 1867): "When numerous workers work side by side, either on one and the same process, or on different but related processes, they are said to cooperate, or to work in cooperation" (Bottomore, 1988, p. 80).

Without going into detail about the issue of value and surplus value, dealt with in depth by the author, Marx added from the definition of cooperation that the collective workforce is more than the sum of the parts. They constitute it, in a vision of totality, which anticipates what would be dealt with in the first half of the following century by gestalt theory and later by general systems theory. In any case, although Marx did not condemn cooperativism, he did not even include cooperativism in the conception of "utopian socialism," most likely because he did not see it as a substitute for a classless society and because he was aware of the deformations already present in existing cooperatives at the time. A contemporary of Marx, John Stuart Mill, the leading name of the so-called ethical-political liberalism, words Antonio Zanotti (2020, p. 15): "Mill was one of the greatest interpreters of the social emancipation movements of the 19th century. He inserts the cooperative enterprise into his model of working-class emancipation." The utilitarian John Stuart Mill has always been a controversial author within liberalism, one of the main defenders of the cooperative system as an overcoming of capitalism, through a hybrid solution, today, certainly, in the eyes of neoliberals, a subversive, although he admitted that capitalism properly reformed still had many advantages. Mill's influence on social reformers extended from the late nineteenth century, with the "Fabianist" movement, to the varied forms of "revisionism" that have derived from the union and political organization of the working class since then. Such movements seek a progressive transformation based on successive reforms of the State and the productive apparatus, to "make democracy more and more solid and broad and replace private property with multiple forms of collective, municipal, state and cooperative property" (Bobbio, 1983, p. 1129).

Currently, there is a wide range of sectors in which the cooperative organization can operate with specificities in different countries, in Brazil, for example, until 2019 its performance was divided into 13 branches of activity: consumption, social, work, education, transport, agriculture, health, credit, housing, production, infrastructure, minerals, and tourism/leisure. Such activities have been grouped into seven branches: agriculture, consumption, credit, infrastructure, health, work, and transport. There are several cooperative arrangements, including self-

[1]It is worth noting that as early as 1895, the ICA (International Cooperative Alliance) was formed, headquartered in Geneva, Switzerland, as a non-governmental entity, bringing together, representing, and providing support to cooperatives and their corresponding organizations throughout the world, following the seven cooperative principles, under the name of Rochdale Principles of Cooperation. Versions of this charter were updated by the ICA in 1966 as the Cooperative Principles and in 1995 as the Declaration on Cooperative Identity.

managed or recovered enterprises (taken over by workers), community associations, production associations, informal productive groups of specific segments, etc. However, the very concepts implied in the principles of cooperativism are very generic, abstract, and not very operational, mostly coming from the fields of philosophy and sociology, thus opening the perspective of encompassing organizations that are not called cooperatives and, otherwise, have a strong identity with such principles. Based on the idea of solidarity in contrast to the individualism that predominates in market societies (Viveret, pp. 169–178), the notion of solidarity economy is more comprehensive and inclusive than that of cooperativism, according to Paul Singer (http://paulsinger.com.br/paul-singer-uma-lucidez-solidaria-e-futurante-por-ruy-namorado/ seen on 03/20/23): "we have a wide and varied set of initiatives still in search of an institutional identity and a consistent legal framework, which do not fit into traditional cooperativism or the so-called third sector. As participants in a new wave of social and economic solidarity, they eventually oppose traditional cooperative forms, already institutionalized in the form of cooperativism." With this critical approach, such initiatives, called Solidarity Economic Enterprises (SEEs), have gone through a process of mutual recognition and unification, gradually aligning themselves with the field called Solidarity Economy.

For the author, with its horizon of overcoming capitalism, the solidary economy implies an unequivocal and unconditional option for democracy, as the founding matrix of any post-capitalist model. In Brazil, the notion of a solidarity economy gained significant relevance in the last decade of the last century, when neoliberalism expressed its dominant thesis at the "World Economic Forum" in Davos, Switzerland. At the time, as a form of resistance to the theses of this dystopian globalism, the "World Social Forum" was born in Porto Alegre (RS) in 2001 to point out alternatives to hegemonic financial capitalism. Among the Brazilian intellectuals who developed the theme of a solidary economy, Paul Singer stands out (Singer, 2002). He became not only a theoretical reference, but he was able to apply such ideas through his political militancy in positions he held. He engaged the public administration of the City of São Paulo as Secretary of Planning in the LuízaErundinaCity-Mayor administration (1989-1992). Eight years later, worked in the Marta SuplicyCity-Mayor government as responsible for the Technological Incubator of Popular Cooperatives. Later on, especially at the federal level, he was the National Secretary for Solidarity Economy (SENAES), in the Ministry of Labor and Employment (MTE) in the first President's Lula governmental term (2003-2006).

2.2 Associated Work Cooperatives and Production Cooperatives

Singer considers production cooperatives to be an elementary modality of solidarity economy, as "an alternative mode of production to capitalism, and not merely interstitial" (Viveret, p. 90). Although production cooperatives are also work cooperatives, a better definition of their specificities is necessary, according to the Dictionary of the Other Economy (Id. pp. 99–105):

> Worker cooperatives and industrial production cooperatives have different definitions, although the terms are used as synonyms. The first refer to the provision of specialized personal services, bringing together, for example, education or health professionals, taxi drivers, and so on. (...) In general, cooperative members use their own instruments, so that the cooperative functions as an intermediary in the acquisition and distribution of services. In the second type of cooperative, the production of goods results from collective work, whose model is the factory. Factories, or production units – offices, workshops – are managed collectively. In any case, although the use of the terms is controversial, sometimes confusing them, the two forms of cooperative are based on work as an element of possession and collective management.

In relation to labor cooperatives, it is worth noting for its definition, the declaration approved in September 2005 by the General Assembly of the International Organization of Cooperatives in Industries and Services (CICOPA)(https://josswinn.org/wp-content/uploads/2014/11/

Declaration_approved_by_ICA_EN-2.pdf seen on 03/20/23).

The document establishes that the purpose of this type of cooperative is the creation and maintenance of sustainable jobs that generate wealth, improve the quality of life and promote community and local development. Reaffirms the classic principles of cooperativism, such as free and voluntary membership in a cooperative relationship, governed by democratically agreed rules and autonomous before the State and third parties in relation to the work and management of the means of production. As can be seen, although anchored in the same original principles of cooperativism since its advent, paradoxically, associated labor cooperativism has never managed to actualize its full potential, perhaps because in the field of service provision there is an abundance of façade cooperatives, thus contributing to demoralize and demobilize the organizational actions of this type of cooperative. One can also hypothesize that more than any other segment, labor cooperativism presupposes the need for collective and coordinated action, which depends much more on the convergence of beliefs and values in a community of interests, which are many difficult times to reconcile. In this way, such cooperatives become more vulnerable to conflicts of interest based much more on competition than on cooperation, often reproducing, as in game theory, those called zero sum, that is, some win and others lose. Such an adverse scenario can only be avoided through profound cultural changes, always in the long term, requiring continued investment in cooperative education and the social and community participation of the cooperative members' families. From this collective and solidary experience, leaders are prepared to unite wills in a project that is greater than individual interests and that results in the possibility of decent work for all, including the eventual need for exclusion through assembly decisions – not always easy to execute – of those who systematically sabotage cooperative principles.

Otherwise, the struggle for power and for ever-increasing chunks of participation in collective results, through various artifices that often reach the verge of mafia behavior, with fraud of all proportions, ends up undermining what is most precious in cooperativism, solidarity and belief in a meaning greater than that of individual progress.

2.3 Authentic Cooperatives and Facade Cooperativism

From what has been exposed so far, it becomes evident the importance of establishing precisely what a true cooperative is, however, such a task is not trivial, in view of the complexity of the relationships that are established between internal and external economic agents to cooperatives, in a wide range of activities in the market economy. In such a scenario, at least we can say that a false cooperative seeks to use the name of cooperative in defiance of cooperative principles. In facade cooperatives there is no free and voluntary membership, management is not democratic, there is no economic participation or autonomy and independence of members, nor educational training, information, or interest in the community, essentially focused on obtaining profit. We can say without fear of being wrong that in the same way that the values bequeathed by the French Revolution and the Enlightenment, in the course of the last two centuries, many times degenerated into corporatism and fascism, without the vigilance of the citizens, cooperativism can also engender monsters, without the surveillance of the cooperates. In the Brazilian reality, "front cooperatives," organized under corporate supervision, taking advantage of loopholes in legislation at various government levels, are forms of outsourcing in the industrial, agricultural, and service sectors. In them, the principle of free adherence is perverted into mandatory adherence to work subordinate to the command of bosses and foremen and derisory payments for extended hours without the right to overtime and unhealthy conditions. This situation generally stems from a huge contingent of unemployed and desperate workers, a condition reminiscent of the dark times of the "Industrial Revolution" in England. In Brazil, the Labor Court, Social Security, the Ministry of Labor and Employment, and the

Public Ministry of Labor are responsible for supervising such situations. In the event of conviction in legal proceedings, the subsidiary liability of service takers takes place. However, even in these cases, in view of the non-accountability in the criminal sphere, the perverse notion often prevails that crime pays, as eloquently asserted by labor judge Valdete Souto Severo (2005, pp. 141–144).

Therefore, the logic of capital is perverse from any angle of analysis. It is perverse when it assigns the burden of the enterprise to the employee. It is perverse when it takes away rights that have already been won, depriving man of his dignity and, therefore, of the condition that differentiates him from animals. It is perverse when it generates a society of survivors who, because they are plundered, do not consume and, by not consuming, do not generate wealth or move the capitalist machine. It must be remembered that the argument that it is better to have a job, even a cooperative one than to be unemployed, has already served to justify the practice of slavery, for several decades, in our country.

2.4 Between the Individual and the Collectivity: A Case and a Parable

The entire history of economics revolves around the tension between collective interest and individual interest, and from such a conflict emerges the phenomenon of inequality ingrained in the very history of Capitalism. As Thomas Piketty well observed in his work published in 2019 "Capital and Ideology" (Piketty, 2020): "Inequality is neither economic nor technological; it is ideological and political." Moreover, the author justifies such an assertion as follows (Id, p. 7):

> In other words, the market and competition, profits and wages, capital and debt, skilled and unskilled workers, natives and aliens, tax havens and competitiveness—none of these things exist as such. All are social and historical constructs, which depend entirely on the legal, fiscal, educational, and political systems that people choose to adopt and the conceptual definitions they choose to work with.

About the conflict between personal/individual desires and the collective interest, in the end, we deal with personal choices, potentialities that are actualized or not, dependent and determinant at the same time of historical conditions, whether or not they are an expression of the collective will. By the way, it is worth highlighting a remarkable case of success in the solidarity economy field: the Mondragón Complex, the largest cooperative conglomerate in the world, which originated in the small Basque town of the same name, in the north of Spain in 1956. the trajectory of José Maria Arizmendiarreta, better known as Father Arizmendi, a Basque cleric, who fought in the Spanish Civil War alongside the republicans, was arrested and sentenced to death, survived to later, in the Francisco Franco period, became a key player in the birth of the Mondragon Cooperative Complex (MCC). There were many Arizmendi adventures in this story, some of them recalled by Paul Singer (Singer, pp. 99–101), which shed light on the apparent paradox of the precedence of individual will over collective will in the birth of cooperation initiatives. The paradox is apparent, as it becomes clear that there is no contradiction between the interest of the individual as a human person and the interest of the collectivity as a community in the field of solidarity economy. It was in 1943 that Arizmendi founded a polytechnic school with democratic management, currently, MondragónUnibertsitatea (University of Mondragón), which played a key role in the resurgence of the global cooperative movement. The institution is a second-degree cooperative formed by three unique cooperatives, each offering one of the three existing courses: Engineering, Economics, and Pedagogy. The successful experience of the University of Mondragón did not happen without the energetic action of Fr Arizmendi, which also culminated in the birth of the first cooperative group, Ularco, formed initially by Ulgor, Arrasate, and Copreci, later also by Ederlan and FagorEletrotécnica. Regarding this assertion, it is worth mentioning Paul Singer, when he observes that shortly after the creation of the first of the Corporation's cooperatives, Arizmendi proposed the creation of a cooperative bank, to guarantee the independence of the cooperatives in relation to the banks, according to the author, Arizmendi (Singer, pp. 99–101):

> "He had to face some resistance from Ulgor's directors, who were engineers and did not dream of becoming bankers. But Arizmendi studied the legislation, wrote the statutes of what would become the CajaLaboral Popular, and forged the minutes of a founding meeting that never took place. Faced with a fait accompli, resistance ceased and Cajacame to light in 1959, becoming the first second-degree cooperative in the Mondragón grouping. From then on, it performed essential functions not only in providing financial services but in forming new cooperatives, functioning as their incubator."

Today, the Mondragon Cooperative Complex is a worldwide success story, made up of 120 cooperative companies, 87 industrial, 1 credit (CajaLaboral), 1 consumer (Eroski), 4 agricultural, 13 research, 6 consultancies, and 8 educational. According to the Financial Cooperativism Internet Site (https://cooperativismodecredito.coop.br/cenario-mundial/expressao-mundial/cooperativismo-de-credito-na-espanha/o-case-de-mondragon-na-espanha/ seen on 03/20/23), MCC is the seventh largest economic group in Spain, with sales of around 13.6 billion euros and a result of 792 million euros in 2007. It is a true work cooperative, as they are "members of the Cooperatives only their workers who currently add up to 93,000 people. In essence, all Mondragón cooperatives are work cooperatives that have different products and services." And this fully respects the basic principles of the International Cooperative Alliance (ICA), however, with a doctrinal explanation of some aspects:

1. Education: to implement the other cooperative principles, people's dedication is fundamental.
2. Sovereignty of work: MCC's experience considers work to be the main transforming factor of nature, society, and human beings themselves.
3. The instrumental and subordinate character of capital: social capital is subordinated to work and is necessary for business development, and its remuneration must be fair, adequate and the minimum value for joining the cooperative cannot be a limiter for free membership.
4. Free membership: as long as the defined principles are respected, there can be no religious, political, ethnic, or gender discrimination, on the occasion of the interest of joining a new member.
5. Participation in management: in addition to being a democratic organization in its essence, cooperatives must seek self-management and the participation of partners in the scope of business management.
6. Retributive solidarity: salaries must be sufficient, according to the reality of the cooperative, they must be equivalent to the salaries of other companies in the region and internally must respect the limit defined between the minimum and maximum remuneration (6x).;
7. Intercooperation: takes place through relations between cooperatives in the same field of activity, gathered in divisions, with sharing of results, and transfer of partner-workers. Intercooperation should also take place with other cooperatives in the MCC group, in the Basque Country, and in the world.
8. Universal character: MCC is in solidarity with all those who work for economic democracy within the scope of the Social Economy, with objectives of Peace, Justice, and Development.

The diagram that represents MCC's corporate management model has at its radiating center, based on the Basic Principle: "It is these people who build a Shared Project and use a Participatory Organization to put it into practice" (Fig. 6.1).

It is noteworthy that MCC, in addition to the university, also maintains primary education schools, as well as research and dissemination centers and training of executives and entrepreneurship in cooperativism. That is, a list of structural conditions that allow and perpetuate a genuinely virtuous circle, in fact, a spiral of growth according to Mikel Lezamiz ((https://

Fig. 6.1 Basic principles of cooperativism (Public Domain)

cooperativismodecredito.coop.br/cenario-mundial/expressao-mundial/cooperativismo-de-credito-na-espanha/o-case-de-mondragon-na-espanha/ seen on 03/20/23), some of the factors that led to MCC's success are as follows:

> "The leadership that Father Arizmendiarrieta had: his vision of the future, the focus on education and formation; The trust Ulgor's founders had in Padre; The performance of the business division of CajaLaboral at the beginning of the group's development; The continuous search for management improvement, employing 4% of the payroll in training expenses; The extraordinary economic growth that Spain had in the '60s; The surplus distribution model that strengthens the group; The expressive contribution of social capital made by the partners; The genuine inter cooperation between the group; The focus on improving the community: generating wealth in the community and not making money; The balance between economic efficiency and social development; Internal stability in social terms (democracy); The pursuit of economic efficiency; The organization of MCC into sectors after 1991."

The MCC model seems exceptional less because of the way it is structured than because of the historical conditions in which it was created, that is, in the Basque Country and the struggle of its people, hence the existence of a strong identity, capable of uniting the members in a community feeling of belonging, which strengthens and feeds back the bonds of solidarity. As we have seen, however, the precondition for the MCC's success was not limited to the feeling of belonging and the historical-cultural ties of a people, but the unequivocal leadership of Arizmendi. Regarding this condition, it is worth highlighting an exciting and influential cinematographic narrative that marked an era, a true parable about the solidarity economy; fiction certainly had an impact on real life, although in a difficult way to measure, either due to the extraordinary success of the public and critics or for the perenniality of what surrounds the drama. It is the film "It's a Wonderful Life" (1946), the favorite work of Frank Capra (1897-1991) and whose protagonist is the legendary George Bailey, played by James Stewart in a great performance. At the center of the plot is the "Bailey Brothers Building & Loan," a community bank in the small "Bedford Falls," an institution that, at the time of the Great Depression, had saved the inhabitants of the community from the clutches of North American commercial banks, in the form of the dreaded mortgages.

The ever-present opponent in the structure of the heroic sagas is Henry F. Potter, an unscrupulous banker eager to absorb the "Building & Loan." Regarded as one of the greatest films of all time, "It's a Wonderful Life" was recognized by the "American Film Institute" (https://www.afi.com/afis-100-years-100-cheers/ seen on 03/20/23) as one of the 100 Greatest American Films and as number 20 in the 2007 Revised List, being number one in the list of films most inspiring Americans of all time. Although considered one of the greatest film directors in the United States, Frank Capra was born in Italy. When he arrived in New York at the age of five and saw the Statue of Liberty with a torch in his hand, he heard from his father: "Look at this! This is the greatest light since the star of Bethlehem! This is the light of freedom!", Capra believed in and worshiped the individual above race, class, gender, group, and social status. Although worshiping the individual, the brilliant film director never lost sight of the value of the community and in "It's a Wonderful Life" he achieved the feat of synthesizing the union of individual values in the heroic and torn figure of George Bailey, with the community of people made up of residents of Bedford Falls. It can be said that perhaps Capra had always intuited that that old lady with the torch of freedom was born and came from the homeland of the "French Revolution," under the motto "Liberty, Equality, Fraternity."

2.5 Financial and Credit Cooperativism

Certainly among the cooperative ventures, the most successful from the point of view of competition in the market is credit (or financial), particularly in central countries, in which this cooperativism was already emerging in a prominent way just 20 years after the emergence of "The Rochdale Society of Equitable Pioneers." However, unlike England, the first cooperative members in Germany were not urban artisans, but rural workers, many of them serfs recently freed from the feudal regime between 1800 and 1848. Heddesdorf was the first rural credit cooperative (RCCs) that appeared in 1864, on the initiative of Friedrich Wilhelm Raiffeisen, with Germany currently in second place in a ranking that includes countries with assets above 100 billion dollars (Port, 2008). The sum of the assets of the top seven in the ranking: France, Germany, Japan, China, Holland, the United States, and Italy is nine trillion dollars. Not only is the enormous volume of assets of financial cooperatives impressive, but the fact that the United States occupies only the sixth position in the ranking, a little ahead of Italy, in addition to the fact that England is placed only in the distant sixteenth position, with 26 billion dollars compared to Brazil with 16 billion dollars in assets.

The sum of the assets of the top seven in the ranking: France, Germany, Japan, China, Holland, the United States, and Italy is nine trillion dollars. Not only is the enormous volume of assets of financial cooperatives impressive, but the fact that the United States occupies only the sixth position in the ranking, a little ahead of Italy, in addition to the fact that England is placed only in the distant sixteenth position, with 26 billion dollars compared to Brazil with 16 billion dollars in assets. Although there are conflicting data on the size of credit unions in the United States, appearing in the financial cooperatives' Internet Site (https://cooperativismodecredito.coop.br/cenario-mundial/expressao-mundial/cooperativas-de-credito-nos-estados-unidos/ seen on 03/20/23) as fourth place, with assets of 1.07 trillion dollars, compared to the size of the country's economy, it would be expected to occupy the first position.

Perhaps this fact is due to the relatively more recent history of this type of cooperative in the United States, since the first credit/financial cooperative, the St. Mary's Bank Credit Union of Manchester will only appear in New Hampshire in 1908, founded by Alphonse Desjardins (founder of the Desjardins Movement). It is also curious that it was precisely in the United States and England in the Reagan and Thatcher era that the so-called neoliberalism was born and flourished, an antipodean doctrine of the solidarity economy which led to the great international financial crisis in 2008, with repercussions that still spread across the world, accentuating social inequalities. At the height of this crisis, cooperative banks provided stability and financial security to millions of people, as their business model, instead of prioritizing profit maximization, emphasizes the development of better products and services for its associated members, with a long-term vision. Cooperative banks do not depend on the financial market to raise their capital. The model is based on democratic governance, proximity, participation, and satisfaction of the interests of its members and clients.

Faced with such success, it is not surprising that financial capitalism wants to interfere in the regulation of cooperatives, introducing true "Trojan horses." Just note that cooperative banks in many countries are doubly regulated, as they also offer services such as savings and loans to non-members, many of whom have shares traded on the capital market, resulting in the fact that they are also controlled by non-members.

Quite different from this contemporary situation of credit unions in typically capitalist countries, in China, which sees itself as a "market socialism" regime, credit unions are more similar to those characteristics that instituted the first cooperative of its kind in the world, in Germany at the end of the nineteenth century. With high capillarity and lending capacity, however, Chinese credit cooperatives suffered from liquidity problems at the turn of the last century. In 2001, default reached 44%, with 53% of Rural Credit Cooperatives (RCCs) with negative equity. That led the Chinese government to impose new rules for the sector, seeking the professionalization and financial viability of RCCs. Currently, China has the strongest credit cooperative movement in the world, according to data from the Financial Cooperativism Internet Site with data updated until 2012 (s://cooperativismodecredito.coop.br/):

> Estimates point that in China 200 million households are members of cooperative financial institutions. For more than 50 years, the system has been the main source of basic financial services for the rural poor in the country. Rural credit cooperatives hold 10% of all bank deposits and represent more than 90% of agricultural loans.

As can be seen, credit unions in China have strong government support and focus on low-income credit and agricultural credit. That is, credit unions make up a state policy, in which any intervention aims to save popular savings and not cooperative companies. Quite the opposite of what happened in the 2008 crisis, in which in the United States, for example, Public resources were used to capitalize private banks, mired in veritable financial pyramids or speculative bubbles, to the detriment of the interests of the vast majority of small investors. Brazil currently occupies the nineteenth position in the ranking of financial cooperative participation (Id), and the first credit cooperative was founded in 1902, in Nova Petrópolis in the State of Rio Grande do Sul. Conceived by Father Theodor Amstad with the support of leaders of the German immigrant community, initially under the name of SparkasseAmstad, 90 years after its creation, it was renamed as Sicred, a brand that consolidated over time into a nationwide cooperative credit system.

The Brazilian Constitution of 1988 was fundamental for the advancement of the credit cooperative sector, insofar as it allowed its operational autonomy and from this achievement, Sicred's growth has been impressive. Currently, the largest cooperative bank in the country and according to data announced by the institution (https://www.sicredi.com.br/site/sobre-nos/ seen on 03/20/23), with over six million members, Sicred is the eighth-largest Brazilian bank in total assets. In 1998, Sicred crossed the country's borders and was affiliated with the World Council of Credit Unions (WOCCU). In 2010, it signed an international partnership with the Rabobank Group, a Dutch cooperative credit system. In December 2019, the Sicred system began the process of joining the "United Nations Global Compact" (UN), committing to adopt corporate social responsibility and sustainability policies. An enterprise initially focused on rural credit, today Sicred is inserted in the entire chain of financial market products, not only in the countryside but also fully inserted in the urban economy of the country. The SICRED is a cooperative venture on an impressive rise that puts it on the heels of the powerful commercial banks in Brazil. Full proof of this fact is that in 2021, Forbes Magazine ranked SICREDI seventh as the best financial institution in Brazil and, at the end of 2022, had its corporate assessment elevated by Moody's Local, to the highest rating of the agency.

3 The Role of the Solidarity Economy in the Movement for Economic Democracy in Brazil

Although they legitimately fit into the concept of solidarity enterprises, countless initiatives of self-organization by society, the cooperative movement, whether due to its very well-defined principles or its global scope and political and institutional organization, has shown itself to be more viable and reproducible. Much more than just a way of organizing work, the cooperative enterprise is, from a formal point of view, synergistic, and it is supported by the legal frameworks of national states. More than that, the cooperative movement has been shown to be the only one capable of confronting capitalism within the framework of the market economy since the advent of the Industrial Revolution. Even in "market socialism," as the Chinese regime calls itself, cooperative ventures have proven essential, even with their autonomy limited by decisive state intervention in a planned economy.

Since its birth, the cooperative movement has never stopped seeing itself as a viable alternative to capitalism, not just an interstitial and peripheral activity in the competitive market. The secular battle continues fiercely with victories and setbacks, but the war was never lost the political field is also the battlefield in the fight for the preservation and the expansion of legal frameworks favorable to cooperativism, conquered with great difficulty. Strengthening and, if possible, expanding such legal boundaries is necessary in the face of antagonistic interests so that the battle takes place with parity of arms in a context of precarious labor relations and an overabundance of "labor" and consequent weakening of union organization. We believe that although different, cooperative and solidary enterprises are synergistic, providing forms of popular participation and economic democracy. Alongside structuring and successful public policies such as the Unified Health System in Brazil (SUS) and the "BolsaFamília" Program, such undertakings act as a kind of catalyst in the process of conquering and evolving economic democracy.

The solidarity economy fulfills essential social functions, for example, providing forms of organization for the poor and unemployed; raising public money for productive investment, offering healthy food to the population, and other aspects necessary for human development. With regard to cooperative ventures in the financial and credit sector, we could see that, as Credit Agricole in France and Rabobank in Holland, cooperative banks in Brazil have also been gaining muscle capable of making them formidable competitors in this market. However, such gigantic institutions have not been able to respond satisfactorily to small borrowers, as is the case of small traders and service providers in urban peripheries and peasant families in the countryside.

To cope with this challenge, several cooperative and solidary enterprises have been multiplying throughout Brazil, such as community banks and creating social currencies for local circulation. Although of different shades, the sum of such experiences impacts not only the present but points to the consolidation of the solidarity economy as a real alternative for developing this country of continental dimensions. A development that points not only to the growth of the economy measured by the Gross Domestic Product (GDP), but also leads to the abolition of the overexploitation of work resulting from endemic unemployment and offers alternatives to monoculture based on the intensive use of pesticides through agroecology.

3.1 Solidarity Economy and Community Banks: The Palmas and Finapop Cases

In the solidarity economy, community development banks gain relevance, among which, in Brazil, Banco Palmas stands out, founded in 1998, in ConjuntoPalmeiras, a neighborhood of 32,000 inhabitants on the outskirts of Fortaleza-CE. A pioneer among 52 other Brazilian community banks, Palmas is managed by the Associação de Moradores do ConjuntoPalmeira (ASMOCONP). The organization defines itself as a bank in the periphery whose objective is "to democratize access to financial and banking services for the population on the outskirts of Fortaleza, with broad participation and social control, mobilization of local associations, seeking the socioeconomic development of neighborhoods and slums" (https://bancopalmas.com/ seen on 03/20/23). Still in the definition of the institution itself, the bank's mission (Id.): "to implement work and income generation projects through solidarity economy systems, where everyone is a producer, consumer, and social actor, focusing primarily on overcoming poverty."

With a staff of primary volunteers, its objective is to guarantee microcredit for local production and consumption with low-interest rates and without the need for registration, proof of income, or a guarantor (the neighbors guarantee the borrower's trustworthiness). In 2000 the bank conceived "Palmas," a social currency that circulates in local commerce and in 2003 created from the experience of ASMOCONP the "Instituto Palmas de Microcredit," a "Civil Society Organization of Public Interest" (OSCIP), with the mission of developing a solidary socio-economy. In 2006, through an agreement with the "National Secretariat for the Solidarity Economy," "Banco Palmas" and a whole network of other community banks began to have access to credit and act as a banking correspondent for "Banco Popular do Brasil." More recently, "CaixaEconômica Federal" and "Banco Nacional de Desenvolvimento" (BNDS) became its main financial partners, in addition to partnerships with international institutions such as "Zurich Brasil," for the development of the new line of products bank insurance.

The theory behind community development banks is that there is no such thing as an inherently economically poor territory, be it a neighborhood, region, or municipality, what impoverishes is the repeated loss of its savings. Thus, despite the level of poverty, a certain territory is capable of achieving economic development, provided that it can be autonomous in its internal economy, in the circulation of the local currency of social circulation, as is the case of "Palma," case otherwise it will not be sustainable. Another interesting experiment in the solidarity economy field that is in full development in Brazil is Finapop, "Popular Financing." Created at the beginning of the Covid-19 pandemic in 2020 on the "Movimento dos TrabalhadoresRuraisSem Terra" (MST) initiative in collaboration with Eduardo Moreira, economist, former banker, and investor, its objective is to raise funds to benefit agriculture peasants. According to Moroni and Pereira Jr. (2022), before the creation of Finapop, cooperatives linked to the MST always faced difficulties in obtaining financing via development agencies such as the "National Bank for Economic and Social Development" (BNDS), because although with similar interest rates:

> (…) "the approval of peasant agriculture projects, via the BNDES, takes much longer, practically 4 to 5 years, compared to Finapop. This is because traditional banks prioritize investments in large farmers' projects. This disparity in the treatment given by the banks to peasant agriculture and large landowners is inconsistent with the reality of food production on the Brazilian table, as 75% of these foods are produced by family farming, peasants, and small scale."

According to an interview given by Eduardo Moreira to "Brasil de Fato" (https://www.brasildefato.com.br/2020/06/03/conheca-o-finapop-um-projeto-inedito-de-financiamento-para-pequenos-agricultores seen on 03/20/23), "Finapop is a popular funding movement. It is not a bank; it is not an investment fund. They are people who are waking up, looking at this situation of injustice in the country, and want to contribute." Still according to Moroni and Pereira Jr.,(Id.), it is a catalytic element of this type of enterprise. It is an instance of a non-representational decision-making system, situated and incorporated, self-organized, and directed by the perception of dynamic and meaningful informational patterns called affordances. In the decision-making process, such an instance incorporates the unconscious associations and logical reasoning made by individuals (in their private mental space): "affective impulses of individuals and collective effects of people in democratic political systems, as well as contingent affordances to specific social and physical interactions.

3.2 Agrarian Cooperatives and the Solidarity Economy: The MST Case

Intimately linked to credit cooperatives in their origins, to the point that some of them have become influential organizations, as is the case of Credit Agricole in France and Rabobank in Holland, agrarian cooperativism has the predominant modality in the world, agricultural service cooperatives. Such cooperatives exist under two forms of denomination and organization: supply cooperatives and market cooperatives. The former has the function of supplying their associated members with inputs for production, such as fertilizers, feed, seeds, machinery, fuel, and others, while the latter organize themselves into processing activities for their agriculture and livestock products, including packaging, distribution, and marketing. Created in Europe in the mid-nineteenth century, the first agricultural cooperatives are currently standing out in emerging and third-world countries, for example, in Colombia, a country where the legal framework (Law No. 454) dates from 1998 and regulates economic solidarity, through the National Administrative Department of Solidarity Economy. An essential distinction exists between agricultural service cooperatives and community land exploitation (or agricultural production) cooperatives, in which land and equipment are shared and members farm communally.

The last form of cooperative organization is quite rare in capitalist countries, examples being the kibbutzim of Israel and in communist countries, collective farms. In Brazil, there is a successful experience with great national and global impact involving solidarity work in the countryside, the "Movimento dos TrabalhadoresRuraisSem Terra" (MST). The MST is an organization born in 1984 from peasant occupations of land in southern Brazil and is currently organized in 24 states in the five regions of the country. In total, there are about 450,000 families settled and, as stated on its website (https://mst.org.br/quem-somos/ seen on 03/20/23), the MST defines itself as "a mass, autonomous, social movement that seeks to articulate and organize rural workers and society to conquer the Agrarian Reform and a Popular Project for Brazil."

A fundamental characteristic of the movement is that even after settling, families remain organized, as the conquest of the land is only the first step towards carrying out a broad agrarian reform that gives the right to ownership of unproductive land and large estates. Still, according to the MST itself, the fight for the transformation of Brazilian society involves the creation of alternatives "in defense of a new development project in which the deconcentration and democratization of land, emancipated work, human beings, and nature are central elements." The work of the MST gave it great worldwide projection, being a prominent member of the international organization "Via Campesina"; it is also one of the founders and animators of the World Social Forum, starting with the first edition in Porto Alegre RS, in 2001. The MST in its first years of existence was strongly influenced by the thought of Karl Johann Kautsky, a Czech-Austrian philosopher, journalist, and Marxist theorist and one of the founders of the social-democratic political-ideological conception, which he saw in workers' unions and in the form of cooperative organization, great virtues in combating forms of capitalist organization of work (Kautsky, 2018, p. 4).

It is irresistible because it is inevitable that the enlarged proletariat puts itself on guard against capitalist exploitation, that it organizes itself in its unions, cooperatives, and political groups, that seeks to conquer better conditions of work and existence and a more considerable political influence.

According to Brenneisen (2002), the MST created a specific structure to promote cooperativism in its settlements, the "Cooperative System of Settlers" (SCA), as an organizational instance composed of a wide range of other instances. However, to the extent that the MST constituted a rigid model for the settlements, imposed from the top down, with priority given to economic viability, ended up accentuating contradictions and reproducing aspects of the capitalism against which it fought. This model was based on mechanized and intensive production, including the use of pesticides, in addition to the division and specialization of labor and scale production. This organization did not consider the specific characteristics of peasants, disregarding the need for autonomy and respect for individual life experiences. That is, the theories of Marx, Lenin, and Kautsky, when transformed into practice by the MST, with few exceptions, did not become successful models to be followed, on the contrary, they created in the settlers a strong feeling of resistance to this cooperative project, in the words of Borsato and Carmo (2013):

> Paradoxically, the MST disseminated in its settlements a model of production that had been the cause of the expropriation of peasants at an earlier time. This model had indeed incorporated new dimensions, such as the collectivization and systematization of work, as well as the social division of capital to overcome previous problems; however, the peasants continued to depend on the market, both for the acquisition of goods for production and for the sale of their goods, a fact that led the settlers to become hostages of situations that were not under their control.

After the failure of this initial experience, the MST creatively sought new theoretical references to base its actions on and build models capable of supporting the organization's ambitious project. Although cooperativism has not ceased to represent a valuable and necessary business model, it is now seen more flexibly and democratically, taking into account local specificities and, currently, there are hundreds of examples of cooperatives linked to the MST that worked, especially in the South and Southeast of Brazil, as is the case of the Regional Commercialization Cooperative of the Far West (Cooperoeste), headquartered in Santa Catarina and the Agricultural Production Cooperative of Settlers and Small Producers of the Northwest Region of the State of São Paulo (Coapar), in Andradina, in the interior of São Paulo. Fundamental to the MST's new directions was the introduction of the theme of Agroecology, which will gain enormous relevance not only in the internal spaces of the Movement but in society as a whole, particularly in sectors of the left in Brazil and in the world. Although best known for land occupations, public or private areas, which do not fulfill their social function, as provided for in the Federal Constitution, with the aim of agrarian reform, the MST is today a showy reality of success. It is also worth remembering that the MST created and maintains the Escola Nacional FlorestanFernandes (ENFF), located in Guararema in São Paulo, with courses at various levels, from literacy to courses in cooperative administration, land pedagogy, community health, agricultural planning, and agroindustrial techniques. The excellent-level courses are administered by professors, many of them linked to partner universities and technical schools, in addition to volunteers.

3.3 Cooperativism and the Solidary Economy: Collectors and Recycling Cooperatives

Certainly one of Brazil's two most significant experiments in social inclusion, alongside the "BolsaFamilia"[2] program, has been the support through public policies for collectors and recycling cooperatives (CCRs). Although these experiences date back to the beginning of the 60 s of the last century, from the engagement of the Catholic Church pastorals, non-governmental organizations, and universities, it was only from the beginning of the present century that an effective emergence of a large-scale process was achieved. A present phenomenon, especially in medium and large Brazilian urban centers, the organization of CCRs has led to the inclusion of large contingents of people in conditions of extreme poverty, victims in their majority of structural racism. It is highly relevant that concomitantly and synergistically, such a process of social inclusion has been taking place simultaneously with the emergence of a consistent environmental sustainability agenda. It was in 2001 with the organization of the National Movement of Recyclable Material Collectors (MNCR), and in 2004 with the founding of the National Street Population Movement (MNPR), that a critical mass was created for a truly virtuous circle in the organization of work. Beyond generating income,

[2]The "BolsaFamilia" program aims to break the generational cycle of poverty through conditional cash transfers, education, and health. It was created in the first Lula government after incorporating the "Zero Hunger" program, instituted 9 months earlier (January 2003) into other income transfer programs. Among the programs incorporated, in addition to "Fome Zero," also the programs established in the previous government (Fernando Henrique Cardoso): Bolsa-Escola, CadÚnico, Bolsa-alimentação, and Auxílio-gás.

it also established a legal framework for the disposal of solid waste in Brazil.

It is worth noting that the phenomenon of social exclusion, with all its ills, was most evidently concentrated around landfills, veritable heaps of unprocessed urban consumption waste discarded in the open by public authorities. In such landfills, miserable populations sought their livelihood, competing with clouds of vultures, stray dogs, pigs, and other animals, until at the beginning of President Lula da Silva's third term (2023-2026), two presidential decrees were signed turning waste pickers into protagonists in the process, in addition to enabling an essential change in the circular economy and reverse logistics model.[3] The first decree institutes the "Programa Diogo Sant'Ana Pró-Catadoras e Catadores para a Reciclagem Popular," based on the former "Pro-Catador" Program, extinguished by the Bolsonaro government (2019-2022). It is also relevant to highlight that the patron chosen by the collectors was a black man of humble origins, DiogoSant'Ana managed to enter the Faculty of Law at USP where he became a student leader in his career, according to the "National Association of Collectors and Collectors of Recyclable Materials" (ANCAT). He was one of: "A brilliant professional and committed to social causes, Sant'Ana worked with the Presidency of the Republic during the Lula and Dilma governments. During this period, he played a key role in building the Regulatory Framework for Civil Society Organizations and several sustainable development policies and support for recyclable material collectors and other social movements" (https://ancat.org.br/nota-de-pesar-diogo-de-santana/ seen on 03/20/23).

The decrees establish the Certificate of Recycling Credit, Structuring and Recycling of Packaging in General, and the Future Mass Credit. The collection of recyclable material, such as cardboard, aluminum, glass, and plastic and metal packaging, depends mainly on these urban workers called collectors. According to the IBGE, from 1989 to 2000, the Brazilian population increased by 16% while the amount of garbage collected in the same period increased by 56%. Data released by IPEA in 2013 (https://ipea.gov.br/portal/busca-geral?q=reciclagem+de+lixo seen on 03/20/23) indicate that collectors pick up 90% of everything that Brazil recycles, although only 7.5% of these workers organize themselves in cooperatives.

Data on the number of these workers is uncertain,[4] but they are estimated to range from 300,000 to 1,000,000. Still, it is undoubtedly a growing number of those who depend on this activity for survival for those who cannot integrate into the formal labor market., such as migrants who have recently arrived in large urban centers, the street population, and other segments of the universe of poverty. According to data released by the MNCR, there are around 35,000 of these workers registered and in four stages of the organization (Source: https://www.mncr.org.br/ seen on 03/20/23):

Situation 1: Group formally organized in an association or cooperative with its press, scales, carts, and shed, with the capacity to expand its physical structure and equipment in order to absorb new collectors and create conditions to implement industrial recycling units. In this situation, the cooperatives are ready to verticalize the production of recyclable materials. Cooperatives in this situation should be seen as important vectors of social inclusion.

[3]Reverse logistics is the mechanism by which the waste producer is responsible for its final destination, and was provided for by Law 12,350/2010.

[4]As of 2013, the collection activity started to be explicitly accounted for. According to information from the Ministry of Social Development (MDS) in June 2015, the CadÚnico lists 49,181 collectors, of which 31,078 receive some type of social benefit. It is important to take into account that the total number of waste pickers in Brazil cannot be estimated only by CadÚnico, since a significant part of the cooperative members and those whose main activity is waste picking, although isolated, is often above the income limit to enter the register. (Catadores de materiais recicláveis: um encontro nacional [Recyclable material collectors: a national meeting] / Bruna Cristina Jaquetto Pereira, Fernanda Lira Goes (organizers) – Rio de Janeiro: Ipea, 2016. 562. p.: il., color maps)

Situation 2: Group formally organized into an association or cooperative, with some equipment, but in need of financial support to acquire other equipment and/or sheds. The cooperatives in this group are in an intermediate phase – lacking some equipment to be able to expand production – in need of infrastructure reinforcement to expand the collection and thus formally include new collectors of recyclable materials.

Situation 3: A group in the organization, with little equipment – some owned by it – needs financial support to acquire almost all the necessary equipment, in addition to its sheds. The formal establishment of your cooperative will mean the inclusion of new jobs for collectors of recyclable materials.

Situation 4: Disorganized group – on the street or in a dump – without equipment, often working in extremely precarious conditions for intermediaries. Financial support is needed for the complete assembly of the infrastructure of buildings and equipment. The formal establishment of your cooperative will mean the inclusion of new jobs for collectors of recyclable materials."

Although still far short of realizing its full potential, the organization of collectors in the form of cooperatives has been taking place quickly, even with a tendency towards verticalization, as observed in situation 1. Verticalization implies an increase in the operational autonomy of cooperatives, in addition to larger sums of financial resources. The growing dominance of the recycling chain by the cooperative movement of collectors is seen as a threat by economic groups, anxious to appropriate the sanitation sector – which includes urban cleaning and the selective collection of recyclable materials. Research carried out by the MNCR estimated that from R$ 168,972,913.12, 39,040 direct jobs could be generated, benefiting around 175,680 people in 199 urban centers in 22 states. The Movement proposes that these resources be transferred directly to the 244 groups and bases of the MNCR, duly registered and spatialized, as this is a previously identified demand. It is also proposed that the "Banco do Brasil" Foundation can make the transfer of this resource operational, given previous extremely successful experience with cooperatives and associations of collectors.

According to the, the generation of jobs in the segment of recyclable material collectors is strategically important due to a set of factors, namely (https://serviosocialemgeral.blogspot.com/2011/04/o-dia-dia-dos-catadores-de-papel.htmlseen on 03/20/23):

1. This is a specific target audience of the ZERO HUNGER Program.
2. The action of the collectors reduces the cost of operating sanitary landfills and conventional garbage collection, increasing the time and useful life of sanitary landfills.
3. The action of the collectors reduces the costs of energy and raw materials through the use of solid residues, reducing the emission of pollutants.
4. The collection activity is relatively rudimentary, without the need for a high degree of technical knowledge. It is a strategic element when faced with a target audience in extreme poverty, with very low levels of education, and who need a short-term response from the point of view of their survival.
5. The collection activity is part of the recycling chain, whose market is dynamic, even in times of macroeconomic constraint, given the extremely high liquidity of recyclable products.
6. The family allowance program includes a significant part of the collectors. The potentization of this economic activity may mean reinforcing the sustainable construction of the program's so-called exit doors.

For all these reasons, these cooperatives are promoted through technical and financial cooperation with the Ministry of Social Development and Fight against Hunger, the Ministry of the Environment, and the General Secretariat of the Presidency of the Republic. The "National Solid

Waste Policy" (PNRS) established by Law No. 12.305/10, covers technical advice, infrastructure, and logistics in waste collection, sorting, processing, and marketing units. The same law also provides for access to lines of credit, literacy, and schooling through professional and technological education, training for self-management, and the institutional strengthening of associative and cooperativity in the recycling production chain, in addition to access to housing, health, and other social rights (Singer et al., 2014). However, even if transformed into public policies, the actions described would not be capable of effectively transforming social reality, without the existence of true cohesion of the social actors directly involved. That is, even with massive financial and logistical investments and political action, without a true, constitutive, and constructive identity of the main social actors involved, the collectors, the program would become just one more among others of a welfare and paternalistic nature. Fortunately, as Sant'Ana and Metello suggest (Pereira & Goes, 2016): "A collector wants to be a collector!" that is, "He/she doesn't want to be anything else that deconfigures the civic affirmation of the collection activity." In addition to this element, the same authors note the existence of a second factor, precisely the form of solidary organization through cooperatives and associations (Id):

> Here, too, there is a fundamental point. The movement's activities are not restricted to defending improvements in working conditions or the rights of its members. In addition, the collectors propose a model of cooperative organization for the economic activity of recycling, considering that the impetus for this activity and the economic benefits of its expansion must be shared fairly, proportionally to the work performed, and solidary way.

These authors also note that "pickers are one of the main political agents in defense of sustainable development and, in terms of popular penetration, the main organized movement" (Ibid). To support this claim, they recall that both, at Rio + 20 in 2012, where they were present at the "Dialogues for Sustainable Development," and at the "Summit of the People" at the "IV National Conference on the Environment in 2013. The presence of waste pickers was fundamental: "These pillars put the movement ahead of its time and allowed the cause to be seen differently by the public authorities. They also allowed public policies to find fertile soil to develop and multiply." In this way, in line with the present time, the Climate Fund (Law 12, 114/2009) and the Amazon Fund could become sources of financing for the recycling chain, primarily contemplating collectors" cooperatives. In addition to these funding sources, Sant'Ana & Metello (Id. 2016, p. 40) observe that mechanisms of the private initiative itself, such as carbon credits, could add resources to entrepreneurs in the recycling sector. They state that expansion of the industrial park would also generate opportunities for the most advanced cooperatives, improving verticalization of the production chain and, in short, they conclude with the following suggestions (Ibid. 2016, pp. 42–43):

(i) To guarantee that cooperatives are contracted by city halls in the formal selective collection system, especially in municipalities Recycling and Social Inclusion in Brazil: balance and challenges that are closing landfills.
(ii) Signing of the Reverse Logistics Sectoral Agreement, especially in the packaging and electronics sector, which may allow for a new leap for recycling in Brazil and an impetus for better organization of cooperatives.
(iii) To stimulate the increase of recycling industrial parks in Brazil, either through investments from different sources already available, such as the Amazon Fund and the Climate Fund, or through the restructuring of taxation of the recycling chain.

With so many promising prospects, we also believe in the ability of these social actors to play a very effective role in the fight for an inclusive and truly sustainable society in Brazil.

4 Final Considerations

We seek to highlight in this chapter how fallacious and malicious the statement that associated cooperative work and human solidarity, in the territory of the market economy, is a mere utopia, and that civilizing projects based on such premises are doomed to failure. The idea that only the pursuit of personal success and the resulting competition can generate economic strength and social enrichment gives the basis to this belief. In the old ideas with new clothes, neoliberalism has shown itself to be the last frontier of orthodoxy and defense of the metaphysical thesis of the "invisible hand of the market," a perverted species of an immanent ubiquity that sanctifies human selfishness as virtuous. The cases dealt with in this section sought to provide an amplified and non-pamphleteer dimension to understanding the anthropological, sociological, economic, and psychological phenomenology of cooperativism and solidarity, as a way of organizing society to face the traditional capitalist model, especially neoliberalism. We could see that among the various forms of self-organization of work in societies and from the legal norms already instituted in the various national states, solidarity work is present and resists all forms of framing it in the old formulas, abounding in number and creativity. Among such forms of organization of society, facing challenges such as the reversal of income concentration and overcoming the alienated work typical of capitalism, the cooperative movement in the world stands out.

The cooperative movement stands out both for its antiquity and resilience in the face of the challenges imposed by the market and for its impressive growth in various sectors of the economy, particularly in the financial and credit sectors, in addition to services, production, and agriculture. Contrary to what could be anticipated in a focus of this nature, we highlight the fundamental importance of the individual as a human person in the process of social self-organization and in the formation of the community in its various forms. These approaches may be the macro dimension, in the nation-state as a nation, or the dimension micro in various forms of solidarity enterprises. As it was well characterized when we dealt with the bioethics of public governance and the strategies for reducing inequity, for the purpose of changing the economic political model, it is very important to promote games with a "positive sum." We believe that solidarity and authentic cooperativism undoubtedly constitute "games" of this nature. We state, however, that the success of solidarity enterprises, in particular, of cooperatives, depends on a continuous exercise of participatory management and surveillance, that is, participatory democracy. It is a fact that the specter of facade cooperativism remains present and is potentiated by perverted legislation on legislative or jurisdictional interpretative bases. Alongside this challenge, the solidarity enterprise faces internal threats, such as the sabotage of individuals or groups that put their interests above collective interests. Thus, for a positive-sum game to work, what the authors Fehr and Gachter (2002) call altruistic punishment is necessary. That is, the cooperative members need to implement a system of "surveillance and punishment of those – the freeloaders – who take advantage of cooperation to exploit others or to obtain undue advantages at the expense of the efforts of the people who are collaborating."

The legislation that regulates cooperativism provides for such mechanisms, through the bylaws (the cooperative's major regulation) and the internal regulations (which regulate the bylaws). The challenge is to ensure that corporatism and cronyism do not overlap with cooperativism and solidarity, because if the profiteers are not identified and punished and, instead, are allocated power and money, the game will become a negative sum, where everyone loses, or even a zero-sum game, where some win and others lose. In the vicissitudes in the work organization process, the continuous conflict between individual interest and collective interest is manifested, a conflict, we would say, constitutive of the human condition itself. However, a dialectical understanding of such phenomenology obliges us to subsume such conflict between individual and society, in a broader dimension and represented by the relationship between the human Person and Community. Through this micro plane from the anthropological point of view, we must move towards sociological and political relations, subsuming the relationship between the human person and community in the broader relationship between the citizen and the nation. In this way, the anthropological conflict between individuals and society, expressed in the field of economics as a conflict between capital and work, can be subsumed in the desirable relationship between solidary enterprise and solidary work. In such a dialectical spiral, there is no place for individualism, nor collectivism, there is something infinitely greater and that implies the contribution of all human power actualized in community relations. At the end of this journey, a new global paradigm vision will fully emerge, as in the extraordinary cooperative experience of Mondragón and its patron Father Arizmendi. But there is no lack of Brazilian examples of this visionary and supportive entrepreneurship in Brazil. They are represented among the human beings who stand out in the country such as the Swiss priest Theodor Amstad, creator of Sicred, the largest cooperative bank in Brazil; the work by Paul Singer, an intellectual and political activist who became the great patron of the solidarity economy in Brazil; and the testimony of Diogo Sant'Ana, the activist who gave his name to the "ProgramaPró-Catadoras e Catadores para a Reciclagem Popular" and more recently Eduardo Moreira, creator of Finapop. They are, in their community works, concrete expressions of a fictional case such as the "Bailey Brothers Building & Loan" and the legendary George Bailey in "It's a Wonderful Life."

References

Bobbio, N. (1983). *Dicionário de política I Norberto Bobbio, Nicola Matteucci e Gianfranco Pasquino; trad. Carmen C, Varriale et ai.; coord. trad. João Ferreira; rev. geral João Ferreira e Luis Guerreiro Pinto Cacais*. Editora Universidade de Brasília.

Borsatto, R. S., & Carmo, M. S. D. (2013). A construção do discurso agroecológico no Movimento dos Trabalhadores Rurais Sem-Terra (MST). *Revista de Economia e Sociologia Rural, 51*, 645–660.

Bottomore, T. (1988). *Dicionário do pensamento marxista*. Editora Schwarcz-Companhia das Letras.

Brenneisen, E. (2002). *Relações de poder, dominação e resistência: o MST e os assentamentos rurais*. Edunioeste.

Dowbor, L. (2017). *A Era do Capital Improdutivo*. Autonomia Literária, 2017.

Faria, M. S. D. (2005). *Autogestão, Cooperativa, Economia Solidária: avatares do trabalho e do capital*.

Fehr, E., & Gachter, S. (2002). Altruistic punishment in humans. *Nature, 415*(6868), 137–140.

Kautsky, K. (2018). El camino del poder. Ed. Alexandria Proletária, Valencia. https://marxists.architexturez.net/espanol/kautsky/1909/1909-caminopoder-kautsky.pdf

Lénine, V. I. (1975). *A transformação socialista da agricultura*. Estampa.

Moroni, J., & Junior, A. P. (2022). A Intuição Ecológica Criativa na Tomada de Decisão Econômica na Democracia: o caso do Banco Finapop. *Acta Scientiarum. Humanand Social Sciences, 44*(1), e62241.

Pereira, B. C. J. O., & Goes, F. L. O. (2016). *Catadores de materiais recicláveis: um encontro nacional*. Ipea. 562p.

Piketty, T. (2014). *Capital in the Twenty-First Century*. Belknap Press.

Piketty, T. (2020). *Capital and ideology, Translated by Arthur Goldhammer*. The Belknap Press of Harvard University Press.

Port, M. (2008). *A expressão do cooperativismo de crédito mundial e seus principais modelos*. Nova Petrópolis, dezembro de.

Severo, V. S. (2005). As Cooperativas de Trabalho. *Revista do Tribunal Regional do Trabalho da 18ª Região, Goiânia, 8*(2005), 141–144.

Singer, P. (2002). *Introdução à Economia Solidária/Paul Singer* (1st ed.). Editora Fundação Perseu Abramo.

Singer, P., Silva, R. M. A., & Schiochet, V. (2014). *Economia Solidária e os desafios da superação da pobreza extrema no Plano Brasil sem Miséria. O Brasil sem miséria*. MDS.

Standing, G. (2012). The Precariat: The New Dangerous Class. *Work, Employment & Society, 26*(4).

Varoufakis, Y. (2020). *Another now: Dispatches from an alternative present*. Bodley Head.

Viveret, P., et al. (2013). In A. D. Cattani, J. L. Coraggio, & J.-L. Laville (Eds.), *Diccionario de la otra economía* (2nd ed.). Universidad Nacional by General Sarmiento. 384 p. (Lecturas de Economia Social).

Zanotti, A. (2020). *John Stuart Mill e l'utopia cooperative*. Ed: Homeless Book.

Enidio Ilario is an associate member of the Brazilian Society of Bioethics, an expert member of GEObs (UNESCO's Global Ethics Observatory), a co-coordinator of the Interdisciplinary Seminars on Economic and Social Inequality, and a member of the Self-Organization Group, both at the Center for Logic, Epistemology, and History da Ciência (CLE) at Universidade Estadual de Campinas (UNICAMP). He is Doctor in Psychology, Master in

Philosophy from Pontificial Catholic University of Campians (PUCCAMP) and Specialist in Collective Health from FCM-UNICAMP and in Internal Medicine from Sociedade Brasileira de Ciências Médicas (SBCM).

Alfredo Pereira Jr is a philosopher of Science, Professor at the Universidade Estadual Paulista Julio de Mesquita Filho (UNESP), Brazil, retired, with degrees in Philosophy and Administration. Currently working in the UNESP Graduate Program in Philosophy. Alfredo has around 300 publications in several areas of knowledge, mostly in the philosophy of neuroscience and the theory of consciousness. Orcid: https://orcid.org/0000-0002-5960-041XAD, Scientific Index: 56th place in Top 100 Latin America Philosophers/Scientists.

Heleno Rodrigues Corrêa Filho is a retired professor from the Department of Collective Health at UNICAMP and a volunteer contributor to the Professional Master's Degree at ESCS-FEPECS and to the University of Brasilia (UnB) FS-DSC Research Group on Disability Assessment.

Solidarity Economy and Self-Management: Theoretical Notes and Practical Experiences of Participatory Democracy and Popular Governance

7

Isabela Aparecida de Oliveira Lussi,
Flávia Sanches de Carvalho,
and Joelson Gonçalves de Carvalho

Abstract

The economic and social development capitalist model highlights the disparity between wealths growth and the rate of poverty and malnutrition. In times of economic and social crisis, this imbalance and, consequently, social inequalities become even more glaring. There is a need for another model of economic and social development, that is more just, more humane, and more supportive. The solidarity economy is a path for this transformation, a concrete possibility for promoting the social and economic inclusion of groups and communities that have been excluded from the world of work. The objective of this chapter is to present the solidarity economy, with self-management as its cornerstone principle, as an economic and social alternative that promotes participatory democracy and popular governance, generating individual and collective emancipation.

I. A. de Oliveira Lussi (✉) · J. G. de Carvalho
Federal University of São Carlos, Sao Paulo, Brazil
e-mail: bellussi@ufscar.br; joelson@ufscar.br

F. S. de Carvalho
ADUFSCAR Syndicate, Federal University of São Carlos (UFSCar), Sao Paulo, Brazil

1 Introduction

Although Brazil is the 12th largest economy in the world, according to data from the International Monetary Fund (IMF) for 2021, it is in the 87th position in the ranking of the Human Development Index (HDI). That is, the growth of wealth in a given economy does not mean a country without poverty, malnutrition, and absence of basic health and education services, because there is no automatic transposition between economic growth and the improvement of the objective conditions of people's lives (Carvalho, 2015). Adding to this more general picture, there is high poverty rate and chronic unemployment, in addition to a contingent of approximately 55% of the country's population with some degree of food insecurity (Penssan, 2021).

In times of economic and social crisis, the solidarity economy gains evidence as a possibility for the promotion of social and economic inclusion of groups and communities that face exclusion or marginalization in the traditional labor market, such as informal workers, the unemployed, rural communities, quilombolas, indigenous people, people with disabilities, people with psychic suffering, homeless people, among others. According to Singer (2002), with self-management, cooperation, collabora-

A. Pereira Jr, F. Sousa (eds.), *Principles for Governance*, Sustainable Development Goals Series,
https://doi.org/10.1007/978-3-031-40025-4_7

tion, democratic participation, and solidarity as its principles, the solidarity economy stands as a form of resistance and social transformation.

It should be noted that the most contemporary Brazilian debate on issues such as economy, democracy, and development, still has no consensus on how to tackle the country's structural problems, owing in large part, to the power that neoliberal thought exerts over people and institutions that insist in denying the State as the privileged entity for overcoming national backwardness. There is an elimination of the debate, notably with regard to the deepening of participatory democracy and popular governance.

For those reasons, the objective of this chapter is to present the solidarity economy, with self-management as its fundamental principle, as an economic and social alternative that promotes participatory democracy and popular governance.

Aiming for that purpose, the chapter will present theoretical notes on the principles and foundations of the solidarity economy and self-management as disruptive empirical elements that contradict the liberal thesis of democracy in which the effective participation of the population, especially the most vulnerable, is inefficient. Having done this, the arguments presented seek to evidence the power of self-management to promote participative democracy and a new type of governance with a popular character, and, therefore, to contribute to emancipatory processes.

It is important to emphasize that this work is based on the hypothesis that, beyond moments of crisis in which the solidarity economy is an alternative to unemployment, it has a structural role and a disruptive power in the modus operandi of the expanded accumulation of capital to the extent that, in its progress, the social subjects that participate in it and build it structure, in an inseparable way, emancipatory processes that strengthen participatory democracy and popular governance.

2 Solidarity Economy Beyond Representation: Theoretical Contributions and Concrete Experiences of Participation and Popular Governance

The concept of governance, as well as others, such as the concept of development, is not fully agreed upon, and there are several understandings and adjectives to qualify it. A single attribute or concept of governance tends to serve as a tool for the construction of a political and normative repertoire that, by disregarding the processualism of the term, seeks to reinforce one paradigm to the detriment of another. In trying to present the importance of popular governance in the construction of a high-intensity type of democracy, it is key to make it clear that, in the construction of the theoretical argument intended here, the tensions arising from the various controversies and interpretations that often oppose each other are not being disregarded.

Throughout history several thinkers have dedicated themselves to understanding and contributing to the political and social processes that drive development, however, the inclusion of equality as one of the fundamental objectives for the democratic exercise, whether social, economic, or political, is still a relatively recent variable in the literature on the subject.

In an attempt to rescue the processualism and present the tensions inherent in the debate, it is worth starting by saying that the liberal view of governance, notably advocated by authors such as Fukuyama (2004), emphasizes the importance of curtailing state power and establishing mechanisms that ensure individual freedom and the protection of individual rights. In this view, governance should be based on a system of clear and predictable rules that encourage private initiative and competition, allowing individuals and companies to exercise their economic activities without excessive interference from the state. In this way, governance, seen from the liberal perspective, would be central in ensuring the creation of a predictable and favorable environment for the exercise of private initiative, the protection of property rights, and the promotion of citizens' welfare through evidence-based public policies and principles of freedom and justice.

By arguing that political power is exercised by a narrow group of individuals or institutions, the liberal vision of governance is conveniently associated with the theory of elites, which, in short, is based on the argument that a privileged minority should exercise political power at the expense of the majority of the population. Remains to popular participation to legitimize the effectiveness of political institutions, through processes structured on the basis of representative participation.

In the early twentieth century, the thought of the so-called Elitists gained dissemination and notoriety, especially the contributions of Schumpeter (1961), especially his criticism of classical democracy, particularly the vision of Rousseau, defender of an ideal of direct and participatory democracy, based on equality among citizens and the alienation of individual sovereignty in favor of the general will of the people around a common good.

Equality was, for Schumpeter, an unattainable ideal in practice, especially in a modern capitalist society. For him, inequality was a natural feature in politics and economic competition and, contrary to what this might suggest, was not an obstacle to more general development.

To reinforce his argument, Schumpeter (1961) asserted that democracy could not guarantee equality of outcome, since most people were unwilling to pay the price necessary to achieve it. He further argued that democracy was founded on competition among political elites and that equality of political opportunity was more important than economic equality. Even though he recognized that inequality could have some negative effect on social and political stability in developing countries, Schumpeter argued that governments should pursue policies that would allow the widest possible access to economic opportunities in order to minimize the negative effects of inequality on social cohesion and political stability, since the people would not be able to perform such actions.

Analyzing the nature of democracy and the conditions necessary for the existence of a democratic political system, Robert A. Dahl, in the 1970s, presented his theory of polyarchy. In this work, as Dahl (1998) argues, the unequal distribution of political resources always remains a threat to political systems in which popular participation is broad and there is a great variety of political interests. In the analysis of these systems, considered by Dahl as more realistic and, in which democracy is limited, there may be polarization between opposing and equally powerful forces and this can lead to hard-to-resolve impasses within the democratic framework (Santos, 2014).

In his analysis, Dahl takes into account variables such as political institutions, popular participation, representation, and political competition in different countries over time. In a comprehensive way, the author seeks to explain the political and social conditions that favor or hinder the existence and consolidation of democracy. Going against the elite theory, Dahl emphasizes the importance of participation and inclusion as important elements of democratic processes.

Following Dahl's theory, democracy is achieved through pluralism in the ownership of resources, which includes popular participation and the extension of suffrage. Polyarchy differs by highlighting the importance of inclusion, allowing citizens to express their preferences and have an active voice in government through the right to vote. Dahl believes that economic development is capable of reducing inequality and that historical sequences are also relevant factors in this process.

Although Dahl's view on democracy has been influential and key for the study of democratic politics, and notwithstanding the criticism made to elites' theory, especially on popular participation, it is a limited view for, among other reasons, having in the extension of suffrage the central element for popular participation, thus reinforcing representative democracy as a sign of effective participation (Miguel, 2017).

By focusing mainly on analyzing electoral participation and party competition as the main indicators of a healthy democracy, the author does not cover direct citizen participation, decentralization of power, and respect for human rights, for example. In summary, it is known that representation does not guarantee the full participation of citizens in the political process (Pateman, 1985).

Regarding the present work, another weakness that can be pointed out in Dahl's vision is that it does not take into account the socioeconomic inequalities that can limit the political participation of certain groups and individuals. Thus, it disregards the fact that democracy should be more comprehensive and encompass not only formal political participation, but also social and economic justice (Fraser, 1992).

As for the Brazilian case, democracy is marked by a series of challenges, such as the haunting shade of dictatorship, social inequality, and political exclusion of certain groups, and the belief in voting as the maximum expression of popular participation in the country's decision-making process. Thinking about this concrete reality in the light of theories of democracy and popular participation brings challenges, but is fundamental to understanding how democracy works – and the obstacles – in the country.

Although forms of popular participation can be traced back to the nineteenth century, it was from the mid-twentieth century onwards that they gained new impetus and became more evident (Valla, 1998), notably from the unions' movements and the movements active in land's conflicts. However, with the coup and the military dictatorship that followed there was a suffocation of movements and their leaderships, as well as the suspension of the mechanisms of representative democracy, especially with the Institutional Act no. 5 that impeded the functioning of the National Congress in 1968.

New political actors emerged in the early 1980s, putting the debate about democracy and popular participation back on the political agenda, namely: the Workers' Party (1980), the Central Workers' Union (1983), and the Landless Rural Workers Movement (1984). These new subjects were fundamental in giving force to the Diretas Já (Direct Elections Now!!) movement that demanded the return of direct elections for president, as well as the installation of the National Constituent Assembly.

The process of elaboration of the new Federal Constitution had, as a backdrop, the great expectation of innumerable popular sectors in Brazil to advance with the effective creation of processes of direct popular participation at the same time that new forms of representative democracy were institutionalized in the country. It is not the object of this chapter to deepen this theme, but it is worth noting that there has been a widening of the institutionality of popular participation, with emphasis on the creation of forums and councils in various sectors, especially health and education. There was also the materialization of a more robust participation, such as the Participatory Budgets. However, agreeing with Miguel (2017, p. 92), "frustration with the institutions of political representation" prevailed, which, in the process of pacts leading to transition to democracy, ended up stifling popular mobilization for the settlements among elites.

Observing how popular participation is built and developed helps to better understand the possibilities and limitations of democracy in Brazil. Among the many experiences and movements that can be analyzed, we have chosen to analyze the solidarity economy because it can be considered both a practice and a movement, since it involves concrete actions of production, consumption, credit, and distribution of goods and services, besides representing a form of social organization that seeks to promote the transformation of economic and social reality (Singer, 2002).

As a practice, the solidarity economy refers to work activities that are carried out through collaborative and self-managed processes, with the aim of creating fairer and more sustainable alternatives to the traditional economy. These activities can be developed through cooperatives, associations, and self-managed companies, among other forms of organization. Being a movement, the solidarity economy represents a political and social struggle for the transformation of the economic development model and of society. It is a movement that seeks to confront social and economic inequalities, to promote social inclusion and the generation of work and income for marginalized groups or those in vulnerable situations, in addition to respecting nature and valuing culture and local knowledge.

In summary, the solidarity economy is a practice that seeks to transform social and economic relations by building fairer and more sustainable alternatives, and a movement that seeks to promote this transformation in a broader and more systemic way, through political actions and social mobilization.

Observing the economic efficiency of solidarity economy enterprises, Gaiger (2007a, b) concluded that it is a direct function of their equity and self-management mechanisms, irreducible to the instrumental logic that presides over ordinary capitalist economic action. The present-day practice of the solidarity economy expresses, according to the author, the impossibility and the refusal to live capitalism and its intrinsic sociability, both in its economic, cultural, and political dimensions. Therefore, the advance of the solidarity economy, as a movement, is the materialization of a historical resistance guided by collective, cooperative, and radically democratic values.

Dimensions	Capitalist Economy	Solidarity Economy
Economics	Based on competition and the use of collusion to favor corporatism and guarantee profits	Based on cooperation; self-management and the suppression of competition as a defining element of the interaction between enterprises in the market
Cultural	Exacerbation of individualism and consumerism as markers of distinction and social status	Valuing collective action and solidarity as ways of strengthening the group and empowering acceprance, inclusion, and autonomy
Politics	Plutocratic, that can even do without democracy as a political model for organizing society	Radically democractic, by demanting the active participation of its members in the deliberative processes

In other words, according to Singer:

> The solidarity economy is the organized response to exclusion by the market, on the part of those who do not want a society driven by competition, from which victors and defeated emerge incessantly. It is above all an ethical, political, and ideological option, which becomes practical when those who opt for it meet those who are in fact excluded and together build productive enterprises, exchange networks, financial institutions, schools, representative bodies, etc., that point toward a society marked by solidarity, from which no one is excluded against their will (Singer, 2005, p. 11).

The solidarity economy in Brazil gained, according to Silva (2017), a larger role from the 1990s, with the encouragement of various initiatives of work organization and income generation for groups excluded from the formal labor market, such as the unemployed, informal workers, and collectors of recyclable materials. Since then, and largely due to its organization as a social movement, the solidarity economy has been the object of public policies and programs aimed at its strengthening. In this sense, it is worth mentioning the creation, in 2003, of the National Secretariat of Solidarity Economy (Senaes in the Portuguese acronym), belonging to the structure of the Ministry of Labor and Employment.

In an even more critical perspective, the work of Boaventura de Sousa Santos (2006), starting from a criticism of liberal democracy and its limitation to popular participation in political decision-making, contributed to the understanding that in democracy citizens must have an active role in political decision-making at all levels. Thus, governance, for this author, must be democratic and participatory, so as to include perspectives from marginalized and oppressed groups. However, to achieve this kind of governance, an expansion of popular participation is needed, which allows citizens to more directly influence public policies and participate more meaningfully in decision-making.

The solidarity economy as a social movement and as a concrete practice becomes, therefore, the generative element of a participatory governance that strengthens the democratic participation and the capacity for action of the social subjects involved.

3 Self-Management and Emancipatory Processes

According to Arruda (2004), human communities emerge from processes that involve the gain and sharing of knowledge among people, the constitution and execution of a set of common goals, and the contract among people of accountability for tasks and contributions and co-responsibility in caring for each other and the common good. Thus, a community develops in a combination of action and reflection, practice and theory. These are requirements that underlie the pedagogical praxis that breaks the dichotomy between theory and practice and gives form and meaning to self-management.

Self-management implies setting up democratic practices in the workplace, but not restricted to it, extending to the territory and the cities (Nascimento, 2020). Self-management refers to the set of social practices characterized by the exercise of democracy in the decision-making process of a collective, providing its autonomy. It is an exercise of shared power marked by social relations of cooperation between people and/or groups in a more horizontal way. Self-management is a model of social management, of business management, but, above all, it stands as a radical criticism proposed by a new form of autonomous management of the social whole (Albuquerque, 2003).

In this sense, the meaning of self-management also refers to other radically new forms of organization, as is the case of education and global politics, besides the economy. Self-management should be understood as:

> [...] a system of organization of social activities, developed through the intentional and convergent action of several people (productive activities, services, administrative activities), where decisions regarding the destinies of the group are directly taken by those who participate, based on the attribution of decision-making power to the collectivities defined by each of the specific structures of activity (company, school, neighborhood, etc.) (Albuquerque, 2003, p. 24).

In this way, the author identifies two essential determinations of the concept of self-management: the first related to overcoming the dichotomy between those who take decisions and those who execute them, and the second related to overcoming the external interference to the collectivity regarding the definition and choice of what to do (Albuquerque, 2003). When understood in this way, it is evident that one of the cornerstones of the solidarity economy is self-management.

The Brazilian debate in the 1990s regarding the generation of work and income, showed a great approximation to the topics of solidarity economy and self-management, assumed as the perspective of emancipatory work for the unemployed population. The solidarity economy, in this context, emerged as a response to the mass unemployment brought about by the neoliberal policies in force in Brazil at the time. However, little by little it also became a form of labor inclusion for people who were completely cut off from the world of work due to their social condition, but also due to a more aggravated situation of vulnerability.

This arrangement of autonomous management of the social whole carries with it positive and negative aspects in the internal daily life of solidarity economic enterprises, which are groups of workers based on the principles of the solidarity economy. The positive aspects are: the increase in the workers' productive capacity due to a more democratic environment; the workers improve the quality of their production because they have a more positive perception of their work; the worker's personal involvement in their colleagues' productivity is greater because the traditional confrontation between employer and employee has disappeared, and the increase in organizational effectiveness favored by the participative environment. However, if on the one hand the democratic environment increases the productive capacity of workers, on the other hand, their participation in management ends up reducing this capacity; depending on how the work is organized, the material stimuli or those of participation open possibilities for a policy of dispute and manipulation (Albuquerque, 2003).

Moreover, the extra effort demanded of workers in order to concern themselves with the general problems of the enterprise can become exhausting, especially when it is necessary to get involved in conflicts, taking sides in favor or against their comrades, and participating in tiresome meetings. This can lead to workers becoming disinterested, refusing the extra effort, and delegating the decisions to the enterprise's management. If the management accepts to decide in the place of the partners, the enterprise goes from self-management to the practice of hetero-management, losing its democratic characteristics and resembling capitalist enterprises in terms of management (Singer, 2002).

Another factor that contributes to the degeneration of the self-management practice is related, most of the time, to the lack of democratic education of the partners. The impositions of the family, school, church, and other institutions make people learn to obey orders and follow patterns. The involvement in emancipatory struggles, such as strikes, protest demonstrations, meetings of grassroot ecclesial communities, land occupations aiming at land reform, among others, causes this alienating charge to be shaken, opening room for democratic practices and excluding authoritarian ones. The practice of political democracy in many countries, as well as the conquests resulting from the feminist revolution and the sexual revolution, make the new generations less repressed and less passive than those of their parents and grandparents, favoring self-management practices (Singer, 2002).

A quite significant roadblock encountered in self-managed enterprises refers to the change of status of the worker, who goes from being a salaried worker to being the owner and responsible for the enterprise. This implies a transition from a worker/boss relationship to one of equality among the partners, which requires a process of learning the new role and the opportunities and responsibilities that the situation of collective property implies. Such a process is even more difficult when the actors involved are people who have suffered extreme forms of social exclusion (Santos & Rodríguez, 2002). To face such difficulties, one alternative is the promotion of recreational, cultural, social, and other activities by the enterprise in order to provide an environment of mutual support among the partners (Santos & Rodríguez, 2002). According to the considerations presented, it can be observed that the practice of self-management enables the incorporation of important attributes by socially excluded people, contributing to their emancipatory processes.

4 The Practice of Self-Management as Promoter of Participatory Democracy and Popular Governance

One of the Brazilian initiatives to promote the solidarity economy is represented by the Technological Incubators of Popular Cooperatives (ITCP in the Portuguese acronym). ITCPs are organizations linked to universities that use their human resources and the knowledge produced to train, qualify, and advise workers on the development of self-managed enterprises (Guimarães, 2003). Thus, one of the main activities developed by ITCPs is the incubation of solidarity economy enterprises.

The experience of the Federal University of São Carlos (UFSCar), for example, is developed through the Multidisciplinary Integrated Nucleus of Studies, Training, and Intervention in Solidarity Economy (NuMI-EcoSol), which accompanies the incubation of solidarity economic enterprises (EES in the Portuguese acronym). Among these enterprises are those constituted by people in situations of social disadvantage,[1] especially people with psychological suffering, on whom this text focuses to explore the power of self-management in the processes of social emancipation. Singer's speech, in this sense, is emblematic: "The solidarity economy and the anti-mental institution movement are born from the same matrix – the fight against social and economic exclusion. Some are excluded because they are insane, others because they are poor" (Singer, 2005, p. 11). Through the observation of the work done by the Center, it is possible to affirm the significant power of the practice of self-management in the emancipatory processes of these people.

It allowed to understand, through the incubation work of NuMI-EcoSol, that education in self-management takes place on a daily basis in the firms. All the activities developed in the firms are based on the principles of self-management, which stimulates the participation of those involved in all the stages of the productive process and commercialization of products, foreseeing processes of decision and collective management, without hierarchy among the participants, with horizontal relations, which contributes, above all, in the process of social emancipation of people.

People with psychic suffering, historically, have been silenced and made invisible, mainly due to the prejudice of society towards the insane. In general, they are offered few possibilities to participate in decisions, be it in the family nucleus or in the larger social nuclei, and they are rarely heard. Thus, they end up presenting difficulties to take initiatives and participate in decision-making processes.

Along with the practice of self-management in the enterprise on a daily basis, participation in decision-making processes becomes part of life, allowing the social subjects involved to take the experience lived in the enterprise to other life contexts. Thus understood, the practice of self-management also promotes a pedagogical process of democratic participation.

Based on the observed results, one can cite the significant increase in the processes of social inclusion through the active participation of people in solidarity economy fairs, in the solidarity economy movement, and the resumption of life projects. These aspects are intrinsically related to social emancipation.

The analysis of the processes of social emancipation that occur in solidarity economy enterprises constituted by socially disadvantaged people is an important example, due to the particularities of the social subjects involved. But there is a huge range of EESs that, in their work processes based on the active participation of their members, build a new type of governance, a popular governance.

Popular participation in decision-making processes at the state level is still residual in Brazil. Through it, there could be an active participation of civil society in the formulation of public policies and implementation of government programs, ensuring the representativeness of the population's interests. Municipal councils of a deliberative nature are good examples to be expanded. Another important example is the experience of participatory budgeting in which, by means of assemblies, meetings, public hearings, and other mechanisms of popular consultation, citizens define the spending priorities for public management. It is a fact that, most of the time, in their governance model, EESs are not able to set the public agenda for the expansion of programs and public policies that could favor mechanisms of participatory democracy within the State. However, through its self-construction, popular governance in the context of EESs, besides being pedagogical, is also disruptive to the extent that it opposes the capitalist model and heterogeneity.

A social subject that, in its daily life, organizes itself politically and economically in a democratic and horizontal way, actively participating

[1] *Persons in social disadvantage* is an expression used to characterize different populations in vulnerable situations such as handicapped, persons with mental health impairments, persons with troubles derived from drinking or using drugs, homeless people, recently freed inmates, young people presenting troubles with the judicial system, among others.

in collective decisions, is able to understand the limits of representative democracy and elections as the maximum expression of popular participation. It is not possible to move towards a high-intensity democracy without social subjects that are willing to build it and actively participate in it. In this sense, the political and economic organization of the solidarity economy is a concrete example of how popular governance applied in EESs can be extended to new ways of thinking about development.

5 Final Considerations

Solidarity economy is a form of social, economic, and political organization of labor that seeks to promote solidarity, self-management, and sustainability through collective and autonomous enterprises, such as cooperatives, associations, and self-managed companies. Also characteristic of the solidarity economy is the collective ownership of the means of production. Being a necessary condition for its existence, the solidarity economy values collective work and self-management, allowing workers to actively participate in decisions that affect their lives and their work, instead of being subjected to a hierarchical and often authoritarian structure. This enables the development of the social subjects involved so that they can act as protagonists of their own lives, through a process that generates individual and collective emancipation.

By valuing solidarity and cooperation among the members of an EES, favorable conditions are created for the construction of support networks and mutual support. This contributes to the development of relationships of trust and mutual respect, which are fundamental for the construction of a more just and egalitarian society.

The solidarity economy praxis is based on the direct participation of workers in decision-making and management, both of the work process and of the firm. In this model, workers assume responsibility for organizing work, for planning and executing activities, for managing resources, and for distributing results. In other words, self-management is inherent to the solidarity economy and disruptive in relation to traditional management models, based on hierarchy and centralized authority. By seeking to promote horizontality in work relations and democracy in the management of EESs, self-management contributes to structuring a different kind of governance, of a popular nature.

The social subjects involved in this popular governance, therefore, challenge those arguments and theories of democracy that tend to disregard social inequalities and asymmetries of political power in decision-making, undermining the capacity of the most vulnerable to actively participate in the decision-making arenas.

Acknowledgments *National Council for Scientific and Technological Development* – CNPq for financial support of this research.

References

Albuquerque, P. P. (2003). Autogestão. In A. D. Cattani (Ed.), *A outra economia* (pp. 20–26). Veraz Editores.

Arruda, M. (2004). Desenvolvimento comunitário e autogestão. In *Brasil. Ministério do Trabalho e Emprego. Secretaria de Políticas Públicas de Emprego. Departamento de Qualificação. Autogestão e economia solidária: uma nova metodologia*. ANTEAG.

Carvalho, J. G. (2015). *Economia Política e Desenvolvimento: um debate teórico*. UFSCar, Coleção Governança e Desenvolvimento, Grupo de Pesquisa de Ideias, Intelectuais e Instituições. Revisto e ampliado em 2017.

Dahl, R. (1998). *On Democracy New Haven*. Yale University Press.

Fraser, N. (1992). Rethinking the public sphere: a contribution to the critique of actually existing democracy. In C. Calhoun (Ed.), *Habermas and the public sphere*. The MIT Press.

Fukuyama, F. (2004). *State-building: governance and world order in the 21st century*. Cornell University Press.

Gaiger, L. (2007a). A outra racionalidade da economia solidária. Conclusões do primeiro Mapeamento Nacional no Brasil. *Revista Crítica de Ciências Sociais, 79*, 57–77.

Gaiger, L. (2007b). La economía solidaria y el capitalismo en la perspectiva de las transiciones historicas. In J. L. Coraggio (Ed.), *La economía social desde la periferia: contribuciones latinoamericanas* (pp. 79–109). Altamira.

Guimarães, G. (2003). Incubadoras tecnológicas de cooperativas populares: contribuição para um modelo alternativo de geração de trabalho e renda. In P. Singer &

A. R. Souza (Eds.), *A economia solidária no Brasil: a autogestão como resposta ao desemprego. Coleção economia* (pp. 111–122). Contexto.

Miguel, L. F. (2017). Resgatar a participação: democracia participativa e representação política no debate contemporâneo. *Lua Nova: Revista De Cultura E Política, 100*, 83–118. https://doi.org/10.1590/0102-083118/100

Nascimento, C. (2020). *A autogestão comunal*. Lutas anticapital.

Santos, B. S. (2006). *Renovar a teoria crítica e reinventar a emancipação social*. Boitempo Editorial.

Santos, B. S., & Rodríguez, C. (2002). Introdução: para ampliar o cânone da produção. In B. S. Santos (Ed.), *Produzir para viver: os caminhos da produção não capitalista* (pp. 23–77). Civilização Brasileira.

Santos, W. G. (2014). Robert A. Dahl, Econômico. *DADOS, Revista de Ciências Sociais, Rio de Janeiro, 57*(2), 289–292. https://doi.org/10.1590/0011-5258201409

Schumpeter, J. A. (1961). *Capitalismo, socialismo e democracia*. (Editado por George Allen e Unwin Ltd., traduzido por Ruy Jungmann). Editora Fundo de Cultura.

Silva, S. P. (2017). *Análise das dimensões socioestruturais dos empreendimentos de economia solidária no Brasil. Texto para discussão. Instituto de Pesquisa Econômica Aplicada*. Ipea.

Singer, P. (2002). *Introdução à Economia Solidária*. Editora Fundação Perseu Abramo.

Singer, P. (2005). Saúde Mental e Economia Solidária. In *Brasil. Ministério da Saúde. Saúde Mental e Economia Solidária: inclusão social pelo trabalho* (pp. 11–12). Editora do Ministério da Saúde. Retrieved at http://bvsms.saude.gov.br/bvs/publicacoes/SAUDE_MENTAL_ECONOMIA_SOLIDARIA.pdf May 24, 2023.

Pateman, C. (1985 [1979]). *The problem of political obligation: A critique of liberal theory Reedição com novo posfácio*. University of California Press.

Penssan, R. (2021). *VIGISAN Inquérito Nacional sobre Insegurança Alimentar no Contexto da Pandemia da Covid-19 no Brasil*.

Valla, V. V. (1998). Sobre participação popular: uma questão de perspectiva. *Cadernos de Saúde Pública, 14*, S07–S18. https://doi.org/10.1590/S0102-311X1998000600002

Isabela Aparecida de Oliveira Lussi holds a Master's degree in Philosophy (UNESP) and a PhD in Sciences, Psychiatric Nursing Program (USP), held a Postdoctoral Fellowship at the Center for Social Studies, University of Coimbra, Portugal. She is a professor at the Department of Occupational Therapy, the Postgraduate Program in Occupational Therapy and the Multidisciplinary Integrated Center for Studies, Training and Intervention in Solidarity Economy (NuMI-EcoSol), Federal University of São Carlos (UFSCar). She is a president of the Brazilian Association of Solidarity Economy Researchers (2018–2021 and 2021–2024).

Flávia Sanches de Carvalho holds a Master's and Doctoral degree in Political Science from the Federal University of São Carlos (UFSCar) and is a researcher at the Ana Lagôa Archive (ALL) of military politics.

Joelson Gonçalves de Carvalho holds a Master's and a PhD in Economic Development from the University of Campinas (Unicamp) and is an associate professor at the Department of Social Sciences at the Federal University of São Carlos (UFSCar) and at the Graduate Program in Agroecology and Rural Development (PPGADR). He is the coordinator of the Núcleo Multidisciplinar Integrado de Estudos, Formação e Intervenção em Economia Solidária (NuMI-EcoSol) and researcher of the Núcleo de Pesquisa e Extensão Rural (Core Group for Rural Research and Outreach – NuPER).

Part III

Philosophy of Public Administration Directed Towards Human Development

Indicative Planning and Government Redistribution Policies: A Systemic Continuous Process

8

Alfredo Maciel da Silveira

Abstract

Projects to reduce social inequalities require a planning methodology that recognizes the interactions of public administrations with the socioeconomic system as a whole, and with the political system where negotiations take place regarding the future that one wants to see built. Such projects must be understood in a set of actions that are consistent with each other, coordinated in time and space, in addition to obeying the restrictions of feasibility. There may even be some room for the "invisible hand of the market", but it is about structural changes in the medium and long term. So you need to have some control and evaluation over goals and objectives. Political negotiation between social groups, which appeals to the present "sacrifices" of some, in exchange for a better future for all, requires the elaboration of "constructed futures", with quantitative and qualitative expression. The analysis of planning then opens up into two combined dimensions: technical and political-institutional, inseparable in their concrete object. The work intends to bring some more general aspects of this methodological question. In particular, it should be noted that redistributive policies, whether income or investment in public or private goods (housing, for example), are mediated in some way by the economic-financial flows of public administrations (fiscal flows, public banks and public companies), thus demanding a careful representation of the relationships between those administrations and the economic system as a whole. It is briefly reviewed the methodological research work in market economies that supports the renewal of indicative planning taking into account the double dimension: technical and political-institutional. Highlight as references (Johansen, Leif, *Lectures on macroeconomic planning* (p. 2v). North-Holland P. Company, 1977-1978; Kerstenetzky, *O Planejamento Econômico e Social em Economias de Mercado: Informações e Compromissos. Debate Econômico* (pp. 17–25). Ed. Fundação João Pinheiro, 1986; Duloy, Sectoral, regional, and project analysis. In H. Chenery, M. Ahluwalia, J. Duloy, et al. (Eds.), *Redistribution with growth*. Oxford University Press, 1976; Bhaduri, Economic policy and the theory of the state. In *Macroeconomics: The dynamics of commodity production* (p. cap.8). Macmillan, 1986; Silveira, *As Administrações Públicas no Planejamento Moderno: Modelização da Estrutura, Conflitos e Estratégias*. 205p. Tese (Doutorado em Economia). IE-UFRJ, Rio de Janeiro, 1993;

A. M. da Silveira (✉)
National Economic Development Bank, Estácio de Sá University (Retired), Rio de Janeiro, Brazil

A. Pereira Jr, F. Sousa (eds.), *Principles for Governance*, Sustainable Development Goals Series,
https://doi.org/10.1007/978-3-031-40025-4_8

Silveira, O Esquecido Artigo 174: repensando o planejamento indicativo no Brasil. In *VI Encontro de Administração Política*. UFF/PPGAd. https://drive.google.com/file/d/0ByYJJ7xV4RaMaXNpbVdIZUFZOGVVQndPMDQ4bVZ4VFRDaXJZ/view, 26 March 2023, 2015). The outlines of an indicative medium-term plan are illustratively exposed, as a guideline for the methodology and interfaces with the political system and society as a whole.

1 Introduction: Redistributive Policies and Indicative Planning

Projects to reduce social inequalities require a planning methodology that recognizes the interactions of public administrations with the socioeconomic system as a whole, and with the political system where negotiations take place regarding the future that one wants to see built. Such projects must be understood in a set of actions that are consistent with each other, coordinated in time and space, in addition to obeying the restrictions of feasibility. There may even be some room for the "invisible hand of the market", but it is about structural changes in the medium and long term. So you need to have some control and evaluation over goals and objectives. And the very political negotiation between social groups, which appeals to the present "sacrifices" of some, in exchange for a better future for all, requires the elaboration of "constructed futures", with quantitative and qualitative expression.

2 First Steps, Negotiation and Concertation

The *production of consistency* in socioeconomic information is a basic input, which is at the beginning of continuous institutional improvement. And the ongoing informational revolution, which impacts culture and everyday life, greatly facilitates citizen participation. It is a whole organizational setting to be developed alongside the instrumental and technical support of this approach to economic and social policy.

The first step is *the simple production of articulated and internally consistent socioeconomic information* (for which the country – Brazil – has the IBGE, IPEA and an entire extensive research network covering civil society) providing the explanation of costs, trade-offs and benefits of the strategies, which in itself would raise the quality of political negotiation. Information is the raw material of negotiation and planning in market economies (Kerstenetzky, 1986, p. 20–25). It should be noted that *this decisive initial step is entirely under the control of the executive branch* and in no way conflicts – quite the contrary – with the "daily conduct" of the set of economic and social policies.

But this informative support must remain in connection with the decision-making process on a social scale and be inserted in the organizational environment proper to this process. The *technical and political-institutional* dimensions of planning then come together, demanding a mutual development, in continuous improvement.

Failing that, the debate on policies and reforms does not leave the sectorial and partial perspective.

In turn, the creation of a "climate", an institutional and macroeconomic environment suitable for private sector confidence, especially for industry, is absolutely crucial also in the context of planning. For the private sector, planning is obviously not mandatory, but *it invites concerted action and alignment of expectations. Therein lies its strength. That concertation is the "heart" of his methodology*. It encompasses a whole, public and private sectors.

In recent decades, the idea of planning has been identified, in many cases with well-founded reasons, with despotism, statism, or technocratic knowledge. The methodological backwardness and the exhaustion of traditional processes that no longer responded to the conditions of today's world put that idea on the defensive. In Brazil, moreover, it is necessary to add the disarticulation of technical teams, the dispersion of specialized staff and the predominance of an immediatist culture in public administration, as factors of absolute emptiness of any specific debate on the subject. The culture of planning was completely lost, with immediacy and improvisation prevailing. And in government guidelines – from the energy "blackout" to the "bullet train" – the superimposition of neoliberalism, voluntarism and "guidance without a plan".

It is true that Brazilians even know how to map the major national issues. But when it comes to pointing out ways of solving things, the ideas of "Plan" and "Strategy", when they appear, do so in a very embryonic way or as a panacea, when not associated with the preconditions of a great "social consensus" and "state reform". It turns out that the methodology must deal, first of all, with *an environment whose rule is competition and conflict*, and from the perspective of a special actor, the government, which takes the initiative for institutional development in its field of action. Planning, then, cannot be placed at the culmination of a "reform" and a "consensus", *but rather at its origins.*

In short, the government is the protagonist. It would suffice for its leadership to understand the conceptual framework of planning and the strategic scope of the initial steps.

3 A Summary of the Conceptual Framework and Methodological Principles

3.1 Theoretical Framework: (More than) Indicative Planning

Meade (1971)

- Indicative planning as an "uncertainty reducer"
- Market uncertainty x environmental uncertainty
- Compilation, processing and distribution of information and proposals that do not imply the commitment of any economic agent. "Pure" form of indicative planning.

Kornai (1970, apud Johansen, 1977, p.44):

- Process of cognition and commitment
- Collecting and evaluating information about the future
- Device for understanding interdependencies and reconciling conflicting interests

Johansen (1977)

- Context: Centralized economic planning with decentralized decision-making power
- Recognition of relevant non-central decision-makers
- Function of decision coordination
- Game structure, strategic interaction
- Cooperative game
- Dual government action:
  - Use of instruments directly controlled by the government
  - Activities aimed at promoting the coordination of decisions and actions to be taken by other agents in the economy.

Kerstenetzky (1986)

- Economic and Social Planning in Market Economies: Information and Commitments.

3.2 Traditional and New Principles

3.2.1 Traditional

Consistency: macroeconomic benchmarking, stages (sectors, regions, projects)

Feasibility: macroeconomic, demographic, natural resources, technology constraints

Temporal chaining: sequences, steps, trajectories.

Integration between design and policy operators

Information and commitments: explanation of trade-offs, opportunity costs, negotiation bases.

3.2.2 New

The *State as an actor (player) alongside the others*. Unlike referee, "neutral"

Cooperative Game: adherence or non-adherence to the Plan

Strategic interaction: non-controllable variables and actors.

Treatment of Uncertainties. "Minimax", adaptive control.

3.3 New Possibilities

Information Technologies

Networks. Interactivity: electronic government, teleconference, public consultation.

3.4 Institutional Evolution

Elaboration of Scenarios

New Decision-Making Process – Institutionalization.

Continuous improvement.

Still on the *new principles*. As for the instrumental and technical support of planning, the quantitative modeling of a *structure* (a set of relatively stable technological, behavioral and institutional relations (evolution of parameters can be simulated) functioning independently of consciences and wills) was segregated more strictly, *leaving aside another representation, that of subjectivities*, of actors, endowed with decision-making autonomy and great economic impact, outside governmental control. For this very reason, scenario methodologies were brought to the field of economic policy and planning.

In the post-war period, that *structure* was represented much more comprehensively. All private agents were included in it. Even private investment decisions were treated as behavioral relationships (like the French INSEE models, say, "of 1500 equations"). Under a systemic approach, the structure was linked to a "control unit" (the government which, once elected, was supposed to be a "neutral" administrator of conflicts of interest, especially those of classes). This was implicit in macroeconomic theory itself, and was intensely debated and revised since the crises of the 1970s and the following decade. A critique of the connection between the vision of cooperative capitalism mediated by a neutral state (post-war Keynesianism), and the conceptual structure of economic policy inspired by Tinbergen (goal-instrument relationship, "trade-offs" between goals) is addressed by Bhaduri (1986, p. 245, 250–256).

Maintaining the systemic approach, the new representation starts to highlight the strategic interaction between central authority and relevant non-central agents – the aforementioned representation of subjectivities or subjects – as well as the recognition of uncertainties now assumed, made explicit and treated as such, as scenario variables. Now, in methodological terms, almost all of this comes from the experience of large multinational corporations (corporate "strategic planning"). But it goes much further, when it comes to coordinating decisions between social or economic agents, negotiating conflicts on a social scale, in short, when it comes to public affairs and State-society and State-market relations.

Thus, the production and circulation of information gains enormous importance. And not just the status quo, the existing information. But mainly prospective information, of "built futures", the object of political negotiation and citizen participation. The characterization of planning by J. Kornai (1970 apud Johansen, 1977, p.44) suits this purpose well:

Planning: a process of cognition and compromise. Planning is an instrument of cognition. The main purposes of planning are the collection and careful evaluation of information about the future. It helps in understanding our own desires, wishes and goals, and helps to confront them with the realities. It is a framework for the exchange of information and the coordination of otherwise independent activities. Since the activities of all participants of the economic system are mutually interdependent planning is a device to understand the interdependencies and reconcile the conflicting interests.

4 Outlines of a Medium-Term Indicative Plan

4.1 Overview

A first characteristic of medium-term planning to be remembered is that the structural and operating conditions of the economy are practically given (horizon of 4 to 7 years). Here we want to draw attention to the fact that, in the initial process of generating alternative trajectories, the degree of freedom that one has to simulate structural changes is significantly smaller than that existing for the long term (10 to 20 years). The same applies with regard to the mobilization of the constellation of potential resources of the country (natural resources, productive capacity, PEA (Economically Active Population), demographic conditions). This does not mean that such factors are fixed, but rather that they will constitute important constraints. Likewise, certain behavioral relationships, such as the relationship between disposable income and personal consumption, in a context of marked inequality in income distribution, suggest that the stratified profile of consumption may not undergo major changes in the medium term. Changes in the country's energy matrix, which would be the object of the plan, would probably only mature beyond the horizon of the plan, given the extended maturation periods, typical of the energy area. Therefore, it will be very important to analyze the terminal conditions of the medium-term plan, which will shed light on the post-horizontal consequences of the plan.

4.2 Some Aspects of the Institutional and Organizational Dimension

Supported by *preliminary projections and simulations* by the Planning Office and the Statistics Office, the central authority (government, executive branch) defines "Major Options" and "Objectives" which are submitted to Parliament for approval. Back in the executive branch, these objectives and options will be converted into the plan itself. At this stage, it is very important to define the *relationship between objectives and instruments*. Here, the Planning Office relies partially on simulation models, whose model structure emphasizes the limitations imposed by the set of instruments actually available, as well as by the behavior of agents outside government control, and by institutions in general. The model generates *formal results, which should be criticized and complemented based on intuitive knowledge, experience and practical judgments.* This means that not all possibilities and constraints should be represented in the model. In this sense, the model should be complex to the point where it produces correct and reliable results. The model must be connected to the Statistical Office's database. This body will also operate models that are closer to an extensive database, with less emphasis on the restrictions imposed by the list of instruments, and *with greater flexibility to explore physically possible results, in a long-term horizon,* which includes the consideration of a demographic module contemplating, in addition to demographic variables, their interaction with the educational system, the supply and demand conditions of the labor market, and the evolution of the living conditions of social groups.

Once the Plan has been drawn up, it must be debated and approved by Parliament. This culminates *in a process that develops in parallel with the elaboration of the Plan, namely, the process of political negotiation within society, between different social groups and economic agents* (Kerstenetzky, 1986, p. 24–25). Particularly in a situation where income and wealth profiles are concentrated, the information set out in the Plan, making trade-offs transparent, particularly those referring to losses and gains of different groups, becomes a fundamental piece in the formation and establishment of commitments around the path or pattern of development to be pursued. As already seen, the role of information in democratic improvement is therefore essential.

During the process described so far, the *Planning Council*, made up of representatives of social institutions, specialists and personalities of notorious knowledge and independence, advises the central authority and the Parliament in an advisory capacity.

During *implementation*, the Planning Office will adjust objective-instrument relationships by formalized and non-formalized means, as mentioned above, based on *new information that is gained during implementation.*

The dissemination of information, the transparency and horizontalization of the decision-making processes, can have highly positive social and political consequences in the sense of the advances of a democratic institutionality. However, based on historical experience, one can argue about the eventual capture of planning by technocratic knowledge and authoritarian statism. There is indeed an undeniable ambivalence and inherent risk in that capture.

It is true that planning has the vocation of creating representations of demands, explaining intentions and revealing conflicts between them and the costs involved, thus dealing with social diversity. It is a fact that in its technocratic version, it does the opposite of this, jettisoning issues, covering up conflicts and supporting "non-decisions".

But the information generated in the planning process can be healthily "dangerous" in the sense of triggering change. A democratic institutionality that makes this process transparent will receive, through the information it produces, a greater impetus for the deepening of democratization.

4.3 Stages and Organizational Structure

The structuring of the Plan in stages aims to study the behavior of a relatively small number of aggregated variables, through an aggregated model whose relationships are independent of the more complex and disaggregated internal structure contained in those variables. It is possible to determine macroeconomic variables and their sectorial or regional composition, but in support of the exchange of information with *operators closer to implementation.*

The aim is to combine the simplicity of analytical planning tasks, for example, generating several tentative calculations at the macro level, with questions concerning operationalization, which requires detailing and disaggregation. Therefore, the stages relate to the *integration between design and policy operators*. For example, certain structural changes in the industry aimed at its competitiveness may show restrictions in the Balance of Payments or a great effort to mobilize domestic savings, which must be analyzed at the macro level, prior to the sectoral disaggregation.

The need for transparency in communication with people with political responsibility at the sectoral, state and municipal levels, and even communication between technicians, recommends that information be arranged in a sequential logic, in which the different results emerge from models chained in stages, instead of a single complex model.

In more aggregated models, the relationships between variables depend on average parameters, whose stability may crucially depend on the composition of the aggregates. For example, the relationship between the mass of wages and the added value of the economy as a whole, which can be traced to an average coefficient, will be sensitive to changes in the sectoral composition of production, assuming relations between the mass of wages and added value, plus stable at the sectoral level. In the context of an inequality in the distribution of income that manifests itself through the functional distribution by sectors, there may be a difficulty here regarding the use of the macro model, aggregated, in the first stage. Analogous example would be the adoption of an average capital-product ratio for the economy as a whole.

However, it has already been pointed out that in the medium term, the impacts on income distribution may not be very significant. But this will obviously also depend on the objectives and the expected trajectory. The medium-term plan links with the short and long terms.

After finding the solution for the higher stage (more aggregated model), the information already contained in this solution is used to obtain the composition of the variables in a disaggregated way. For example, the breakdown of workers' consumption according to different products, according to the mass of wages, given the average propensities to consume broken down by product.

Once disaggregated values are obtained, the average parameters implicit in these results can be calculated ex-post and thus compared with the average parameters adopted in the most aggregated stage. Very pronounced deviations will be corrected by an iterative process, where the two stages will be repeated.

John Duloy draws attention to the two possible forms of connection between the two levels of models: formal connection (decomposition of models and solution by precisely defined algorithms); and informal linkage, where information flows boil down to a small number of key variables. In the latter case, he states: [...] "The analyst must use his judgment concerning the information and relationships which can be disregarded. Further, in an informal linkage, unlike the formal, it is not necessary to iterate amongst models until the system converges upon a solution" (Duloy, 1976, in Chenery et al., p. 196).

The conception in stages does not imply, in itself, any form of organizational setting of the planning. However, one can think of an organizational parallelism, making each stage correspond to an organizational instance in a hierarchical chain. But in principle, all stages could be developed within the same organization or institution. Attention should be drawn here to the work of *Sectorial Groups* – by the way well known in the Brazilian experience – in the sectoral elaboration phase of the plan, when there is an exchange of information between the Planning Office and the Groups around aggregated and sectoral information.

4.4 Market Uncertainty and Coordination of the Capital Formation Process

"Market uncertainty", concerns "[...]things which some people know for certain and others do not know at all" – distinctly from "environmental uncertainty", [...] "about things which nobody knows" (Meade, 1971, p. 150). Reiterating and relativizing the definition, as Meade himself does, market uncertainties then also concern facts about which some can at least form an authoritative opinion, or even facts "[...] about which some agents can form a much better opinion than others" (ibid, loc.cit.). A typical case of these uncertainties arises from the interdependence between projects of different activities. Then the various investments should be coordinated in order *to reduce or eliminate such uncertainties*. The interdependence can be:

(a) In quantities: supply and demand linkages, identifiable from detailed knowledge of *input-output* relationships. They are relevant especially when substitution possibilities are limited; have implications for decisions about the scale of projects.
(b) In prices: problem of economies of scale. Implicit coordination in price signals is not enough in these cases. Need to coordinate the schedule of the:

1. Timing, of investments, schedule in time and synchronization;
2. Location of projects;
3. Scales;
4. Choice of processes, technology.

When it comes to promoting structural changes, there is no forecast of *future market prices*, which can only be considered *through the coordination of interdependent investments*.

In an iterative process, the sphere of global and multisectoral planning informs the specific, sectorial instance about the "shadow prices" of resources (investment funds, foreign exchange, labor, main inputs, exogenous to the sector) and about aggregate projections by sector (production, consumption, investment, budget allocations). Structural changes can provide new dynamic comparative advantages that are difficult to assess based on current market prices.

In turn, the sectoral instance will provide the central instance with elements for updating technological coefficients and for new aggregations, based on sectoral studies. With this, the sensitivity of the global and multisectoral plan to the choices of processes within the sector can be tested.

4.5 The Plan and the Agents Operating the Policies

There is also interactivity between the sectoral and project levels. In a first iteration, sectoral planning attempts to isolate combinations of projects that appear to be the most attractive, so that engineering studies focus on these projects (*engineering*). The evaluation of these projects at a microanalytical level, carried out by engineers and specialists in the sector, can however reveal certain factors that were omitted in the sectoral planning, whose inclusion would alter the "optimal program" initially generated at the sectoral level. As a rule, planners do not realize some relevant factor until they are confronted with the implications of their omission.

By analogy, one can think of spending on social infrastructure, budgeted at the macro level but defined at the base by municipalities and communities. Interactivity then takes place according to a spatial disaggregation of the stages. Based on models about schools, creches, kindergartens and urban equipment, expenditure can be budgeted at a macro level. But the solutions, of course, come from the base. And a mayor (or a participatory budgeting council member) would simulate, on his notebook or smartphone, the budgetary impact of the macro scenario.

References

Bhaduri, A. (1986). Economic policy and the theory of the state. In *Macroeconomics: The dynamics of commodity production* (p. cap.8). Macmillan.

Duloy, J. H. (1976). Sectoral, regional, and project analysis. In H. Chenery, M. Ahluwalia, J. Duloy, et al. (Eds.), *Redistribution with growth.* Oxford University Press.

Johansen, L. (1977-1978). *Lectures on macroeconomic planning* (p. 2v). North-Holland P. Company.

Kerstenetzky, I. (1986, December). *O Planejamento Econômico e Social em Economias de Mercado: Informações e Compromissos. Debate Econômico* (pp. 17–25). Ed. Fundação João Pinheiro.

Meade, J. E. (1971). *The controlled economy. Principles of political economy.* George Allen & Unwin Ltd.

Silveira, A. M. (1993). *As Administrações Públicas no Planejamento Moderno: Modelização da Estrutura, Conflitos e Estratégias.* 205p. Tese (Doutorado em Economia). IE-UFRJ, Rio de Janeiro, 1993.

Silveira, A. M. (2015). O Esquecido Artigo 174: repensando o planejamento indicativo no Brasil. In *VI Encontro de Administração Política.* UFF/PPGAd. https://drive.google.com/file/d/0ByYJJ7xV4RaMaXNpbVdIZUFZOGVVQndPMDQ4bVZ4VFRDaXJZ/view, 26 March 2023.

Alfredo Maciel da Silveira is an electric engineer, EEUFRJ, 1970. He holds Master in Production Engineering, Economic Systems – COPPE/UFRJ, 1974. He has done his Doctorate in Economic in Economics, area of planning, IE/UFRJ, 1993. He was an engineeer (1974–1975) and former intern (1968–1971) at the National Development Bank (BNDE) on financing of industrial and infrastructure projects. He was an engineer at ELETROBRAS on sectorial and economic-financial planning, development of models and decision support systems, scenario studies and training (1975–1991). He is a member of the BNDES/PETROBRAS/ELETROBRAS Working Group (1986). He was a professor at Estácio de Sá University (1995–2007) for courses in Business Administration and Economics and a coordinator of the Masters in Business Administration and Development (1996–2000). He is the author of 'Planning in Brazil: A Historical-Institutional Approach' for the Working Group 'National Strategic Plan' (2015–2016).

The Semiotics of Money: Towards the Governance of Development

9

Luís Otávio Bau Macedo

Abstract

Development is a concept pervaded by diverse foundations, such as technologies, institutions, and cultural traits. My contribution aims to support the claim that semiotics is the science of the analysis of these broad alignments assembled in the study of money in capitalism. For this purpose, I make use of Charles Sanders Peirce's semiotic formulation and John Searle's analysis of social reality, by which development trajectories are depicted as specific compounds of institutions (signs), technologies (objects), and markets (interpretants) that create meaningful properties depending on symbolical forms. In this view, development is socially structured and, ontologically, a observer-relative phenomenon. Semiosis (meaning creation) depends upon the symbolic powers of money that structure and assemble collective intentionality in capitalism. In this process, I advance two critical conditions of development. The first is mentioned as semiotic intra-coherence which is related to the bridging of dispositional functions in a coordinated way in networks of artefacts and users of money. The second is termed as inter-coherence and takes place as the supervenient causality of social structures and the performative character of habitual patterns of monetary behaviour. Both are intertwined in syntax and semantics forms that evolve the formal causality to development.

1 Introduction

The evolutionary and institutional fields of economic inquiry have advanced several important aspects of the analysis of development. Evolutionary thinkers tend to stress that development is the outcome of economic change which is emergent from behavioural proclivities such as routines and habits of taught (Hodgson & Knudson, 2010; Nelson, 2008; Sidney, 1984). Consequently, technology and productivity are depicted as the long-term drivers of income growth and better living conditions, and both are consequential of systems of knowledge and information (Dopfer & Potts, 2004; Nelson, 2008; Nelson & Winter, 1982). On the other hand, an institutional analysis highlights the role of social coordination in economic efficiency in the form of normative instances such as the "rule of law", "property rights, and "individual liberties". All these instances are perceived as

L. O. B. Macedo (✉)
Rondonopolis Federal University,
Rondonópolis, MT, Brazil
e-mail: luis.otavio@ufr.edu.br

A. Pereira Jr, F. Sousa (eds.), *Principles for Governance*, Sustainable Development Goals Series,
https://doi.org/10.1007/978-3-031-40025-4_9

necessary outcomes of "inclusive societies" (Acemoglu & Robinson, 2012, 2019; North, 1990).

Indeed, both perspectives have correct epistemic claims and are well established in terms of the convergence and mutual reinforcement mechanisms that are supported by empirical scrutiny (Hodgson, 2007; Nelson, 2002; Potts, 2007; Vromen, 1995). On the grounds of the ontological foundations of economics, however, the divergence is much greater. Evolutionary and institutional theoretical analysis is based on two non-congruent approaches of prior causation. First, evolutionary analysis is undergirded by biologic metaphors such as mutation, selection, and inheritance, which are questioned in terms of their adequacy to represent social facts (Foster, 1997; Herrmann-Pillath, 2001; Witt, 1997, 1999; for a counter perspective, see Hodgson, 2002). Second, the institutional field nowadays is still ontologically related, with a semi-detached version of *Homo economicus,* to the principles of bounded rationality and individualistic causation that is not well-fitted to non-equilibrium dynamics (Furubotn & Richter, 2005; Ménard & Shirley, 2008; Williamson, 1985).

The outcome is that the lack of common foundations is a serious problem for development studies. The selection of specific mechanisms related to epistemic causation requires a previous specification of the ontological design of social interactions. For instance, a common trait of the evolutionary and institutional perspectives is the claim that there are no ex ante recipes to be followed. Evolving social structures are co-emergent and entangled with power structures and historical contexts.

Thus, social facts are heavily ingrained in the endogenous cultural and historical trends. Path dependence is, conversely, consequential and dependent upon "mental models", a term coined by Douglass C. North (1995), that assemble collective modes of perception in social interactions. The most important signalling mechanism in capitalism is the institution of money.

In the face of this analytical grounding, this article aims at supporting the view that semiotic emergence is the key ontological property to development studies. Semiotics is the science of causally integrating signs, objects, and meanings. It is enmeshed with the configuration of physical entanglements evolving from symbolical processes in economic facts. This perspective is adjacent to the contributions of neuroscience and biosemiotic approaches that support the view that human civilisation and culture are emergent from the non-dualism of mind and body processes (see Damasio, 2018; Dennett, 2017).

Additionally, following a social ontology perspective, monetary functions are structured by the syntax forms in which information and knowledge are decoded according to modes of understanding. This is necessary in the semantics of behavioural proclivities, such as consumer trends, organisational decisions, and investment strategies. Thus, symbolical mechanisms are subject-dependent and constituted by functions that encapsulate mutual understanding that elicits epistemic evaluation according to path-dependent economic trajectories.

In these terms, economic transactions are mechanisms by which knowledge and information are physically transmitted by signs of exchange spread along with market interactions. These forms, such as prices, currency exchanges, and interest rates, are the vehicles by which intersubjectivity is shared along with the diverse sets of economic institutions. The semiotic mechanisms enable the emergence of economies of scale and scope in the form of specialised functions. Therefore, economic systems are evolving structured modes of interaction, intertwining networks of interpreters that constantly reassess the instrumentality of the behavioural proclivities forged mainly by the monetary mechanisms of exchange and value creation.

The semiotic ontology proposed in this chapter is based on Charles Sanders Peirce's approach and the appliances to dialogic systems in economics (Herrmann-Pillath, 2013; Macedo & Herrmann-Pillath, 2019; Marrais, 2019) and to social sciences more generally (Radviir & Cobley, 2009).

Also, this perspective is similar to the "distributed cognition" framework of the relationship between the neuronal functions and artefacts.

Following these general settings, the article is structured in the following order: Section 2 discusses the basic semiotic model and the outcomes of the ontological conception of money and development. Section 3 develops a blending articulation of the semiotic perspective in networks of monetary and techno users. Section 4 concludes with an outlook of the monetary semiotic implications to developments studies and the articulation of possible alternatives to the contemporary crisis of capitalism.

2 Semiotics and the Emergence of Monetary Systems

2.1 A Contextualisation of Charles Sanders Peirce's Ontology

Peircean ontology is the foundation of the pragmatist philosophical school and extension of the evolutionary perspectives of early institutional thinkers. The characteristic trait of pragmatism is the contextual standpoint of working out of the human mind and the emergence of behavioural proclivities. This relationship is clear-cut in the works of Charles Sanders Peirce, William James, and John Dewey regarding the pragmatist maxim and logical postulates of knowledge not mirroring factual reality. However, they diverge in terms of the true implications of knowledge.

James stressed a subjective and individualistic perspective to truth. On the other hand, Dewey postulated an approach based on well-verified social claims. Differently, Peirce validated truth claims solely as the outcome of agreements in the scientific community. In this realistic and materialistic approach, truth is not a common belief. A concept is assigned as true when validated by the empirical assurance of science. Thus, beliefs are the outcome of the tendency towards the emergence of laws in nature.

The influence of Peircean pragmatism was profound in early institutionalism. Pragmatic reasoning may be perceived, for instance, in the works of Thorstein Veblen and John Commons. Both applied Peirce's ideas to economics, especially when making use of the evolutionary concepts of habits and the symbolical traits of social interaction (Brier, 2008).

Admittedly, the evolutionary approach of Veblen is elaborated in the constitutive role of the relationship between institutions and technology in the formation and sedimentation of habits of taught (Veblen, 1898). In his works, the importance of symbolical contexts is a medium of status and meaning in social structures (Veblen, 1994[1899]).

Also, Commons, who was later received as an important source for the new institutionalism, was greatly influenced by Peirce. Commons' concept of *collective action* is ingrained in the pragmatist idea of the individual action emerging as the outcome of social coordination. For instance, in the analysis of common law, normativity is conveyed as evolving from habitual patterns of behaviour (Defalvard, 2005; Grinberg, 2001).

Nonetheless, of this clear inheritance, Veblen and Commons did not leave sequiturs, and Peircean ontology was forgotten by later economic thinkers. In sum, the biological metaphor embedded in "generalised Darwinism" took the role of the main "open system" approach in economics. Indeed, the use of biological terms to explain social evolutionary traits of human interaction requires a proper analytical formulation (for these approaches, see Foster, 1997; Herrmann-Pillath, 2001; Ramstad, 1994; Witt, 1999).

The application of "biologic and natural metaphors" to social facts creates an assemblage of very eclectic analytical inputs, such as thermodynamics, biology, systems theory, complexity theory, cognitive science, and neuroscience. All this diversity makes it difficult to understand the ontological meanings of these terms in the context of economics. Essentially, the depiction and understanding of brain functions are not enough for the conceptualisation of the interactive mechanisms of sociability in economic transactions. Salient epistemic grounds are related to the social semiotic networks of "extended minds" in the anthropic systems of knowledge and information.

Hodgson (2002), for instance, when advocating the view of a generalised Darwinian ontology for economics, rejected the idea of it being a form of biological reductionism. He correctly linked this approach to the long tradition of pragmatist philosophers, such as Peirce and James. He even admitted the need for a precise ontology of social facts in the economic framework:

> It is important to re-emphasise that devotees of Universal Darwinism do not attempt to explain everything in biological terms. The alleged universality of Darwinian mechanisms does not mean that the process involved is always that of genetic variation and selection. Furthermore, when genetic evolution does exist, this does not rule out additional evolutionary processes, acting on different entities, at additional ontological levels. (Hodgson, 2002: 271)

Markedly, Peircean ontology is grounded in the attributes put forth by Hodgson. It is a non-reductionist evolutionary ontology well suited to human-acquired functions in social systems with several levels of emergence. Most importantly, signs are analytical tools for the interfaces between physical artefacts and symbolical performative exchanges in complex networks of users (similar to the view of Latour, 2005).

A discussion on how common perceptions of the world are created, likewise, is suitable for the institutional framings of investigation. Douglass North pointed this out in a discussion of the subject-dependent attributes upon which assessments of reality are built. The "mental models" are culturally forged on sets of meaningful criteria:

> Individuals possess mental models to interpret the world around them. These are in part culturally derived – that is, produced by the intergenerational transfer of knowledge, value, and norms which vary radically among different ethnic groups and societies [...]. Consequently, there is immense variation in mental models, and as a result different perceptions of the world and the way it 'works'. (North, 1995: 18)

For those reasons, I think semiotics is the best way to understand the dispositions in the diversity of perceptions of social interactions. On the most general grounds, semiotics is embedded in the roots of the institutional and evolutionary schools of economic thought. At the same time, neuroscience, genetics, and biology are increasingly taking root in the attributes of complex systems and providing important clues to a materialistic account of selective designs in human society.

To that extent, I think it is not advantageous to claim that the quest for a biophysical explanation of the constitutive roles of the human mind is well grounded in early pragmatism. Therefore, following this orientation, in the next section, I will advance the theoretical perspective of semiotic interaction in the economic system.

2.2 The Scheme of the Semiotic Analysis

Now turning to the core of the claim regarding a "semiotic turn" in the economic analysis of money, my contribution is similar to the distributed cognition framework (Damasio, 2018; Dennett, 2017) and profoundly connected with the theory of social ontology (Searle, 1995). Briefly, Peircean semiosis is based on three ontological categories: firstness, secondness, and thirdness. Firstness is related to the sensing experiences of quality, feeling, spontaneity, and diversity. Secondness is the category of opposition exercised by discreteness, characterised by the "brutal" facts of the nature of the self. Thirdness evolves from the universal proclivity towards homogeneity and lawfulness, both in inanimate objects and in living beings, creating habits (Brier, 2008; Ransdell, 1977; Short, 2007).

The emergence of habits according to Peirce is not deterministic. The three categories are related to variety and uncertainty – tychism – a property in which facts are open to chance. In this sense, the evolutionary perspective of Peirce is based on the law of large numbers and evolves from empirical tendencies that warrant scientific generalisation (Reynalds, 2002). Each choice of action tends to take place with habitual reinforcement and thus increases homogeneity. Therefore, the evolutionary approach of Peirce is more acquainted with the Lamarckian perspective of acquired characteristics in time – synechism.

In a schematic description of the semiotic ontology of mind, three points are crucial.

First, Peirce's metaphysical portrayal of human consciousness is related to symbolical operation – semiosis – which amalgamates, non-dualistically, external inputs into inferences, by the working out of signs in pursuit of responses as functions. Most importantly, according to Peirce, all human proclivities are targeted at some intentionality, in his words, to "action". The semiotic formulation is the entanglement of the three sets of causality embedded in the basic categories: causal, formal, and final.

Second, the semiosis by which meaning and action are performed depends upon the operation of the integrated relationship between the sign and the object. The sign by the mechanism of formal causality induces the symbolical codification of inference that is stimulated by the interaction with the object (efficient causality).

Third, final causality, or the actual meaningful response performed by an interpretant, is the functional mechanism accredited by the interaction between the object and the sign, which is in the advent of the semiosis. Final causality is the response in which the interpretant includes the repetition of mental proclivities. Most importantly, the selection of the appropriate response functions is worked out by physical responses qua intentionality in the flows of processed information (Fig. 9.1).

Let us think now about the semiotic perspective relative to economic facts. Technical knowledge in this assessment is due to the scientific knowledge incorporated in the production system that makes available to human use specific physical properties. The generalisation of innovative gadgets, for instance, in compounded functional mechanisms that encapsulate technological knowledge and information, provides novel modes of response according to feedback signalling in the networks of users (for a detailed description, see Herrmann-Pillath, 2013: Ch. 4).

Institutions are the signs that mediate technology and the network of users by the working out of syntax forms (codes of language). Indeed, in this sense, the normativity of monetary arrangements have the role of enabling mutual understanding and social coordination. Think, for instance, of the sets of users intertwined via institutions in supply chains performing functions, such as consumers, suppliers, and investors. They are *techno users* exchanging information and performing the constitutive roles of money in social structures: consuming, investing, employing, and so on.

Most importantly still, social actors are conveyed by information flows emerging from signs, such as prices, interest rates, exchange rates, to mention just a few, that are dispositional to selective drives of monetary behavioural responses. These connections are forged in interactive networks. The performed practices accomplished in the markets are the combined sets of these patterns of signalling mechanisms. In this sense, the design is emergent in the functions in which actors are engaged by the medium of money in

Fig. 9.1 Semiosis and function. (Source: Macedo & Herrmann-Pillath, 2019)

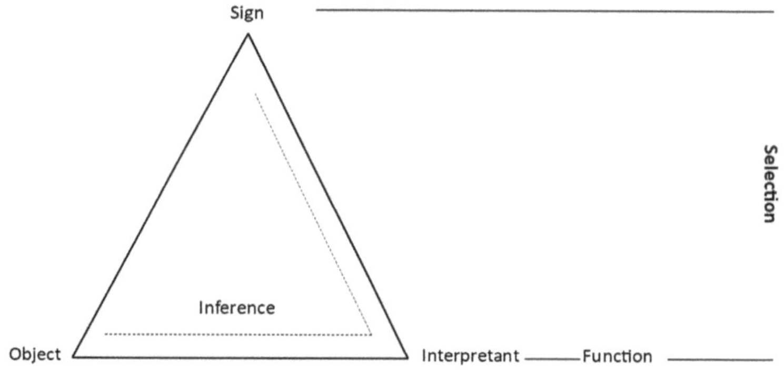

the creation and selection of techno properties assembled in goods and services along the exchanges occurring in the networks.

Semiotics, along this exchange, is the mechanism which operates on the performance of response functions (causal and formal causalities) towards some goal qua final causality. Hence, with respect to these criteria, inferential assessments collapse into behavioural dispositions, a logical process termed "abduction" by Peirce. The inferences creating behavioural dispositions depend upon physical artefacts of syntax forms: the institutions. Thus, final causality is the resultant semantic form related to the process of meaning-making. The key point is that semiotics conveys the assemblage of information criteria according to the cognitive distributed input drivers of selection in which money is the critical meta-sign of capitalism that coordinates market exchanges.

Furthermore, the semiotic perspective provides analytical tools for understanding learning processes, such as technological development, and the institutionalisation of behavioural patterns. Both are key determinants to economic growth and the functioning of markets. Accordingly, my further objective in the next section is to systematise the semiotics of the combined performance of money, technology, and markets.

3 Money and Development in the Semiotics Analysis

Development is conceived as the improvements in living conditions in the long run, measured by indicators such as income, life expectancy, educational achievements, and individual liberties. Indeed, development is *path-dependent* in terms of historical trajectories and the relative assessment of the diverse sets of human intentionality. Markedly, following the ontology of social facts (Searle, 1995), two points are key to the semiotics of development.

First, development is an ontological observer-relative fact. What is considered to be "developed" depends upon human reasoning. Nowadays, there is extensive discussion on how development should be measured, whether solely by per capita income and a limited number of education and health indicators or also by "sustainability", "happiness", and "spiritual well-being" proxies of targeted features (McGillivra & White, 1993; Meier & Stiglitz, 2000; Sen, 1999). In these terms, development studies require addressing the interdependence between observers and social reality.

Let's make one drastic example to make the point clearer. The Kamayurá is an indigenous people in the Brazilian Amazon forest that has the tradition of committing infanticide of the disabled offspring of single mothers and twins. In the relative assessment of that particular culture, infanticide is a legitimate practice to guarantee group survival (Oliveira, 2018). In a milder instance, the same ontological subjective attribute is operating in the distinct assessments of the trade-off between inequality and pro-market policies in countries such as Germany and the USA.

Second, development is an intrinsic observer-invariant epistemic fact. Social facts assign causal powers to structure human actions. These positional attributes enable scientific assessment in social sciences and humanities. For instance, the Kamayurá are linked together by their social structures, and the practice of infanticide is causally imposed on individuals. This is only possible because collective intentionality assigns functions that are performed according to accepted claims: to the Kamayuará, infanticide is necessary for group survival. Similarly, diverse sets of entanglements of pro-market policies and inequalities are expedient for scientific appraisal concerning their causal effects on market economies.

The social ontology point is that the constitutive attributes of social facts, such as development, are subject-relative phenomena. Therefore, in this framing, they vary from observers' mappings of reality. In a nutshell, human intentionality designates which facts are perceived as development. Think about the diversity of perceptions in several branches of today's societies regarding conditions for development. To what degree should their views related to a "better life" be considered as the foundation of economic systems? This idea has strong implications, which I will work on later, related to the degree by which common concepts of development can be coherently shared inside and among networks.

Accordingly, my claim is that the bridging of the relative character of social facts and the emergence of social structures are conducive to development and operate according to the working out of specific symbolic representations. Cognitive distributed systems rely on external artefacts to gather information and amplify processing capacity in the face of environmental stimuli. This is possible at the expense of the semiotic functions of the social structures.

At the core of the semiotic proposition lies the fact that economic systems evolve from information and knowledge flows of exchange that require modes of integration. So then the medium of social facts is the operation of signs in the selection and accomplishment of strands of collective intentionality. In this sense, money is the prime mechanism of economic integration and provides important clues to the understanding of the capitalist intercourse. In a nutshell, money is the language of capitalism. Following Noam Chomsky's (1965) theory of linguistics, the performativity of language is assembled in the rational adequacy of the (i) descriptive and (ii) explanatory functions of the syntax forms. Any ordinary and healthy human being can make use of it. Conversely, Wittgenstein's prescription of a "language of users" requires the matching of the syntax forms to the user (a person or a technological device) as a *formula* of representation. The point is that the stimulation of the input data is contextual to the collective *praxis* of the language. Hence, the explanatory adequacy is not independent of the criteria used by the user system: meaning and action are interrelated.

In applying this perspective to a semiotic system of meanings and economic practices, the closure is that signs and technology are intertwined in the mechanisms of production and transaction. Both aspects are internalised in the strands of collective intentionality that emerge from the operation of symbolic modes of interaction.

I summarise this analysis in Fig. 9.2.

First, formal causality is the structure of language forms that enable the categorisation of input data. In this sense, the underpinning of causal powers depends upon the prior design of the operative rules related to decoding symbolic functions. Think, for instance, about the encoded signals shared by credit-rating agencies on standardised metrics of default risk. The criteria used by the ratings guide financial analysts in their

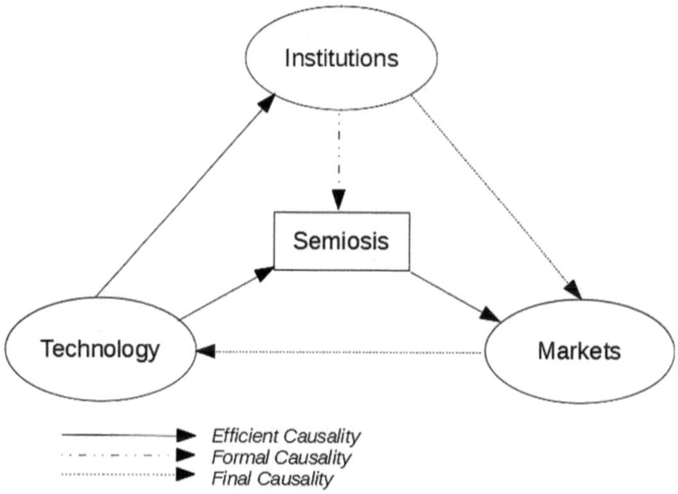

Fig. 9.2 The semiosis of the economic system. (Source: Elaborated by the author (based on Herrmann-Pillath, 2013: 41))

operations, which are gathered in labels embedded in specific modes of response, such as investment grades or speculative assets.

Most importantly, different syntax forms are designed according to distinctive systemic functions that are not directly congruent between specific niche applications. Consider, for instance, the non-compatibility between Mac, Windows, or Linux operating systems and the consequent adaptive requirements for the use of common applications.

In the network of users, in the most general terms, technology is related to patterns of knowledge and information along with diffusion processes. These are similar not only to the energy sources by which mechanisms operate (e.g. fossil fuels, electric energy, and chemicals) but also to the syntax information transmitted in social interactions. Therefore, any system of applied knowledge requires the compatibilisation of physical properties with the symbolical codification of technological functions.

Think of the flows of energy and information that are exchanged in global value chains. For each economic transaction performed, there is a binding relationship between physical and signalling properties. For instance, when a good is sold, an intricate array of physical processes take place simultaneously along with diverse sets of firms in the value chain, such as suppliers, transporters, and financial intermediaries. Each of these functions is intertwined with the formal rules of interplay that elicit the exchange of meaningful information that enables the *performa*nce of functions in the form of transactions, such as accounting methods, software applications, and, most basically, language. Therefore, technologies are embedded in symbolical phenomena.

Second, the role of institutions is related to the operation of signs that enforce the coordination of human intentionalities by creating meaningful modes of interplay and understanding. It is important to make clear that signs are physical phenomena that are the part piece of technology. They are the normativity that makes meaningful decisions possible. By this, I mean that final causality, understood as the patterns of performed functions, evolves from the proclivity of responses shared collectively in the face of environmental stimuli. Innovation and behavioural divergence, conversely, are reinforced or not, according to instrumental performativity. In this sense, new functions derive from ongoing social beliefs that are grounded in justified claims.

Considering the criteria taken as the basis of desired outcomes, the justification of claims depends upon forging socially congruent logical assessments that abide in semantic forms. In this sense, institutions are the common grounds by which decisions (final causality) are forged and socialised in the network of techno users. For instance, according to the later examples, transactions are consequent to instrumental cognitive mappings existent in accounting rules, programming languages, financial indicators, and grammar.

In the network of users, each individual has their proclivities, which are in connection with other individuals and networks eliciting drives of mutual acquaintance or disengagement. In this sense, signs are functional representations elicited in the operation of institutions (rules and norms) forging behavioural proclivities (habits). The outcomes of these common meaningful acquaintances are perceived as feedback effects that are channelled into organisations as market trends.

Now, to take stock, semiosis is the evolutionary entanglement of institutions, technology, and markets. This interplay is an open-system phenomenon in which uncertainty (tychism) and in-process causality (synechism) are at the core of path dependence in economic systems. Each of the three categories operates endogenously to each other and therefore the semiotic causality is a continuous contextual framing of the evaluation of economic development.

Thereupon, the ontological foundation of these mechanisms is evolutionary but without the use of metaphors more acquainted to other strands of inquiry, such as biology. The vector of relations which forge semiosis are subject-relative attributed, and the outcome of their intrinsic attributes is consequential of collective acceptance of syntax and semantic forms.

Likewise, change and disruptions are embedded in behavioural patterns that depend upon the convergence of specific conditions to be unbound and emerge. More precisely, in the semiotic perspective, development has emergent properties that are not reducible to the lower levels of the three categories. Semiotic convergence is the conduit by which the supervenient downward causation arises in the form of systemic coordination. These are critical conditions for development studies that deserve further detailed analysis in the next section.

4 Implications of the Semiotics of Money for Development

Greater income and better living conditions are the outcomes of increasing market extension and the specialisation of labour. Since Adam Smith, productivity has been pointed out as the driving force of development, and it is correctly perceived as the bullseye of economic policies. According to the semiotic prescription, economic activities are emergent functions of the operation of monetary functions. In this sense, markets are expressions of money's signalling mechanisms that depend upon the commonalities of meaningful behavioural proclivities.

Remembering Adam Smith's pin factory example, all the described outcomes are derived from the division of labour – specialisation and mechanisation – and may be outlined as emergent attributes of the working out of semiotic interactions. The advancement of the factory as an innovative organisational design in the early industrial revolution enhanced productivity by applying novel technological mechanisms and structuring social relations.

More than that, in today's integrated economic system, increasing coordination and strengthening knowledge and information flows are even more crucial to the performance of the extended global semiosis. Indeed, the role of integrating final causalities is the prime assessing function of monetary transactions. This point is gravely confirmed by the different degrees of the adequacy of the international community when addressing problems such as a financial crisis or the increasing economic inequality.

Therefore, I think that the analysis of the attributes of semiotic interactions is approachable to frame the convergence of development studies. In this regard, the ontological constituents of the alignments that operate in the monetary semiosis enable the extension of markets and feedback adaptations that induce greater productivity.

The first alignment, the "*intra-coherence*", is related to the formal mechanisms – norms and rules – that pervade and delimit the scope by which knowledge and information are shared and apprehended in the monetary system. It pervades the relationships in which syntax modes operate in technological mechanisms via artefacts that are shared and enforced by institutions, such as commercial banks, stock and commodity exchanges, and the economic authorities. Accordingly, intra-coherence is internalised in networks of users and creates proclivities delivering coordination and mutual acquaintance along the supply chains.

The second alignment is the "*inter-coherence*" in the modus operandi of performative behaviours among techno users. It is worked out by the collective endorsement of meaningful purposes according to different degrees of congruence. Thereupon, inter-coherence is causally supervenient to networks and is operated by modes of semantic understanding. The feedback of the semantic functions of inter-coherence is responsive to contextual conditions that are internalised into techno and institutional design as market conditions in the "real" economy.

Both alignments are responsible for the path dependence in development trajectories, which are the dynamic forces behind the operation of the semiotic intercourse. Next, I will describe the details of the functions working on each alignment.

4.1 Intra-coherence

Intra-coherence enables the assembling of different systems for the exchange of data in a coordinated way by the operation of the monetary system. This is an essential attribute of semiosis in the sense that it enables greater specialisation and mutual dependence along with meaning creation. Each network of techno users relies on specific modes of interplay between artefacts and coordinating mechanisms that require proper compatibilisation to amplify systemic instrumentality.

An exemplary case of this category are the challenges related to the diffusion of technological innovations along distinctive sectors of the economy. The degree of novelty of each niche appliance depends upon the adaptation of the old criteria in place into new modes of interaction, such as endorsed by venture and equity capital, and baseline incentives provided by development banks. The specialised functions of money are required to deal with the patterns in which the interplay of the novel technological mechanisms and the old behavioural proclivities is structured.

Think, for instance, of the case of the adoption of new agricultural practices in traditional communities in underdeveloped countries. Usually, new farming techniques are developed and applied first in developed countries. Hence, how novelties are applied in agriculture will profoundly differ in the grounding contextual conditions, normally related to public financing with lower interest rates and longer installments. The signs which in charge of combining and transmitting practices to techno users are in the form of specific patterns of applications. As the new technology is implemented, new mechanisms of exchange are required to be performed in place of the old ones. However, the challenge, of course, is not the *translation* of the knowledge and information embedded in the new technology.

The real task is that monetary arrangements need to assert meaningful behaviours for users. Even if a government agency freely provides the new technology, without *first* advancing the ways by which the new mechanisms are to be implemented by the current functions of the productive system, the initiative will not succeed. In this case, the enlargement of the procedural mappings with the novel applications requires the matching of old and new efficient modes of causality to local conditions.

For instance, organisations, like extension agencies or universities, must provide an environment in which farmers can find ways to participate in the process of the diffusion of technology. Simultaneously, the assimilation and the assembling of symbolical functions *is not an individually observer-relative phenomen*on. The acceptance of the meaning of a monetary coordination arrangement is dependent upon the relative assessments that operate in the networks. Signalling functions are enforced by the collective acceptance that is emergent at the higher level of normative causality. In the first place, although requiring individual agreement, the matching of new functions is constitutive of the prior assigned powers emanating from the structured relations.

Things become even more complex when technological functions occur in common assembling paradigms in use among diverse sets of applications. In these situations, monetary signs need to deal with the non-compatibilities manifest in several user networks operating simultaneously. This founding condition requires taking stock of contextual biases that influence the frames by which interactions are configured.

On these general grounds, semiotic intra-coherence of money is outlined by five basic properties: (i) categorisation, (ii) congruence, (iii) mapping, (iv) symbiosis, and (v) redundancy. These five properties are interactive in the workings of the sign mechanisms operated by the interfaces embedded in the institutional arrangements and the techno functions. In a directive manner, semiotic intra-coherence operates in the flows of exchange from (1) formal, (2) to efficient, and (3) to final causality.

4.1.1 Categorisation

Signs group phenomena by *labels* that systematise meaningful information regarded as constitutive. Categorisation enables functions to be assessed according to some indicators of qualia that match the elicited responses of the techno users. In this sense, goods and services, for instance, are classified according to criteria that provide mappings of attributes in the face of perceptual stimuli, such as the design of a car or the texture of cloth. This is possible, instead, only if the symbolical interactions are developed on the grounds of the properties in use by the efficient mechanisms in place.

For instance, in the former example, the modernisation of practices in agriculture requires that the labels in use by the techno users (farmers) be able to coherently systematise functional mechanisms according to their appliance. Prices such as of fertilisers, livestock, farming inputs, and pesticides are grouped according to the compound of practices in place. More clearly, each input data, such as a raw material of a specific chemical compound, is required to be categorised by the functions it performs in the "technological package" the combines several inputs all together (i.e. fertiliser, seeds, pesticides).

In this manner, the compound of the signalling inputs is the vector of combined knowledge and information that is applied in the monetary functions in use by the techno users (i.e. prices, contracts, financing). The characteristics of each categorisation in play influence the ways in which the meaningful criteria of the functions will be performed. Therefore, the criteria depend upon the systemic design that is contextual to the actual appliance by transactions.

4.1.2 Congruence

The functions assembled in the categories of use need to be congruent with each other's workings. The monetary design requires properly identifying to users the mechanisms in which mutual entanglements occur and the ways by which properties are reinforcing or not. If successful, signs are operational in the ways of delivering synergic endowments by the assembling of the system's properties.

For instance, congruence is the key characteristic of innovative networks in which actors are combined with well-coordinated arrangements, such as among research centres, universities, venture capitalists, and entrepreneurs. However, one design is not well-fitted to all circumstances of congruence. Corporations' vertical integration and government planning are in principle suitable ways to deal with the compatibility of formal requirements. Signs in the form of user guides, business plans, company amendments, government plans, and so on are instances of the ways to coherently deal with diverse sets of group functions and coordination.

In developing countries, increasing network synergies tends to be particularly difficult in the face of the superposition of conflicting intentionalities, such as from diverse ethnicities, high inequality, and the conflicts between the enforcement of law and traditional norms. The key point is that intra-coherence pervades the ontological property of delivering a congruent meaningful criterion to the entanglements of functions and the coordination of collective efforts among different niches of applications.

4.1.3 Mapping

The provision of money in networks allow for the emergence of recurrent patterns of social positionings. As these recurrences are sanctioned, they are reinforced and institutionalised. The mappings, likewise, are the structures embedded in the mental models according to the plethora of normative causal powers in the rules of the network. Decisions and ultimately instrumental behaviours are framed according to the mappings which users endorse as legitimate.

In the example of the challenges related to the modernisation of traditional agriculture, rearranging the meaningful instances of the productive practices requires the matching of old and the new modes of symbolical functions. The modernisation takes place in rebuilding cognitive mappings that guide social interactions. For instance, social positions related to traditional beliefs must be downgraded in favour of technical powers and scientific scrutiny. According to the same criteria, the expansion of formal education and the specialisation of the labour market change the roles of kinship and proximity. The consequence is that reliance on personal relations based on affection is reconfigured into mappings based on abstract and impersonal normative instances.

Comparatively, as the integration of the global supply chain increases in capitalism, the cognitive mappings become less diverse in terms of how networks are configured. On the one hand, greater procedural homogeneity induced by markets facilitates communication and coordination that increases the scale and scope of the emergence of mutualities. But, on the other hand, the decrease in the stock of diversity of communal mappings in use reduces the available degrees of change and novelty in the face of new drives for selection.

4.1.4 Symbiosis

Distributed cognition systems are at the core of the human coordination skills and are integrated by designs that blend biological and techno attributes. Technology is pervasive in users' functions and is conditioned by the design of artefacts. In this sense, symbiosis relies on the interfaces of organic and inorganic functions that are conducive to the intentionality that frames social structures. Consequently, the performance of the mechanisms elicited by the symbolical exchange sanctions causal powers.

To some extent, agency is not solely a human property because the endogenous matching of functions of artefacts and the brain is profoundly enmeshed. Neuronal plasticity is the vehicle by which the combination of the semiotic applications and the scope of the techno design is accomplished. Learning in this perspective is a key function of the operation of symbolical interfaces and is not performed exclusively by humans. The symbiotic performance is carried out by combined organic and inorganic functions, which are evaluated according to bimodal instrumental criteria.

Think, for instance, that electronic transactions require early development of digital learning and the requirements assembled to assess the quality of services among consumers. Internet relations are structured by the inducement of technological acquaintance into common patterns of social behaviour. Nevertheless, the outcome of habitual apprenticeship is the extension of plastic neuronal applications in networks of techno users. The assessment of quality in digital learning is based on resolving practical "problems" and "tests" by consumers, enabling their ability to develop instrumental knowledge with the use of electronic artefacts. Consumers are required to be proficient in the use of signs, such as icons, in applications related to functions regarded as mandatory by the digital corporations. In the same way, artefacts are reassembled and redesigned in this process, such as by the constant redevelopment of software applications and cell phones which are used in economic online applications.

4.1.5 Redundancy

The semiotic system must endow degrees of freedom to the interfaces between users, artefacts, and monetary functions. Redundancy emerges from the multiple structural configurations that can support the same compounded use of information. Greater efficiency and flexibility are outcomes of redundancy. The selection of distinct modes of interaction requires the prior existence of a plurality of cognitive channels of intercourse. As environmental conditions evolve, the system tends to use the excess of the potential diversity to modulate the flows of information according to the best available outcome.

For this reason, the bimodal channels – symbolic and functional – can operate interchangeably as long as the stabilisation of the features of the economic system is not complete. As the network becomes more homogeneous in the scope of alignments, the transactions face a selective strain in terms of the degeneracy of inefficient routes. Otherwise, some degree of redundancy is maintained to foster the continuous cleavage of network conditions still unknown to markets.

In the case of development policies, redundancy may be a source of opportunities for "leaping forward" as new technological paradigms emerge. Traits of the social structures perceived as dysfunctional, when paradigmatic conditions change, may provide newly acquainted modes of interaction. For instance, multicultural societies embedded in the diversity of cultural inheritances may be better equipped to face the environment of greater global interconnection. In these terms, the redundancy of transactions' attributes that pervades the monetary system is better suited to delve into more complex routes of exchange and coordination enabling greater innovation and value creation.

In this case, more interconnected and diverse systems of exchange must provide flexibility to diverse sets of use to economic actor. Think, for instance, the use of cell phones in poor communities to deliver goods and services in areas not reached by the supply chains.

4.2 Inter-coherence

Inter-coherence provides the basis for mutual understanding and performative action in the networks. These collective instances of behavioural cleavages evolve from semantic forms that are pervasive in the contextual conditions of clusters of artefacts and signs. In particular, I mean that monetary semiosis is how the final economic causality collapses into actual performative behaviour.

Habits are the instances in which behaviours are continuously assessed and vindicated on the grounds of instrumental scrutiny. In the most general terms, inter-coherence is delimited by the performative instances of the decision-making of economic actors, such as consumers, workers, public officials, and businessmen. All these social positions are entitled to rights and obligations and are committed with entangled functions in the networks. The semantic understandings pervading the fluxes of economic exchanges are the backup for the specific performances required for each social position. Most importantly, performative behaviours are the feedback inputs for the technological design and the mechanisms in charge of efficient causality in the monetary arena.

Development, in this portrayal, is the combined alignment of performances in the pursuit of the specific outcomes of a flourishing life. Better living conditions, freedom, and social justice are mediated conditions by which actors in networks are embedded. In this sense, inter-coherence depends upon the workings of beliefs that are the outcome of the human reflexive endowments. Decisions are nurtured in the semiotic operation of artefacts in the money's symbiotic relationship with the human cognitive functions. In these terms, the agency is a bimodal construct developed by the organic and inorganic features of the economic networks.

Think, for instance, of how the patterns of sanctioned behaviours are interlocked with the mechanisms of artefacts. The ways in which users engage with technology delimit how techno mechanisms are shifted towards actual performance. Likewise, phone calls and internet applications are devices, when in use by local communities, that can amplify the scope of exchanges already in place, such as the commercialisation of staples or the search for job opportunities in specific ways of learning by doing. Consequently, the technological appliances are assembled along with the practices interrelated with the contextual outcomes and the modes of exchange evolving continuously among economic actors.

Additionally, the feedback of responses from the users provides the basis for the design of new mechanisms and the refitment of the modes of market exchange. In the previous case of the modernisation of agriculture, how new inputs are combined and how technologies are applied are intertwined with the implementation of farmers' practices in the field according to traits that are assessed instrumentally.

Conversely, the roles of the social positionings that mediate the adoption of practices with the patterns of behaviour are at the core of economic performance. Design is undergirded by the evaluation of the mechanisms in which technology is assembled according to the flows of value creation delivered. Hence, the important consequence of this perspective is that the economic path-dependence is integrated into users' modes of monetary interplay.

Based on this scheme, semiotic inter-coherence is arranged according to five properties: (i) performativity, (ii) assertivity, (iii) identity, (iv) interactivity, and (v) reflexivity. These five properties are assigned according to the performance mediated by users in networks according to the semiotic assemblage. In this sense, inter-coherence is defined by the flows of intercourse from (1) final, (2) to efficient, and (3) to formal causality.

4.2.1 Performativity

Actors in the economic networks deploy modes of mutual commitment to accomplish some nurtured causalities. Functions are performed inducing the emergence of causal powers according to these collective agreements. Performativity in that sense is an outcome of a system of interpretation that is operational at endorsing beliefs in the contextual conditions of the monetary transactions.

Users' performance and the techno design are interlocked in that constant interplay. In this way, performance may be contextualised as an attribute of "habits of taught", in the Venblenian context of the continuous assurance of the modus operandi of the practices. In a grand perspective, features such as the endorsement of common beliefs in the credibility of the property rights are dependent on the actual performance of the actors in the economic networks.

In a more parochial framing, innovation and niche market segmentation rely on features of use that are not always incorporated initially by design. For instance, software applications are the outcome of the features of how users engage with the properties of *systemic* programming. Advertising on Facebook was deployed initially by small businesses as a free tool to advance commercial opportunities that were not screened initially by the internet industry. By this way of interaction, online communities gathered via social media in settings of common features that induced the blending of core interests and practices matching the economic interests of sellers and buyers. In this process, the role of traditional ways of marketing tends to be reassessed and transformed into different monetary relationships. In sum, performance and design tend to be conjoined within novel modes of governance of the economic transactions.

4.2.2 Assertiveness

As economic networks engage in pursuing objectives regarded as desirable, some guidance to action is required. In this sense, the collective intentionality encapsulates common grounds of market outcomes that are framed as *conditions of felicity*. Narratives have an important role as the guidance to coordinate consumer performance and to provide simplified schemes of causal relationships in a capitalist directive way.

Therefore, assertiveness is the property of the semantic forms to assemble coherently distinctive roles in time and scope according to sets of economic outcomes. Broad narratives are instances to lead the way from measures to results. In sum, assertiveness bridges the gap between means and ends.

From a semiotic perspective, the task of deploying a guiding scheme is to achieve common grounds of acceptance and to guarantee features of adjustment perceived as legitimate according to property rights. This is only possible if the assigned roles are active in delivering behavioural proclivities that match with the planned schemes previously assigned in the transactions.

From a general perspective, development is a grand narrative that features collective action in the direction of enlarging economic opportunities to deliver social inclusion and economic prosperity. Any successful attempt to implement policies to revamp governance to foster economic productivity and greater income equality requires the assignment of assertive thrusting utterances. Deng Xiaoping's "reform and opening up" in China, Bismarck's "social reforms" in Germany, and Roosevelt's "new deal" in the USA are examples of how assertiveness plays a crucial part in gathering collective acceptance and commitment to economic planning.

4.2.3 Identity

Actors in networks are entitled to roles that are assigned with defined rights and obligations. Identity is the outcome of the internalisation by the individual of a position in the social structure. First, the attributes of the position enable the networking of resources and the causal powers to which the economic actor is disposed. Second, the combination of different identities in the pursuit of a common objective provides the specialisation of productive capabilities and the extent of the synergies achieved with market intercourse.

Conversely, identity in the networks is intertwined with the artefacts. For instance, a government official is required to deal with functions, such as financial reports, budget plans, and market surveillance. Each of these assignments is combined with a collection of artefacts to which the specific identity is applied to perform.

As a result, artefacts are combined with the identities and the functions performed to deliver causal powers. Equally, the role of this kind of symbolical reification is in the charge of political leaders and business entrepreneurs during political crises and business imbalances. Therefore, performance is the outcome of the internalisation of these applications among actors, with the use of signs of status, to foster coordination and social order.

Think about the role of monetary authorities' communication during an impending banking run. The specific terms and gestures emitted by the Central bank are perceived according to inbreed roles that are reinforced by market operators and analysts. Accordingly, any deviation from the expected measures is counter to investment decisions taken, eliciting dissatisfaction and criticism by economists of financial institutions.

4.2.4 Interactivity

Interaction is the basis upon which the social intercourse of functions is conducted in a coordinated way. Performance is a plural endeavour even if realised solely. Indeed, the mechanisms that make interactive behaviours effective are intertwined in the routines that are shared in the economic network. For instance, organisations are bestowed with detailed arrangements to guarantee that contracts are performed according to accepted degrees of deviance. Nonetheless, innovation and non-conformance are required attributes of interaction concerning the emergence of new proclivities and adaptive measures of flexibility in transactions.

Additionally, interactivity must deal with the patterns of information shared among actors in the economic system by the delimitation of the criteria of privacy and openness of disclosure in monetary exchanges. Cultural and technological biases are constantly reinforced or taken aside by the modus operandi of the practices related to the pooling of information in the network.

Therefore, characteristics such as how actors engage with others' monetary functions and performances depend on the contextual conditions framed in the intersection of the techno properties and the cultural endowments. For instance, organisational cultures and national or regional *ways of doing* are spread from more-or-less rigid acquaintance with rules and norms and are determinant of the binding commitment to common objectives. Each niche market is embedded in a specific business ethics that provides guidance to the credibility of market players.

All these phenomena are consequential of a plethora of signs that are transmitted by formal causality. Body language, norms of etiquette, and memes in general represent important devices to integrate individual performance into the borderline fringes of collective intentionality.

4.2.5 Reflexivity

Reflexivity is the unique attribute of human beings that abides in the constant flows of reassuring proclivities. It is a cognitive attribute of forging and abandoning habits of thought. The basis for the setting aside of behavioural patterns entrenched in the economic structures requires a reassessment of the performed actions. Consequently, openness to scrutiny and the exchange of diverse points of view qualify as the emergence of conditions for change.

Reflexivity also is a property of stability in terms of the instrumentality of the acquired modes of monetary positioning. This means that the conditions of exchange in the networks are reinforced when the techno properties of artefacts are stable. Technology is the main efficient drive of causality that operates in the framing by which actors measure the degrees of the adequacy of the semiotic modes of interplay.

For this reason, the development of new techno paradigms increases the incentives for the reconfiguration of monetary functions in capitalism. As the pressure to abandon the "old ways of doing" mounts, new attributes are meanwhile assured by reflexivity. In this perspective, closed networks are not amid environments for change because the conduit of self-reflection is not linked with the free exchange of modes of connectivity within players.

In these terms, semiotic reflexivity is operated by the workings of habits and routines intertwined with the structure of organisation positionings. For instance, the advancement of scientific knowledge is not a clear-cut indicator of innovation in productive systems. The exchange of information and the internalisation of economic gains increase the incentives for the appliance of research and development in the supply chains. Therefore, the scope of the operation of monetary functions and the abandonment of stringent proclivities are magnified in capitalism. The measures of applied reflexive means abound, such as the recall campaigns, the sorting of group segments for advertisement, and the data analysis of audiences. All these devices are ways by which organisations put forward reflexive forms of systemic scrutiny.

4.3 The Semiotic Integration of the Modes of Development

The semiotic scheme can be summed up as the interplay of syntax and semantic forms that structure monetary niche dispositions into collective performance (see Fig. 9.3). First, signs – prices, labels, contracts, and so on – are the systems of information and knowledge that underpin the design of artefacts according to behavioural proclivities. Second, networks are in the role of the assemblage of the niche dispositions via normative systems of property rights and obligations.

In this perspective, intra-coherence is the causality of the symbolical attributes of the normative instances in the social structure. Thus, the meanings that abide in the syntax interplay are at the core of the collective performance. These are the properties in the scope of the behavioural individual dispositions encapsulated in the semantic forms.

Wherefore, inter-coherence is the operation of the performative feedback inputs that drive the channels of path dependence in the semiotic monetary exchanges. The selection of instrumental responses is adaptive and built upon the attributes of the mutual reasonings elicited by the semantic forms of the system. In sum, the actual performance is consequential of the continuous assessment via the symbolical features of the monetary functions and the techno paradigm.

Thereupon, development in the sense of *life flourishment* is the integrated bimodal entanglement of the intra- and inter-scopes of the economic instrumental and final causalities, elicited by the semiotic interplay. A key outcome is that selection and adaptive capabilities are framed in the specific *conditions of felicity* existent in the networks. Underdevelopment in these terms is a collective observer emergent state that pervades how actors are assembled with the workings of the semantic properties of the monetary functions. Most importantly still, the challenges to improve living conditions are deeply rooted in the attributes of normativity in the governance structures that induce greater inclusion and prosperity into the monetary system.

5 Final Remarks

The current global economic system is endowed with greater interconnection and diversity that require the strengthening of collective synergies. The task of delivering coordination via bimodal semiotic coherence abides in the structure of the signs that encapsulate dispositions and performances by monetary functions. In this perspective, I think three aspects of these entanglements are outstanding in development studies:

(i) Dual societies that have a large degree of inequality are poorly equipped to face the challenges in inducing the convergence of the modes of technological advancement and institutional reform. Policies that deliver income fairness are intertwined with the evocation of common grounds of dispositions. This is a reason why triggering monetary policies to improve income distribution is at the core of the collective performance.
(ii) Multiculturalism in a semiotic perspective is not just the side effect of the economic networks. Embedded diversity requires the emergence of a blended assemblage of systemic

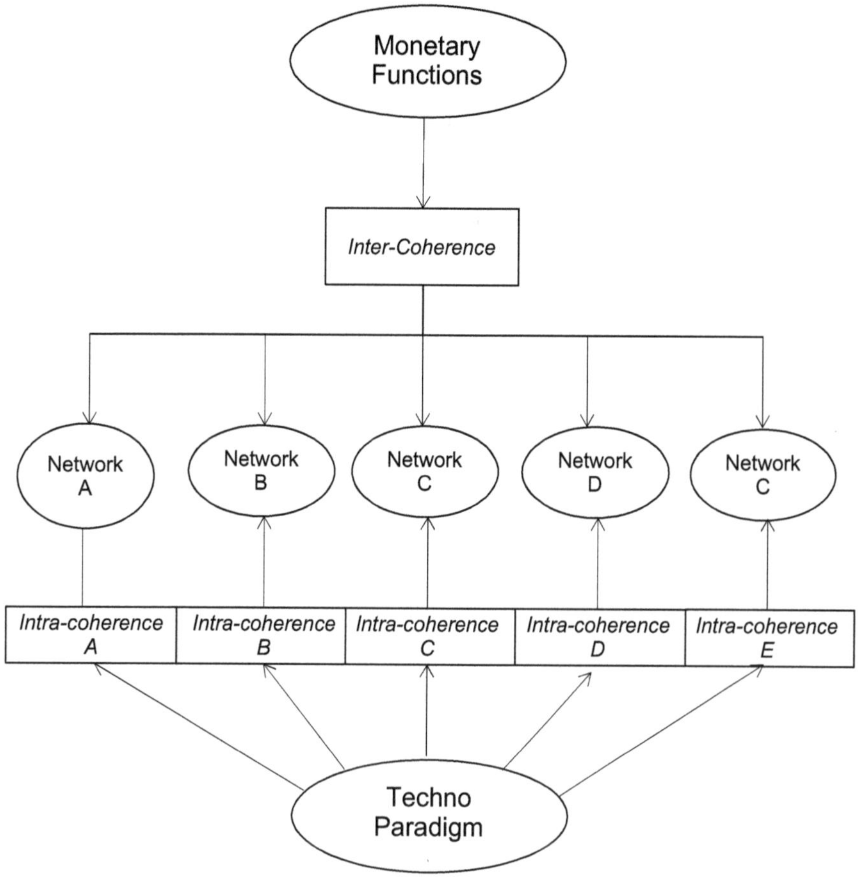

Fig. 9.3 The semiotic system of intercourse. (Source: Elaborated by the author)

multicultural structures that refit the fabrics of the normative positionings in transactions. Monetary efficacy should not be assessed only by a restrictive collection of indicators pervasive in the niche networks of the financial system, such as public debt, long-term yield curve, or inflationary expectations. Development in these terms is only possible when monetary functions reach the requirements of the inclusiveness in the fluxes of knowledge, information, and investment enabling synergies among supply chains.

(iii) Environmental concerns and the challenges of the climate crisis are of great importance to the drivers of change in semiotic causation. Habits and behavioural propensities are learned and performed by artefacts that reconfigure already entrenched practices. In this way, revolutionising public "green funding" to adaptation and mitigation to climate change are combined part pieces in the process of delivering sustainable attributes to new patterns of production.

In all these challenges, the development of new modes of semiotic intercourse requires that both monetary and technological coherence to be reconfigured simultaneously. Governance arrangements and the relations among the dispositions in the economic networks must be structured in parallel with the advancement and reconfiguration of the geopolitical arena.

Post-industrial societies and the current digital connective era have provided increasing challenges and opportunities for the amalgam of greater diversity and complexity into the monetary system. In this perspective, the semiotics of money and technology is at the core of the issue of delivering performances that enhance human potential and the scope of collective endurance in development. The great task is to develop new approaches to enlarge the scope of the monetary policy according with the interests of the most vulnerable in society.

References

Acemoglu, D., & Robinson, J. A. (2012). *Why nations fail: the origins of power, prosperity, and poverty*. New York: Crown Business.

Acemoglu, D., & Robinson, J. A. (2019). *The narrow corridor: states, societies, and the fate of liberty*. London: Penguin Press.

Brier, S. (2008). The paradigm of Peircean biosemiotics. *Signs, 2*(1), 20–81.

Chomsky, N. (1965). *Aspects of the theory of syntax*. MIT Press.

Damasio, A. (2018). *The strange order of things: Life, feeling, and the making of cultures*. Patheon.

de Oliveira, C.. (2018). The right to kill: Should Brazil keep its Amazon tribes from taking the lives of their children? *Foreign Affairs*, 9th April. https://foreignpolicy.com/2018/04/09/the-right-to-kill-brazil-infanticide/

Defalvard, H. (2005). Pragmatisme et institutionnalisme en économie une voie outliée. *Revue de Métaphysique et de Morale, 47*(3), 375–389.

Dennett, D. C. (2017). *From bacteria to Bach and back: The evolution of minds*. W.W. Norton & Company.

Dopfer, K., & Potts, J. (2004). Evolutionary realism: A new ontology for economics. *Journal of Economic Methodology, 11*(2), 195–212.

Foster, J. (1997). The analytical foundations of evolutionary economics: From biological analogy to economic self-organisation. *Structural Change and Economic Dynamics, 8*, 427–451.

Furubotn, E. G., & Richter, R. (2005). *Institutions and economic theory: The contribution of the new institutional economics* (2nd ed.). University of Michigan Press.

Grinberg, M. (2001). Un cheval est-il un cheval? Les mots, les faits, le capitalism et le droit. *Cahiers d'économie politique, 40*, 177–189.

Herrmann-Pillath, C. (2001). On the ontological foundations of evolutionary economics. In K. Dopfer (Ed.), *Evolutionary economics: Program and scope (Recent economic thought series 74)* (pp. 89–139). Springer.

Herrmann-Pillath, C. (2013). *Foundations of economic evolution*. Edward Elgar Publishing.

Hodgson, G. M. (2002). Darwinism in economics: From analogy to ontology. *Journal of Evolutionary Economics, 12*, 259–281.

Hodgson, G. M. (2007). Evolutionary and institutional economics as the new mainstream? *Evolutionary and Institutional Economic Review, 4*(1), 7–25.

Hodgson, G. M., & Knudsen, T. (2010). *Darwin's conjecture: The search for general principles of social and economic evolution*. University of Chicago Press.

Latour, B. (2005). *Reassembling the social: An introduction to actor-network-theory*. Oxford University Press.

Macedo, L. O. B., & Herrmann-Pillath, C. (2019). Towards a semiotic theory of the corporation. *Social Semiotics, 31*(2), 282–304.

Marrais, J. (2019). What development stand for? A socio-semiotic conceptualization. *Social Semiotics, 29*(1), 15–28.

McGillivra, M., & White, H. (1993). Measuring development? The UNDP's human development index. *Journal of International Development, 5*(2), 183–192.

Meier, G. M., & Stiglitz, J. E. (Eds.). (2000). *Frontiers of development economics: The future in perspective*. World Bank and Oxford University Press.

Ménard, C., & Shirley, M. M. (Eds.). (2008). *Handbook of new institutional economics*. Springer.

Nelson, R. R. (2002). Bringing institutions into evolutionary growth theory. In V. FitzGerald (Ed.), *Social institutions and economic development*. Springer.

Nelson, R. R. (2008). Economic development from the perspective of evolutionary economic theory. *Oxford Development Studies, 36*(1), 9–21.

Nelson, R. R., & Winter, S. G. (1982). *Evolutionary theory of economic change*. Harvard University Press.

North, D. (1990). *Institutions, Institutional change and economic performance*. New York: Cambridge University Press.

North, D. (1995). The new institutional economics and third world development. In J. Harriss, J. Hunter, & C. M. Lewis (Eds.), *The new institutional economics and the third world development* (pp. 17–26). Routledge.

Potts, J. (2007). Evolutionary institutional economics. *Journal of Economic Issues, 41*(2), 341–350.

Ramstad, Y. (1994). On the nature of economic evolution: John R. Commons and the metaphor of artificial selection. In L. Magnusson (Ed.), *Evolutionary and neo-Schumpeterian approaches to economics* (pp. 65–121). Kluwer.

Randviir, A., & Cobley, P. (2009). Sociosemiotics. In P. Cobley (Ed.), *The Routledge companion to semiotics* (pp. 140–156). Routledge.

Ransdell, J. (1977). Some leading ideas of Peirce's semiotics. *Semiotica, 19*(1), 157–178.

Reynalds, A. (2002). *Peirce's scientific metaphysics: The philosophy of chance, law, and evolution*. Vanderbilt University Press.

Searle, J. R. (1995). *The construction of social reality*. New York: The Free Press.

Sen, A. (1999). *Development as freedom*. Alfred Knopf.

Short, T. L. (2007). *Peirce's theory of signs*. Cambridge University Press.

Sidney, W. (1984). Schumpeterian competition in alternative technological regimes. *Journal of Economic Behavior & Organization, 5*(3–4), 287–320.

Veblen, T. B. (1898). Why is economics not an evolutionary science? *Quarterly Journal of Economics, 12*(3), 373–397.

Veblen, T. B. (1994 [1899]). *The theory of the leisure class*. Penguin Books.

Vromen, J. J. (1995). *Economic evolution: An enquiry into the foundations of new institutional economics*. Routledge.

Williamson, O. (1985). *The economic institutions of capitalism: Firms, markets, relational contracting*. The Free Press.

Witt, U. (1997). Self-organisation and economics – What is new? *Structural Change and Economic Dynamics, 8*, 489–507.

Witt, U. (1999). Bioeconomics as economics from a Darwinian perspective. *Journal of Bioeconomics, 1*(1), 19–34.

Luís Otávio Bau Macedo holds PhD in Economics from the University of São Paulo (USP) and MSc in Economics from Kingston University. He is a professor in the Federal University of Rondonopolis (UFR).

A Public/Private Partnership for the Implementation of a Job Guarantee Program

10

Ricardo Ribeiro Gudwin

Abstract

This chapter contains a preliminary study about the possibility of creation of a public-private partnership between the municipal government and businessmen who pay taxes in a given city/town, aiming to create conditions so that unemployed citizens of this city/town might have their labor force in useful services in the private sector, providing a reduction in the tax burden due to the municipal treasury and ending unemployment at the city/town. The proposal is based on the creation of a program to offer positions of work, from the private sector, with a wage equivalent to a minimum salary, which may be fully paid by the city/town government, using a special account to be managed by its city hall or company designated by it for this purpose, equivalent to a restaurant ticket or similar. In this way, any private company established in the given city/town (or even a domestic employer residing in it) will be able to apply in this program, offering a job position, to be disputed by the unemployed interested in the program. The service will be fully managed by the private sector, but will be fully paid by the municipality, in the form of a credit in the worker's account in a municipal ticket mechanism. That is, instead of receiving his/her payment in cash, the worker receives a credit equivalent to a minimum wage, in his/her municipal ticket account.

R. R. Gudwin (✉)
University of Campinas – School of Electrical and Computer Engineering, Campinas, SP, Brazil
e-mail: gudwin@unicamp.br

1 Introduction

This chapter presents a proposal for a utility service designed to guarantee to every citizen of a target town a source of income to support his/her basic needs, serving to alleviate unemployment in the target town. It proposes the creation of a public-private partnership between the municipal government and taxpayers who pay taxes in the envisaged town, aiming to create conditions for local unemployed citizens to have their workforce used as labor in useful services, in the private initiative, providing a reduction in the tax burden owed to the municipal treasury.

The proposal is based on the creation of a program to offer job positions, by the private sector, with a salary equivalent to the minimum wage, which may be fully paid by the city hall, through the creation of a municipal ticket, a special account to be managed by the city hall, or the company designated by it, for this purpose, equivalent to a food bond or similar. In this way, any private company established in the city (or even a domestic employer residing in the city) can enroll in this program, offering a job posi-

A. Pereira Jr, F. Sousa (eds.), *Principles for Governance*, Sustainable Development Goals Series,
https://doi.org/10.1007/978-3-031-40025-4_10

tion, to be disputed by unemployed people interested in the program. The service will be fully managed by the private sector, in terms of its duties and work to be carried out, but will be fully paid by the municipality treasure, in the form of a credit to the worker's account in the municipal ticket mechanism. That is, instead of receiving payment in currency, the worker will receive a credit equivalent to one minimum wage in his/her account at the municipal ticket account.

These credits in the municipal ticket account can be used by local taxpayers, for the payment of their taxes owed to the city hall, with discounts. The idea is for merchants in the city to start accepting municipal ticket credits as a form of payment for their products and services, as they will have the advantage of being able to pay their municipal taxes with discounts. In this way, small businesses, or even markets and supermarkets that pay taxes to the city hall, will be able to enjoy discounts on their taxes paid, if they do so through credits on municipal ticket program, instead of in cash. In this way, workers who will receive their payment in municipal ticket credits will be able to spend their resources directly at those commercial establishments that accept these credits as a form of payment.

2 Job Creation Mechanism

The works to be offered in this program must follow some preconditions. The idea of the program is to offer economic relief to the mass of unemployed people in the envisaged city, at the same time as investing in professional training for these people. In this way, the jobs to be offered must demand a maximum of 4 h a day, for 5 days a week, guaranteeing a break on weekends, thus integrating 20 h of work per week. The idea is that program participants can guarantee a source of income, allowing enough time for them to also study and qualify for more suitable jobs in the future. The idea is that these jobs are occupied on a temporary basis, while these workers are unable to occupy a more permanent position in the labor market. Likewise, payment in amounts equivalent to a minimum wage should make this type of work the equivalent of an insurance providing a guaranteed income in times of despair, and not a professional career to be pursued. This is exactly why the proposal sets this part-time limit, so that the program participant has the possibility of guaranteeing at least the other part-time period as a time in which he/she can study and improve his/her professional qualification.

The advantage for those who offer these jobs is to get a workforce fully financed by the municipal government, to collaborate in their work demands, in a commercial, industrial, or domestic environment. Thus, in principle, any type of job can be offered within the scope of this program, as long as it is limited within this context to 20 h a week and payment equivalent to the minimum wage. The management and monitoring of the activities to be carried out by the program participant will be a full responsibility of the employer (who offers the job). Employees who do not fulfill their assigned duties may be dismissed, and the employer may re-offer the job position again under the program. In this way, workers are expected to do their best within the program, since if their employers are not satisfied with the work performed, they can fire them.

On the other hand, the choice of jobs that will be used in the program is completely up to the person who will carry out the work. Thus, offering a job within the program is not a guarantee that this job will be used. When someone interested in participating in the program as a worker registers for it, he/she will be offered all open job positions available in the city hall system. In the description of the jobs, the workplace, the activity to be carried out, and the worker's responsibilities must be clear, so that he/she can choose from among the job positions the one that he/she feels most interested in performing. In this way, there will be a natural competition between those who will offer the jobs, so that they have better working conditions, since only the best offers will actually be taken advantage of, receiving public funding. In this way, a healthy competition will be created, ensuring that hard-working participants have the best job opportunities at their disposal.

Employers offering job positions will not be obliged, however, to accept anyone who is interested. The negotiation will take place in an analogous way to a traditional job, in which both employer and worker come to a common understanding so that the position is actually occupied. If a worker is not accepted for a position he/she is applying for, he/she may choose another position among those offered and engender a new negotiation with other potential employers. Thus, in the same way that the employer must seek to offer the best working conditions, the employee must seek to present himself/herself in the best possible way so that he/she is accepted by the employer, as it would be in a conventional labor negotiation. The only difference, in this case, is that the salary will be paid by the city hall, and not by the employer, if there is an agreement between the program participant and the employer.

The participant must agree, in order to participate in the program, that his/her salary will be paid through municipal ticket credits, and not through currency. In this way, program participants previously accept that the use of resources received as compensation for their work is limited to establishments that accept them as a form of payment.

3 The Municipal Ticket Management Mechanism

The salary of the worker participating in the program will be credited in the form of credits equivalent to one minimum wage, in a special municipal ticket account, to be managed by the city hall. To use their credits, workers will receive a magnetic card, which they can use to make payments at local establishments that accept this payment mechanism. Likewise, the city hall may provide an application for smartphones or the Internet to transfer municipal ticket credits to other municipal ticket account holders. In this way, merchants or other local tax payers can request the opening of an account in the municipal ticket program to receive credits, and later use these credits to pay municipal taxes at a discount.

In principle, any individual (with a valid CPF[1]) or company (with a valid CNPJ[2]) may request to open an account at the municipal ticket program, and once they have funds in their account, they may transfer these credits to another account holder, or to pay taxes. The creation of new credits, however, should only occur as a form of payment to workers participating in the job-guaranteed employment program. If it is in the interest of workers and taxpayers with municipal ticket accounts, they will be able to freely transact their credits. In this way, it is thought that these credits can be traded on the free market, by those who own them and those interested in paying their municipal taxes.

4 Advantages of the Present Program

The main advantage of this program is the reduction of unemployment rates in the city where this program is supposed to be applied. An additional advantage, however, is the expected boost in the municipal economy, since a portion of the population of the municipality that ordinarily would not be able to consume now will have resources that will be directed to consumption, improving the municipal economy, mainly businesses related to food and other basic necessities, as we estimate that municipal ticket credits will be used, mainly, to guarantee the sustenance of program participants. This boost in the economy may even increase the collection of municipal taxes, in some situations. In the same way, new businesses can be created, aiming to meet the consumption profile of those benefited by the program.

Another advantage is that the cost of the program is diluted over time. Usually, a city hall is only able to spend money after receiving it as taxes are paid. In the current program, the city hall can spend the money before actually receiv-

[1]CPF is an official ID that every taxpayer (as a person) in Brazil is required to have, in order to pay its taxes.

[2]CNPJ is an official ID that every company in Brazil is required to have, in order to pay its taxes.

ing it, because municipal ticket credits are virtually created on the fly, when required. In fact, they are just a number at the user system's accounts. Also, if municipal ticket credits circulate in the market, they will in fact work as a parallel currency, which does not impact the finances of the city hall, except at the final moment when they are used by taxpayers as payment of due taxes.

The mechanism would work as a tax "overbooking" mechanism, in which one can literally spend more than is collected, since the consumption of these credits will be distributed over time. In the same way that a bank can have more credits added among all its account holders than the actual reserves of money they have, the municipal ticket will be able to move a set of resources greater than the equivalent amount of taxes that would be collected if the municipal ticket credits did not existed. In practice, this would be a money creation mechanism similar to those that banks use to earn profit.

From the point of view of unemployment, the job-guaranteed employment mechanism would allow many types of work that are currently unpaid, such as mothers who take care of the house or children, to be recognized in the form of remuneration through the program. Likewise, various types of social work that are currently not possible due to lack of resources could be made possible, with the provision of jobs in charitable or social associations. Public utility or social projects may become more attractive as jobs than jobs in private companies.

Finally, an advantage of this policy is that it would be a democratic and distributed incentive policy, as those who benefit from it, namely, those who are benefiting from the work being performed by the workers in the program, will be democratically chosen by the workers themselves. We can understand that this subsidy would be justified by the relevance of the social benefit that would be the reduction of unemployment, practically to zero in the local municipality. In the same way, the workers themselves who will receive from the city hall can be seen as benefiting, although less than the employers, since in practice they will be receiving for the work performed. However, they are indirectly benefited, because in the absence of the program, they would remain unemployed. In this way, we understand that they will also benefit from the program. With this, we can understand that the same financial resource benefits triply: It benefits the employers, who will be receiving the work without having to bear its costs; the workers, who gain the opportunity to have a job, when it would not normally be available in the labor market; and finally, it benefits the city's economy, which will receive consumption that would not exist if the program did not exist.

5 Disadvantages and Risks

Like everything, there are not only advantages in this program. First, there is indeed a cost. This cost would be in the reduction of taxes received, since potentially a large number of taxpayers would use municipal tickets to pay their municipal taxes, at least, those who had the means to obtain them. Taxes collected in this way would not, in fact, add to the municipal public treasury. In practice, they would be like the postponed liquidation of a resource that has already been used in advance, when payment is made to a worker within the program. The advantage is that the amount of existing credits in the form of municipal tickets can be controlled and limited to the value of one minimum wage multiplied by the number of unemployed people currently in the local municipality, adding the discount that would be given for the payment of taxes through municipal ticket credits.

There are still some risks that would need to be mitigated for the plan to be successful. The first risk would be that fictitious job positions would be advertised and taken over by individuals in collusion to circumvent the system. Thus, two individuals A and B who wanted to circumvent the system could both create fictitious work positions, with B taking the position of A and A of B, and the two would just pretend to work, both being paid by the city hall. Therefore, it would be necessary for the city hall to somehow supervise the jobs, periodically visiting program workers at their jobs, in order to verify that these jobs are legitimate work and not a mechanism to circumvent the system. Another possible situation that could occur would be the individual who creates jobs where there is no real work, proposing the worker to share the resources obtained with the payment from the city hall. This situation could also be detected and avoided with periodic visits to the work locations, but it would be less interesting for the worker, as he would receive less value than he could receive if he occupied a legitimate job at this place. One way to avoid this situation could be to prevent people who already have some other formal job from enrolling in the program. Thus, if the employee attempted such collusion, he would not be able to perform any other formal work. We must remember, however, that as the employee is paid in the form of municipal ticket credits and not in currency, it would be much more difficult for agreements of this type to be engendered, to circumvent the system, since differently from ordinary currency, which cannot be tracked, all transactions would be registered and controlled by municipal ticket management. With that, cases of corruption and misuse of the resource could be much more easily tracked and the corrupt punished. Certainly the ability to create corrupting mechanisms for the system is subject to the imagination of the miscreants, and we cannot, at this time, conceive of all the means by which the system could be corrupted and misused. However, the same is true of several other incentive mechanisms managed by the government. Only time would bring us a real perspective of the corruption possibilities of the present mechanism.

Another risk would be that many employers who currently employ employees paying the minimum wage could consider dismissing their employees, seeking the possibility of having these same jobs paid by the city hall, without having to assume these costs. Although, at first, the offer of a job does not necessarily imply that it is filled, we could imagine a situation in which an employer colludes with an employee, dismissing him and offering his position within the program, being that the employee could register and take the same vacancy that was his before, receiving some advantage from the former employer, or even no advantages, just to not lose his/her source of income. A situation like this could be prevented, for example, by checking the candidate's last job and ensuring that he was not rehired by the same employer.

In fact, a possible side effect that this plan could cause would be the reverse of that. Many workers who currently work receiving minimum wage could decide to leave their current jobs to try another job, with better conditions, within the program. Why continue working in the old job, with not so good working conditions, if it is now possible to look for a job within the program, where employers compete with each other so that their job offers are chosen by program workers? In this way, it may be that the current jobs that pay the minimum wage need to increase the wages paid, in order to keep their workers. Otherwise, one could expect an increased number of layoffs for current minimum wage workers. This side effect could cause a possible increase in demand for the program greater than expected (the size of the current unemployed population in the municipality) and would need to be better analyzed, so that the possible costs involved are better estimated.

6 Precautions in the Implementation of the Present Proposal

The present proposal demands that a series of precautions be taken for the project to have a successful operation. In the first place, before a municipal ticket program is effectively implemented and made available to workers, it is necessary that the local market, where these workers will spend their credits in the program, has been previously alerted to the creation of the program, and that a reasonable number of companies have been previously registered in the program to accept the payment of expenses in the form of municipal ticket credits. This precondition is very important, because if workers do not have a reasonable number of places where they can use their credits, the program will have operational difficulties, which can lead to the failure of the program as a whole. It is necessary that both workers and businessmen who will accept municipal ticket credits be prepared and organized to receive these consumers who will use municipal ticket credits, instead of currency in the purchase of goods and services.

Therefore, it is necessary that businessmen clearly understand the advantage of accepting municipal ticket credits as payment for their products or services. As it may happen that the amount of goods or services that will be acquired through municipal ticket credits is greater than the taxes that the company will have to pay to the city hall, it is necessary that the mechanisms for outsourcing the excess credits are already properly implemented, in such a way that the entrepreneur who remains with a positive balance in credits from the program, have the possibility of selling their surplus credits to other municipal taxpayers who do not have the means to receive these credits directly. It is important that the credits in the city hall's account have credibility, so that account holders feel free to keep these credits in their account, for future use, if necessary. If the impression remains that such credits have no real value, that is, that there are difficulties in converting them into real value for companies or workers, this could also compromise the good operation of the program.

In this way, before the program is effectively put into operation, it is necessary to carry out a prior work to clarify public opinion and a prior registration of entrepreneurs who wish to participate in the program, as well as the construction of the computational infrastructure for accounting for these credits: the user accounts related to CPFs or CNPJs of each participant and agile mechanisms through which it is possible to consult the current balance and obtain statements of movement of these accounts and tools for transferring credits between account holders. Likewise, it is necessary to provide the necessary information technology infrastructure, so that the magnetic cards of the account holders can be used in the participating companies that will accept these credits, in order to make their payments, with the same level of ease with which one makes a payment with credit card, debit card, or food tickets.

The program can only be started if this previous preparation has been completed satisfactorily. Otherwise, all sorts of problems can arise and make the good development of the program unfeasible.

7 Similarity with Other Proposals and Inspirations

The present proposal is based on a series of other proposals that have been circulating in different places and streams of thought. The idea of a job-guaranteed employment program is similar, to a certain extent, to the job-guaranteed employment proposal designed by supporters of the MMT – *Modern Monetary Theory* (Wray, 1998). The MMT proposal, however, provides for the offering of employment positions in the form of public jobs. In our view, our proposal is more interesting than that of the MMT because one of the difficulties in absorbing this mass of workers in public jobs would be to create the positions and description of the work to be performed by the workers. There would always be a criticism that the positions would not generate work that is really useful or necessary, in addition to a greater difficulty in controlling whether the work being carried out is being done satisfactorily, if the worker is doing his/her best work, and who should oversee its proper development. Who would be responsible for monitoring the worker? What to do in the case the worker does not fulfill his/her duties well? How to rearrange the same in a situation like this? Delegating the determination of the job position to the private initiative itself, delegates the responsibility for creating job positions that are really useful, and the supervision of the quality of work carried out by the worker, to whoever is requesting this job position. With this, we understand that our proposal is better than a proposal making positions available in the public service. Another difference between the present proposal and that of the MMT is that in the MMT proposal, the project would be financed through the issuance of currency by the federal government. As this project is municipal in nature, this possibility would not be operational. For this purpose, the creation of virtual municipal ticket credits is proposed instead. The project is also inspired by the mechanism by which banks can virtually create money by holding banking accounts with values that do not really exist, because they are lent to other clients. Banks usually multiply the real money deposited in their accounts by assuming all their clients are not going to withdraw it at the same time. If all bank customers decide to withdraw the money they supposedly have at their bank accounts, the bank will not be able to honor the full amounts, as they do not really exist. The current proposal rely on a similar presumption, regarding an expectation of taxes that are not yet paid. The idea of a municipal ticket was also inspired by the Mumbuca Project (de Maricá, 2013), developed by the City Hall of Maricá – RJ. In the city of Maricá – RJ, the municipal government created a social currency, called Mumbuca, which was made available to needy citizens, in the form of a universal basic income. This currency was made available through a magnetic card, with a monthly credit to program participants, who can pay their expenses at establishments in the city of Maricá registered in the program, which were prepared to receive payments in that currency. Unlike municipal tickets, however, the Municipality of Maricá guarantees a direct convertibility between mumbucas and currency, which can only be carried out by registered merchants. In this way, these merchants collect payments in mumbuca, using the program's card, and monthly convert the amounts received into currency. This payment is financed, in the case of Maricá, by royalties from the pre-salt layer, received by the Municipality of Maricá. Thus, we see that the idea of a municipal ticket is fundamentally different in several points. First, that in the case of the Mumbuca Project, resources were made available to all citizens participating in the program, not conditioned to any type of work carried out by them. Therefore, it is not a job-guaranteed employment program, but a universal basic income program. Second, that the municipal ticket model does not provide for direct convertibility of credits into currency. In this way, the city hall will not need to previously have the resources available to make the expenditures with the payments, and these may remain circulating in the market for a certain time, until they are actually used, for the payment of taxes due. The inspiration for the municipal ticket model also comes from the MMT, which says that state currencies are consolidated as currency, due to the fact that it is only possible to pay taxes in the form of state currency. Thus, the municipal ticket program was adapted, in an attempt to create something similar to a municipal currency. It is not possible to guarantee that it will actually be used as a currency, that is, as a store of value, until the program is put into practice. But there is such a possibility, according to the theory of MMT. If that happens, the conversion of municipal ticket credits into taxes could be postponed, and thus, while this convertibility does not occur, there would be no burden for the city hall. With this, the city hall would function, virtually, as an issuer of a currency of local circulation. One of the findings of the Maricá experiment is that there has been a strengthening of local commerce in the city since Mumbuca was instituted. As users of the social currency can only use it within the scope of the city, this brought a whole boost to the economy of Maricá, which had a previously known demand, which could be considered guaranteed. With this, it was possible to establish a consumption profile for these credits, and encourage the development of new specialized commercial establishments to meet this demand. Merchants were interested in receiving payments in Mumbucas, as only those merchants who were accepting Mumbucas were able to receive these customers and do business with them. In this sense, it is expected that something similar will happen with municipal ticket credits, once the program is installed in the city.

8 Final Remarks

The present proposal is just an initial idea, which still requires several additional points to be better analyzed and established, such as, for example, a total estimate of investment expected in the program, the percentage of discount on tax rebates that will be proposed upon payment of taxes through municipal ticket credits and perhaps a more careful analysis of possible mechanisms of fraud or circumvention of the program's original objectives, which would need to be analyzed in more detail. The present text only aims to present a first vision of the proposal, with the intention of attracting possible supporters to it, so that it can, later on, be implemented in real towns. Any comments, criticisms, suggestions, or opinions regarding them are welcome by the author of this proposal.

References

de Maricá, C. M. (2013). *Mumbuca social currency law – Law No. 2.448 of June 26, 2013*. Municipality of Maricá-RJ. http://www.institutobancopalmas.org/wp-content/uploads/lei-moeda-social-mumbuca.pdf . Accessed 16 Oct 2021

Wray, L. (1998). *Understanding modern money: The key to full employment and price stability*. Edward Elgar.

Ricardo Ribeiro Gudwin is an Associate Professor at the School of Electrical and Computer Engineering, University of Campinas – Brazil. He was born in Campinas-SP, Brazil, in 1967, and received the B.S. degree in Electrical Engineering in 1989, the M.S. degree in Electrical Engineering in 1992 and the Ph.D. in Electrical Engineering in 1996, all of them from the Faculty of Electrical and Computer Engineering, State University of Campinas – Brazil.

Part IV

Ecological Sustainability and Social Consciousness

A Circular Approach to a Sustainable Economy

11

Inês Botão

Abstract

Our planet and resources are increasingly finite. The aim of this chapter is to introduce the concept of circular economy as a solution toward reducing global waste, as well as raise awareness of its opportunities and challenges. To this end, we provide some examples across industries and markets, such as the reuse of plastics for paving highways, air-conditioning "Cooling as a Service" business model, and IT solutions that reduce waste and minimize the environmental impact of IT equipment. We will also comment on the Europe Union's current Action Plan as the institutional framework for the circular economy. Further, we will highlight strategies for enterprises and how they can position their businesses in the circular context.

1 Introduction

Our planet and resources are increasingly finite. Since 1900, the global population has quadrupled to 8 billion in 2022 (see https://www.worldometers.info/world-population/). As the global population continues to grow, so does the amount of waste we produce. Humanity's consumption has been unsustainable for the past five decades. Our global economy has been increasingly consuming more than Earth's natural resources each year. According to the Global Footprint Network, we currently would need 1.75 Earths to sustain our current level of consumption (see https://data.footprintnetwork.org/#/countryTrends?cn=5001&type=earth).

In addition, the global waste we generate is expected to reach 3.4 billion tons by 2050, which is more than double the rate of population growth over the same period. The average rate of recycling and composting waste is only 19% globally. While there has been some progress in increasing recycling and composting efforts, the reality is that most of the waste is not being properly managed with the majority being sent to landfills, burned, or left as litter (Kaza et al., 2018).

To address this problem, the concept of circular economy has emerged as a solution to not only reduce waste but also to keep resources in use for as long as possible. It aims to create a closed loop of materials and products, where waste is minimized and resources are conserved. The circular economy approach is not only beneficial for the environment but also for businesses and society. It also creates new opportunities for innovation, job creation, and economic growth.

I. Botão (✉)
Kyndryl, Chartered Institute of Management Accountants (Accreditation), London, UK

A. Pereira Jr, F. Sousa (eds.), *Principles for Governance*, Sustainable Development Goals Series,
https://doi.org/10.1007/978-3-031-40025-4_11

2 What Is a Circular Economy?

A circular economy is an economic system in which resources are kept in use for as long as possible, and waste and pollution are minimized. It is based on the principles of designing out waste and pollution, keeping products and materials in use, and regenerating natural systems. In a circular economy, growth is decoupled from the consumption of finite resources, and economic activity is regenerative by design. It is an alternative to the traditional linear economy, in which resources are extracted, used, and then discarded as waste.

According to the directions of the Ellen MacArthur Foundation, recognized as a leading authority on the topic, the simplest way to understand circular economy is to think about the difference between a straight line and a circle. When we consider how we utilize materials such as metals and plastics in our current economy, they will eventually become waste. However, in a circular economy, this linear process is transformed into a closed-loop system. In such a system, all the resources used in the economy circulate continuously. Waste is minimized, and resources are maximized as materials are fed back into the system. Food waste can be broken down and converted into fertilizer, metals can be recycled and utilized in the creation of the next product, and so on (see https://ellenmacarthurfoundation.org/topics/circular-economy-introduction/overview).

3 What Is the McArthur Foundation?

The Ellen MacArthur Foundation is a UK-based charity that was established in 2010 with a mission to accelerate the transition to a circular economy. The foundation was founded by Dame Ellen MacArthur, and environmental activist and retired British sailor who previously held the world record for the fastest solo circumnavigation of the globe (see https://ellenmacarthurfoundation.org/about-us/ellens-story). Since its establishment, the foundation has become a leading authority on the circular economy, working with businesses, governments, and academics to develop to promote circular economy principles and practices. The foundation also conducts research and publishes reports on various aspects of the circular economy, such as the economic benefits and business models for circularity. It has become a global thought leader in the circular economy space and has played a significant role in raising awareness and driving action toward a more sustainable and regenerative economy.

4 Recycling and Circular Economy

Recycling refers to the process of taking materials that have been used and turning them into new products that would otherwise be discarded as waste. Recycling can be considered a reactive measure because it is primarily focused on managing waste after it has been produced. This helps to reduce the amount of waste sent to landfills and conserve natural resources, but it does not address the root causes of waste production.

On the other hand, the circular economy can be considered a preventive measure because it aims to address the root causes of waste production and resource depletion. It is a systemic approach to economic development that seeks to keep resources in use for as long as possible, extracting the maximum value from them before recovering or regenerating them. This can involve a wide range of strategies, including recycling, but also includes other practices such as product design for longevity and repairability, as well as creating closed-loop systems where waste is minimized and resources are continuously cycled. The ultimate goal of a circular economy is to create a sustainable and regenerative economic system that minimizes waste and reduces dependence on finite resources.

While recycled products are often associated with lower quality, products designed for circular economy are often associated with higher quality because they are designed with circular principles in mind, which include durability, repairability, and the use of sustainable materials.

In summary, recycling is a specific aspect of the circular economy that focuses on the end-of-life of products and materials, while the circular economy is a comprehensive approach to designing and managing resources for a more sustainable future.

5 Bioeconomy and Circular Economy

Bugge et al. (2016) suggest that the concept of the bioeconomy is complex and multifaceted and different visions of the bioeconomy coexist. For the International Advisory Council on Global Bioeconomy (IACGB, 2020), "The bioeconomy is the production, utilization, conservation, and regeneration of biological resources, including related knowledge, science, technology, and innovation, to provide sustainable solutions (information, products, processes and services) within and across all economic sectors and enable a transformation to a sustainable economy."

Bioeconomy and circular economy are two related but distinct concepts. While both circular economy and bioeconomy promote sustainability and the efficient use of resources, the circular economy has a broader scope and encompasses all types of resources. In contrast, bioeconomy specifically focuses on the use of biological resources such as plants, animals, and microorganisms.

Bioeconomy can be considered more circular by nature because it involves the use of renewable biological resources that can be replenished over time, unlike nonrenewable resources such as fossil fuels, which are finite and depletable. However, it is important to note that bio-based products still must be designed to be biodegradable or compostable to support circularity by returning nutrients to the soil. Additionally, the production of bio-based products can require significant amounts of energy, water, and other resources and may compete with food production which could further lead to deforestation and other negative environmental impacts.

In practice, bioeconomy and circular economy can complement each other. Bioeconomy can provide renewable and biodegradable alternatives to nonrenewable and nonbiodegradable materials, such as plastics and fossil fuels, while circular economy can ensure that these materials are reused, repaired, or recycled, reducing the need for virgin resources, and promoting a closed-loop system.

6 Circular Economy Business Model

Have you ever stopped to consider how long your first mobile phone or fridge lasted compared to the products you have now? And once these items are no longer useful, what happens to them and who is responsible for their disposal?

In the past, products were designed with longevity in mind and seen as valuable investments that could be passed down from generation to generation. Additionally, resources were scarce, and manufacturing was more costly, so companies had to ensure that their products had a longer lifespan to maximize their investments. However, as time has passed, there has been a shift in business models. Products are now designed to have a shorter lifespan, with the primary objective being to increase profits by selling as much as possible (Achterberg et al., 2016). This linear model follows a "take-make-dispose" system where companies extract raw materials to produce products that are sold to customers and eventually discarded as waste at the end of their useful life. Consequently, in this system, the responsibility for end-of-life product disposal rests mostly on the customers and municipalities rather than the companies that produced them.

According to the new EU Circular Economy Action Plan, 80% of a product's environmental impact is determined during its design phase. However, many products are designed with features that prevent easy reuse, repair, or recycling, leading to quick breakdown and making them suitable only for single use, and companies are not provided with enough incentives to make their products more circular.

Circular business models, on the other hand, strive to extend the life of resources and reduce waste. Atasu et al. (2021) discuss three primary strategies for creating circular business models:

1. *Retain product ownership* (*RPO*) by renting or leasing products is essential for circular economy business models because it shifts the responsibility of a product's end-of-life phase from the consumer to the producer. By retaining ownership of their products, companies can take back and manage the products at the end of their useful life. This approach also encourages companies to design products that are more durable and modular, making it easier to replace parts and upgrade components instead of disposing of the entire product. RPO can also create opportunities for new revenue streams and value creation by offering product-as-a-service models, which can help offset the costs of refurbishing and remanufacturing products.
2. *Product life extension* (*PLE*) involves designing products to last longer and extending their lifespan. When a company retains product ownership, it is in their best interest to keep the product in use for as long as possible and prevent the need for costly replacements. When products are designed to last longer, they need to be replaced less frequently, which means that fewer resources are required to produce new products. This, in turn, reduces the environmental impact of manufacturing and helps to conserve natural resources. Moreover, PLE can also create new business opportunities for companies that focus on refurbishing, repairing, and remanufacturing products. By extending the life of products, companies can create a new market for used goods, which can be sold at a lower price point than new products, providing affordable options for customers. Additionally, it can help companies differentiate themselves from competitors by offering high-quality, durable products that last longer than products from other companies.
3. *Design for recycling* (*DFR*) implies designing products and manufacturing processes to maximize recoverability of materials for use in new products. By designing products with materials that can be easily disassembled and recycled, companies can reduce the amount of waste that is generated when products reach the end of their useful lives. This, in turn, reduces the need for virgin materials to be extracted from the environment, which reduces the overall environmental impact of production.

7 Institutional Framework, Government, and Policy

The circular economy is gaining popularity among businesses, policymakers, and investors as a way to address climate change and global challenges while delivering long-term growth. Many companies are adopting circular economy principles to generate new revenues, reduce costs, and spur innovation. The finance sector is also responding to growing demand for sustainable investment strategies.

Public policies are essential for enabling circular economy solutions because they can provide incentives for businesses, promote innovation, create a level playing field, and raise awareness among the public.

7.1 Europe Union Common Charger

In 2022, the European Parliament approved the common charger directive, which requires electronic device manufacturers to provide a universal charger that can be used with all devices. Starting from 2024, the use of USB-C port will be mandatory for a wide range of electronic devices such as mobile phones, tablets, digital cameras, headphones, headsets, portable speakers, handheld videogame consoles, e-readers, earbuds, keyboards, mice, and portable navigation systems, while laptops will be required to have USB-C port as of 2026.

The Commission argues that the current situation of multiple chargers and cable types leads to unnecessary waste, as consumers often need to purchase new chargers with each new device they buy. By standardizing the charging port and connector type, the new regulation is estimated to reduce electronic waste by 980 tons per year, lower greenhouse gas emissions, and save consumers at least €250 million a year on unnecessary charger purchases (see https://single-market-economy.ec.europa.eu/sectors/electrical-and-electronic-engineering-industries-eei/radio-equipment-directive-red/one-common-charging-solution-all_en).

7.2 Europe Union Single-Use Plastic

Plastic waste has become an enormous threat to the environment, and most of the waste is not being recycled but rather incinerated or discarded. An OECD study in 2022 revealed that the world generates double the amount of plastic waste today than it did 20 years ago, with most of it being either buried in landfills, burned, or released into the environment, and just 9% being properly recycled.

The European Union (EU) single-use plastic legislation is a set of regulations aimed at reducing the impact of plastic waste on the environment. The legislation was adopted in 2019 and came into effect in 2021.

Under the legislation, certain single-use plastic products, including straws, plates, cutlery, and cotton buds, have been banned throughout the EU. Additionally, Member States are required to take measures to reduce the use of other single-use plastics, such as food containers and cups, by promoting reusable alternatives.

The EU single-use plastic legislation also requires producers to cover the cost of waste management and cleanup for their products. This means that producers need to contribute to the cost of collecting and recycling the plastic products they place on the market.

The goal of the EU single-use plastic legislation is to reduce marine litter and the impact of plastic waste on the environment. By reducing the use of single-use plastics and ensuring that producers cover the cost of managing the waste they produce, the legislation aims to promote a more sustainable and circular economy (see https://ec.europa.eu/environment/topics/plastics/single-use-plastics_en).

7.3 Amazon Fund

The need to protect natural resources is a key driver behind the creation of a circular economy. The Amazon Fund is a financial mechanism created by the Brazilian government in 2008 to finance projects aimed at reducing deforestation and promoting sustainable development in the Brazilian Amazon region.

The fund provides financial support for a range of projects, including those related to forest conservation, sustainable agriculture, and indigenous peoples' rights. It also supported projects focused on monitoring and combating illegal deforestation activities.

The fund receives donations from international donors, including Norway and Germany, as well as contributions from the Brazilian government. These countries have committed to provide financial support to Brazil to help reduce greenhouse gas emissions from deforestation in the Amazon region. The donations are made based on Brazil's successful reduction of deforestation rates and its compliance with environmental safeguards.

Since its inception, the Amazon Fund has supported over 100 projects in the Brazilian Amazon region and has disbursed more than $1 billion in funds. However, the fund has faced challenges in recent years due to changes in Brazilian environmental policies and governance, which have led to concerns about the fund's effectiveness and transparency (see https://climatefundsupdate.org/the-funds/amazon-fund/).

8 Financing Circular Economy

Circular practices often require significant upfront investment, which can be a barrier to adoption for many businesses and organizations. Financing solutions play a crucial role in overcoming these barriers by providing the necessary capital to fund it.

8.1 BlackRock

In October 2019, BlackRock, the world's largest asset manager based in New York, launched a circular economy public equity fund that has grown to almost USD 2 billion assets under management as of February 2023. The fund aims to drive investments in publicly listed companies already contributing to, or benefiting from, the transition to a circular economy.

Investing in the circular economy not only contributes to a more resilient and regenerative society but also offers the opportunity to generate better financial and competitive returns. In addition, BlackRock's circular economy fund serves as a strong signal to other investors and companies that the transition to a circular economy is underway (see https://ellenmacarthurfoundation.org/circular-examples/the-worlds-largest-investor-embraces-the-circular-economy-blackrock).

8.2 RobecoSAM Global Sustainable Water Fund

The RobecoSAM Global Sustainable Water Fund is a mutual fund that invests in companies that are involved in providing sustainable water solutions. The fund was launched in 2001 by RobecoSAM, a Zurich-based investment firm that specializes in sustainable investing.

The Global Sustainable Water Fund invests in companies across the water value chain, including those involved in water supply and treatment, water infrastructure, and water technology. The fund focuses on companies that have demonstrated a commitment to sustainable practices, such as reducing water consumption, managing water quality, and promoting water conservation.

The RobecoSAM Global Sustainable Water Fund is designed for investors who want to support the transition to a more sustainable global water system while seeking long-term capital appreciation. As of March 2023, the fund had assets under management of over USD 3 billion (see https://www.robeco.com/doca/CGF_WATEE_D-fact-202302-profgloben.pdf?t=1679866917548).

9 Visionary Business Models

9.1 Headphones as a Service

Gerrard Street is a Dutch startup that offers a circular subscription-based model for high-quality headphones.

The company has pioneered and developed a unique business model based on subscription service for its modular headphones. By offering a subscription, the company can collect and refurbish used headphones, reusing up to 85% of their components. This allows customers to receive free replacements for any damaged parts, while the monthly fee provides an affordable way for users to access high-quality headphones without making a significant upfront investment. This approach helps Gerrard Street to attract a wider market segment and provides flexible access to users who desire excellent audio experience (see https://repeat.audio/en/).

9.2 Plastic Highway

India has been leading the world in exploring the use of plastic waste in road construction to address its plastic waste management challenges and improve the quality of its roads.

The idea was first introduced in India in 2001 by a chemistry professor named Rajagopalan Vasudevan, who developed this process to create a stronger and more durable form of road. This process involves the melting of plastic waste along with bitumen, a sticky black liquid that is used as a binding agent in road construction. The resulting mix is then combined with gravel to lay the road surface. The plastic-bitumen mix is claimed to be more resistant to wear and tear and to reduce the amount of bitumen required for the road, which is a nonrenewable resource (Thiagarajan, 2018).

As of May 2019, more than 2500 km (1560 miles) of India's national highways and rural roads have been made with the use of plastic (Lee, 2021).

This solution does not follow the typical footprint of circular economy of designing a product with its circularity in mind, but it serves as a practical example on how to repurpose and convert waste into an asset and serves as a foundation for other innovative solutions.

PlasticRoad BV, a company based in the Netherlands, has managed to further develop this circular concept. The Dutch concept utilizes prefabricated modular products from 100% municipal waste recycled plastics for the construction of roads. These elements are designed to be durable, weather-resistant, and have a longer lifespan than traditional road construction materials. Each component is detachable and can be repurposed and reused at a new site, preventing valuable raw materials from going to waste and reducing environmental impact (see https://plasticroad.com/en/).

9.3 Cooling as a Service

Air-conditioning units and electric fans currently consume around 10% of the world's total electricity and contribute to approximately a fifth of the electricity consumed in buildings globally. According to the International Energy Agency (IEA, 2018), the use of air conditioners is expected to increase in the future as economic and demographic growth becomes concentrated in hotter countries. With rising incomes and living standards, more people are likely to purchase and use air conditioners as a means of keeping cool. If not properly managed, demand for space cooling could more than triple by 2050.

Kaer is a Singapore-based air-conditioning product and service provider that has been offering air-conditioning systems to commercial and industrial properties throughout Asia for 70 years. In 2013, the company shifted its business model to Cooling as a Service (CaaS), which provides several benefits. Rather than having to pay for the capital expenditures associated with building equipment, customers are charged a fixed pay-as-you-use rate based on their consumption. Kaer uses a modular design approach that allows for the addition or removal of air-conditioning units as needed. By using modular design, building owners can free up space and avoid paying for unnecessary equipment. In addition, Kaer utilizes real-time building monitoring and control systems to supervise and optimize energy efficiency, ensuring the systems are running at their most efficient levels. Kaer solutions have successfully reduced cooling energy consumption by 20% and up to 70% in some cases (see https://www.kaer.com/).

9.4 Circular Laptops

Every day 160 thousand "old" laptops are disposed of in the EU alone, while 70% of those laptops could be reused.

CircularComputing is a UK-based company that provides sustainable IT solutions. The company has developed a unique and much needed solution of reducing the unnecessary disposal of a functional laptop, which would have otherwise contributed to e-waste.

The process involves taking used IT equipment and refurbishing it to a like-new condition. Partnering with HP, Dell, and Lenovo, the process includes cleaning, testing, and replacing any parts that are not functioning properly. The company then installs new software and ensures that the equipment is fully functional before it is resold, offering up to 3 years of warranty.

The company's refurbished IT equipment is sold at a lower cost than new equipment, making it also an affordable option for businesses and individuals.

Overall, CircularComputing's business model is based on providing sustainable IT solutions that reduce waste and minimize the environmental impact of IT equipment. Like other cases, the company does not follow the typical footprint of circular economy of owning the product and designing it with circularity in mind, but it serves as an innovative example on how to repurpose and convert waste into an asset and also as a foundation for other future innovative solutions(seehttps://circularcomputing.com/what-is-sustainable-it/).

10 Concluding Remarks

In conclusion, it is evident that the current level of global consumption is unsustainable and that the traditional linear economic model is not equipped to address the resulting resource consumption inefficiencies. The circular economy concept has emerged as a viable solution to this challenge, emphasizing the importance of reducing, reusing, and recycling resources to create a more sustainable economic system. As an emerging concept, the circular economy movement is gaining momentum, with increasing public awareness, as well as support from governments and financial markets. Companies adopting creative solutions in support of a circular economy are emerging as leaders in this field. This demonstrates that it is possible to create profitable businesses while contributing to a more sustainable future. However, much work is yet to be done to fully realize the potential of the circular economy and to address the systemic issues that underpin our current economic practices.

References

Achterberg, E., Hinfelaar, J., & Bocken, N. (2016). *Master circular business models with the Value Hill*. Circle Economy. https://v.fastcdn.co/u/bcd99f8d/48799295-0-finance-white-paper-.pdf

Atasu, A., Dumas, C., & Van Wassenhove, L. N. (2021). The circular business model. *Harvard Business Review*. https://hbr.org/2021/07/the-circular-business-model

Bugge, M. M., Hansen, T., & Klitkou, A. (2016). What is the bioeconomy? *A Review of the Literature. Sustainability, 8*(7), 691. https://doi.org/10.3390/su8070691

IEA. (2018). *The future of cooling*. IEA. License: CC BY 4.0. https://www.iea.org/reports/the-future-of-cooling

International Advisory Council on Global Bioeconomy (IACGB). (2020). *Expanding the sustainable bioeconomy – Vision and way forward*. Communiqué of the Global Bioeconomy Summit 2020. https://gbs2020.net/wp-content/uploads/2020/11/GBS2020_IACGB-Communique.pdf

Kaza, S., Yao, L. C., Bhada-Tata, P., & Van Woerden, F. (2018). *What a waste 2.0: A global snapshot of solid waste management to 2050. Urban development*. World Bank. http://hdl.handle.net/10986/30317

Lee, C. (2021, March 3). *Could plastic roads make for a smoother ride?* BBC. https://www.bbc.com/future/article/20210302-could-plastic-roads-make-for-a-smoother-ride

OECD. (2022). *Global plastics outlook: Economic drivers, environmental impacts and policy options*. OECD Publishing. https://www.oecd.org/environment/plastic-pollution-is-growing-relentlessly-as-waste-management-and-recycling-fall-short.htm

Thiagarajan, K. (2018, July 9). The man who paves India's roads with old plastic. *The Guardian*. https://www.theguardian.com/world/2018/jul/09/the-man-who-paves-indias-roads-with-old-plastic

Inês Botão is a finance professional with over 10 years of experience in accounting, financial analysis and business advisory covering both developed and emerging economies. In addition, Inês is deeply passionate about raising awareness of the urgent need to address climate change. Inês is currently working for Kyndryl, while pursuing the Chartered Global Management Accountant accreditation with Chartered Institute of Management Accountants (CIMA – UK). Further, Inês finance background includes working with Deloitte, IBM and Limiar Capital Management, a hedge fund company based in the USA.

Metabolic Imperative: Deep Gaps in Western Culture Concerning Our Dependence on the Environment

12

Romeu Cardoso Guimarães

Practice proceeds stepwise; theory comprehends the whole journey.

Referred to Bertolt Brecht (Concílio & Koudela, 2019; free translation)

Abstract

Ecological aspects are fundamental to understand various facets of the contemporary global crisis. Climate changes caused widespread productivity modifications and losses, which destabilized nearly all aspects of ecosystems and social systems. We were surprised to witness the absence of the concept of the environment as a bona fide entity for more than 2 millennia, from the Ancient Greece sixth century before zero (-sixth) to the nineteenth century. The gap spans the interval between Thales' Four Essences (earth, water, air, fire) and the maturation of the evolutionary ideas, when the correlations between characters of environments and of living beings could be rationalized. It took more than another century for humans to feel impacted by the consequences of the cognitive deficiency. Our essential dependence on the contexts also became clear. It is a learning process for philosophers and scientists that our predecessors were blind and insensitive to the metabolic imperative and its corollaries: all livings beings are essentially degraders of their environments.

1 Introduction

How could the environment be neglected in theories of economic and human development, if it is an essential initial and boundary condition for the productive system working properly, and for the quality of life of the human population? Similar cognitive abysses are identified in many areas of science. The present ecological crisis may have roots in such ignorance in relation to the environment, possibly reflecting human arrogance in face of natural sustainment. Present cultural scenarios are plentiful with spiritualistic and conspiratorial explanations. They attribute a tyrannical attitude to science that would intend to be normative in all realms. Do we run this risk? How to react and justify? The composition of the interactive amalgam of reason, intuition, and sensibility can be approached aiming at changes that are necessary to overcome the crisis, but it is still hard to envisage the limits that will be reached by the hyper-technological and hyper-informational proposals, if

R. C. Guimarães (✉)
Laboratory of Biodiversity and Molecular Evolution, Department of Genetics UFMG, Institute of Biological Sciences, Federal University of Minas Gerais (UFMG), Belo Horizonte, Brazil
e-mail: romeucg@ufmg.br

A. Pereira Jr, F. Sousa (eds.), *Principles for Governance*, Sustainable Development Goals Series,
https://doi.org/10.1007/978-3-031-40025-4_12

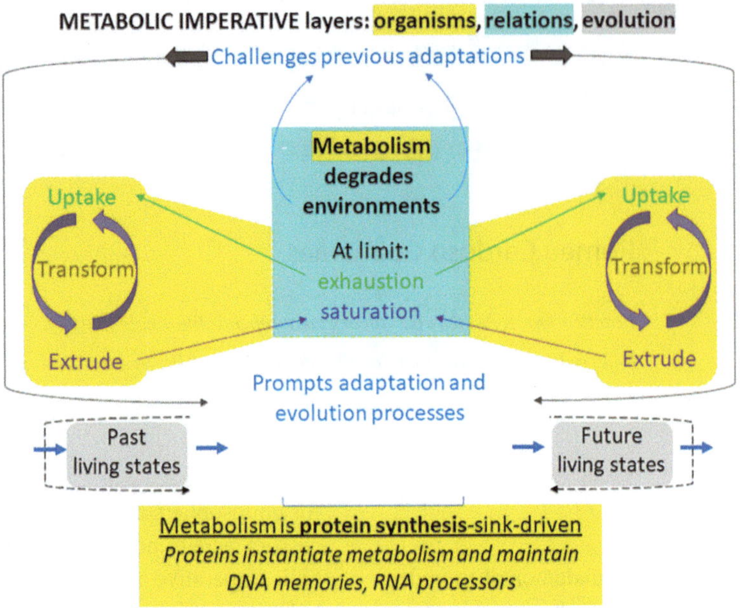

Fig. 12.1 The metabolic imperative is structured in three layers. (1) Living systems' metabolic activities degrade their environments via (a) depletion of matter (masses and energies) that is taken up. After (b) internal processing, the transformations produce (c) waste matter that is extruded into the environments. (2) The previous adapted state is challenged to (3) (re)adapt to the new contexts, at both ontogenetic and phylogenetic levels. The cycle is repeated indefinitely, through the reproductive lineages, the ecosystems, and the biodiversity

the dependence seems to be absolute on the airs, waters, soils, and energies.

Why is it that there is resistance to norms offered by reason, for example, science, while yielding to the norms offered by religious beliefs? What are the hindrances to material and practical knowledge? Science should be proactive and attentive to embrace the wide scope of applications of its developments, including the techniques and tools that may be useful in some but harmful in other aspects of society. Consequences of metabolic activities, if let run free in the limited space, are exhaustion of resources and saturation of the sink of waste. The essential metabolic contradiction shows up flourished: organisms degrade the environments on which they depend. The challenge is solved by organisms that are capable of adaptive changes (at least, replenishment – *anaplerosis* – of nutrients, and care for accumulated waste) and evolution. At evolution (Brentari, 2015; Darwin, 1859), some previous identities may be lost, others modified, which is the nonconservative character of the process: in the case of survival, the rule is impermanence and generalized diversity. In this context, information and its processing into the organization of systems increase. Our species is relatively new and still learning how to develop sustainability in ecosystems and social systems. This is our quest: modulate and moderate growth with focus in systemic qualities, since quantities have reached saturation (Fig. 12.1).

2 Crisis

The contemporary crisis is global, but I am cautious in referring to the Western culture only. This window may seem narrow, but its consequences are wide and strong. Western contributions have spread and affected all areas of the globe, mostly in consequence of science and technology. The crisis seems to be fundamentally ecologic, that is, related to the interactions of living beings with environments. These interactions are characterized as metabolic, in the broad sense. The essence of metabolism is contradictory: living beings modify, transform, and degrade the environment, which is, at the same time, their support. They take nutrients up from environments and extrude derivatives that may be vastly different and new to the exterior. In consequence, adaptations and evolution are mandatory to the life process.

Humans are not different, but only now they take notice of it, in view of the consequences of excesses. These derive from population sizes and from the amplified scale of technological-industrial devastation, associated with the accumulation of waste. The regenerative capacity of natural processes is much lower than the industrial harvesting. Most notorious among the waste materials are the greenhouse effects of gases and vapor that are gravitationally retained in between the tropo-stratospheric layers that surround the planet. The alarm sounds heavily in the form of global warming and climate changes that accelerate the previous damages in a domino/cascade fashion. In consequence, widespread loss of productivity ensues, which trigger political and cultural consequences, including war. The model applied to the present crisis enchains zoonoses, economic recession, climate change, war, biodiversity collapse, and much else.

Most evident is the cognitive aspect. I used to enjoy the notion that scientific curiosity would encompass the widest range possible, but now I perceive its shortsightedness and other limitations. How could it be that the environment was neglected and spoiled for so long? I recall that Thales of Miletus (Ancient Greece, sixth to seventh centuries) said that the basic stuff of the universe being composed of earth-water-air-fire. Although not sophisticated as the modern periodic table to describe chemical elements, the pre-Socratic preoccupation with the *arché* (the elements all things are made of) reveals a degree of environmental consciousness that was absent until recently.

The thesis of the great blackout with respect to the environment, inside the range of interests of Western culture, will, very auspiciously, be challenged along the history of punctuations traversing the vast interval between the periods of Classical Greece and the nineteenth-century Europe. Attention may be given to two rebirths of the thinking on "the human place in the universe": one in the realms of rationality and humanism, which is Lucretius (Pompei -99 – Rome -55), and the other in the Christian spiritualist tradition, Francis of Assisi (1181–1226). It is amazing to witness the vastness of historical time between the clarifying episodes. One part of the rarefied cultural periods may be justified by the lack of writing, drawing, and printing of documents, meaning that oral traditions may not propitiate much fruitful conceptual buildup.

The vagaries and slowness in the progress of reason are, for example, consequent to difficulties in the definition of causes of events. These are frequently unrepeatable, because of dependency on historical contexts. Forces are sometimes highly abstract, difficult to exemplify with empirical and handy demonstrations. In spite of all that, it is required that no gaps are left behind; we are continuously attempting to reach explicit causes. Discussions inside science may sometimes reach the absurdities and stalemates or filigrees of oppositions, when the solution may be so easy as to place one of the opponents as a special case of the other. We may hear now of physicists placing Newton's inside Einstein's. I wonder if it is possible to consider the flat Earth real, when looking at the shape of recent morphologies where the North-South diameter is smaller than the East-West. The difference becomes just quantitative, in this sense. The roundness vs flatness may also be measured by the proportional invariance of the intensity of gravitational acceleration all around the various points of measurements. Or by showing that abrupt borders are never found, all roads are smooth around the ellipse. The discussion gets clarified, meaning that what is in question is not in the geophysical order but in entirely different realms that are not comparable – not commensurable.

Lucretius' writings – the long poem *De Rerum Natura, On the Nature of Things* – conveys the explanations of Epicurean physics (341–270) to people of Roman times to the Middle Ages.

Written fragments were discovered in the ninth century, in full only in 1417, therewith making part of proto-Renaissance. Very interestingly, both highlight the addition of indeterminacy to Democritus' (460–370) atomism, which means their liberation from rigidity and acquisition of possibly "free will" – imagined as an atomic "dance" or swerve, also a God-like character. Atomic interactions result from their intrinsic properties, elaborated by Fortuna, which would embody the concepts of chance and probability. Such thinking has been classified as proto-atheist and might also mark the realm of proto-self-organization, inside natural philosophy. Deities would reside in their own peaceful – paradisiac – realm, not interfering with the earthly battles. It is reported that the school building of Epicurus – called The Garden – stayed right across the road from Plato's Academy, in Athens.

Francis of Assisi is the highly successful *poverello* (the little poor, beggar) whose life was dignified by the creation of an Order in Roman Catholicism. Pope John Paul II declared him the Saint Patron of Ecology (1979), which followed the growth of the leading discipline of contemporary science. The youth period of present-day third-agers was highly influenced by the rise and wideness of spread of the ecological thinking, elected by, for example, nothing less than Aldous Huxley's *Brave New World* (1932). Such influential role gave rise also to the choice of the present Pope Francis' patron (2013; the Argentine-born Bergoglio was born in 1936). Part of the essential message in the Assisian may be obtained from the Canticle of the Sun, also known as *Laudes Creaturarum* (Praise of the Creatures). He would say of nature as the mirror of God, all of it being sisters and brothers to any and all of us. Science's aim is to describe God's creation. Pope Francis encyclical of 2015 adopts the Saint's title "Laudato Si', Mi' Signore," in the name of our sister and mother Earth. Humanity, which used to be nature's Lord and Queen, is now the guardian of creation, and this, our common home.

Humanity and its home, the Earth, are all together sharing the same destiny. Ecological thought is long term and encompasses the inter-specific collectives of ecosystems, multi-lineage and multi-generational, overcoming and including all individualities, while respecting them as components of the big networks. The good to the human equals the good to the whole collectives of species that compose ecosystems. These are embedded one inside others – ecosystem Earth is like the onion bulb, a collective of sub-ecosystems, each skin participating in the composition of the big whole (Fig. 12.2). One may answer respectfully to religious questionings

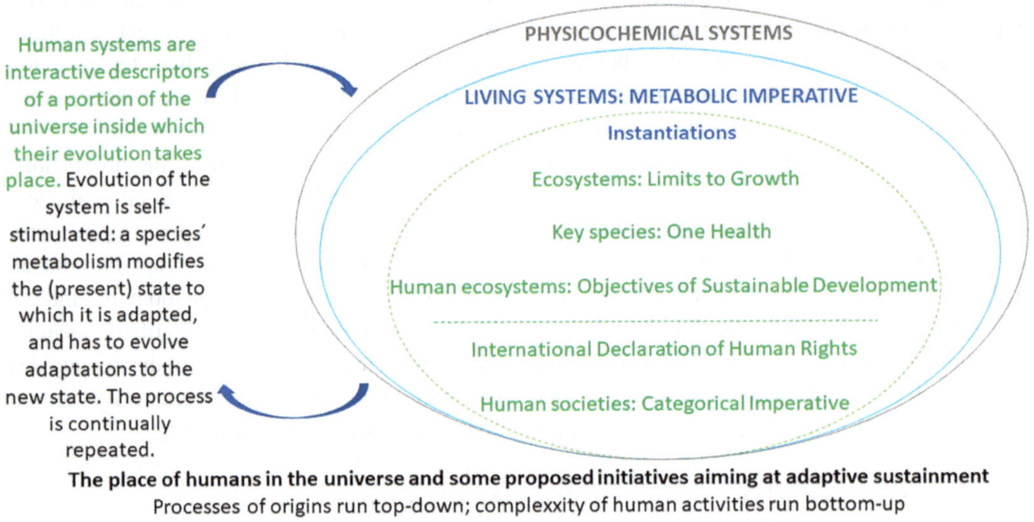

Fig. 12.2 The place of humans in the universe

that, in the case there is God, He/She/It will be praised the same by our caring for the creation, so that we will not be excluded from any post-mortem eventual benefits.

Evolutionary thought showed up in thermodynamics, together with the industrial revolution, in paleontology and then Darwin (1859). In my own medical background, the situation is also drastically deficient. How come nobody talked about the environment as protagonist, with its proper identity and prominence? It is commented only punctually in some instances such as pollution, miasmas, and toxicity. Our umwelt would be somewhat hidden from observation, for example, inside the digestive tract and the airways. Materials ingested only become part of our internal milieu after mucosal uptake and alveolar absorption, respectively.

I recall other environmental crises along the history of life, at least two of them being caused by living beings: the rise of atmospheric oxygen (from 2 Ga) and the rise of rooted plants (~365 Ma).

The description of the world stayed static (as in the *Scala Naturae*; Lovejoy, 1964), around and below mankind for very long, for example, the well-known "know thyself," from the Greeks, and the "world and nature are here to serve us," among the Abrahamic monotheisms. The real things are the Platonic transcendental, metaphysical, the superlative infinities and, again, in the monotheisms, while our terrestrial realm is just a temporary and imperfect "shade in the cave." The Cartesian "I think, so I am," is as much hubric and arrogant. In an opposing humility, Darwin's "I think" is just a proposition in the form of a sketched small phylogenetic tree.

We are in a dynamic and evolutionary world. Perplexity rules, attempting to learn from the crisis, imagining futures, and searching for meanings and senses to the life process. It is probably established that individualisms may have the necessary qualities of objectivism, but they are not sufficient for description and understanding the dynamics. It is mandatory, aiming at completeness, to account for the collectivities, such as populations, lineages, ecosystems, and societies, in spite of these being fuzzy systems, and always under continuous construction and reconstruction. In the face of the obligatory death of organisms, and of their shorter ontogenetic adaptive intervals, what may remain to accomplish the long-term expectancies of the living is the phylogenetic biodiversity.

In spite of the limited reach and uncertainties, we should express concern about the cognitive gap referring to the lack of attention given to, or intentional neglect of, the environments. The gap contributes to widen the feeling of surprise with respect to the "novelty" that the environment is hitting us too hard, in relation to the prevailing concept that it should be passive, assuming the role of "being here to serve us humans." Such feeling of novelty and surprise may add to the backing of the denialist (deeper yet, escapist) kinds of reaction and even the attacks that science, culture, and the universities are being subjected to.

It is also consequent to deliberate "selective blindness" (recalling José Saramago, 1995) toward disturbing facts and advice that have been offered to us since the early twentieth century, on the piling up of waste, desiccation and hydrological crises, loss of habitats, and loss of biodiversity. Generalized slowed global productivity is reported mainly from the 1970s. Many of the alerts from the Club of Rome's *Limits to Growth* (1972) are coming to light.

The cultural gap refers to the late (re)discovery of the concept of environment. The fault is very wide, between Ancient Greece and the nineteenth century, when the notion of evolution showed up more or less simultaneously in the distant branches of science such as physics, paleontology, and biology. We are not alone in noticing gaps in our cultural background. It also springs up in other areas, such as in studies on the attribute of consciousness (Horgan, 2016). I got used to nourishing the concept that the curiosity of scholars, from scientists to philosophers, was necessarily wide to the point of reaching nearly all corners. Ancient Greece attests to this point. Then I came to notice that the environment was not talked about or studied throughout the almost entire span of our culture. It was a deception to note my naïve expectancies.

3 Environment

The environment is now a fashionable concept, environmentalists and denialists clashing about it; great cultural changes are expected. Humankind might now find its wider home, birthplace, and cosmic integration. Can we grow to the multirooted human, *Homo radiculatus* (Fig. 12.3)?

I would expect this to follow from the fixation of the notions that: (1) Living beings are the entities – cells, bodies, organisms – that instantiate the (2) life process. There is no life outside of the living beings, dispersed in the universe. (3) Living beings are differentiations – biochemical systems – from geochemical systems. The latter are, therefore, our mothers and wombs and, by extension, also the Earth: Pachamama, Mother Earth, Gaia, Terra, and Tellus. (4) The environment is our origin, and we have never liberated from it – we are not "autonomous."

The correct context says that environment and the living body are interpenetrated. No demarcation is clear-cut; limits are fuzzy. Organisms are embedded in the environment and this belongs also to the inside of them. Water molecules may have the same structure in and out, but organisms transform them into derivatives that may be useful, such as the hydrogen, utilized for the reduction of organic compounds, or need to be handled with much care, such as the oxygen, that may be necessary for respiration but may be turned into toxic reactive species. Some organic compounds may leak out of their sites of origin and accumulate in the atmosphere layers.

This means that the environment, at least its portion in our immediate neighborhood, that feeds us and exchanges with our metabolism, our "umwelt" (Brentari, 2015) and root, is a functional part of our bodies (or vice versa, we are part of it). This conclusion enhances the magnitude of the present cognitive crisis, with respect to the contrast that becomes apparent only now, confronting the previous view of a clear-cut separation or demarcation. Societies are suddenly facing a large challenge, coming from a previously unsuspected source. The source is diffuse, innominate, and not personalized, which may be causing the reactionaries to get angry and nervous generally, also diffusely against all evolutionary propositions. The escapist attitudes are unfocused, directed against everything, from the shape of celestial objects – the flat Earth – to the indigenous inhabitants of all remote cone-tips of Earth's crust and ocean – which would "own too large territories." Where are the gases in the sky that I don't see and that scientists blame so much for being accumulating?

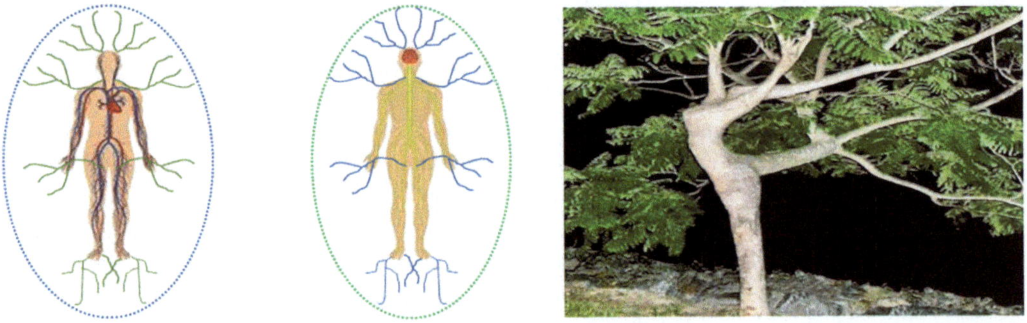

Fig. 12.3 *The fully rooted Homo radiculatus* senses, feels, and acts with both "hearts and nerves" (left and center, respectively). It is embedded in and integrated to the entities and history of the ecosystem and social networks, in consonance with the principles of both their conservativeness (as memories) and adaptivity/evolvability (from relational plasticity). The *Femina radiculata* (right) may be an inspiration, from the plant world

4 Science Communication

All of a sudden, people are told of the existence of not just one but multiple gravitational ceilings above the air that we breath. This resides in a thin layer surrounding the planet. We can only see, in daylight, the water vapor accumulated in clouds, dispersed amid the blue vastness; open space, no ceiling. We can have a faint notion of layers when thinking of the rarefaction of the air in the high altitudes of mountains – a few to various kilometers above ground and sea level – together with the nearness of the clouds. We can sense it when riding airplanes, above the cloud layer and inside a tight bubble holding adequate pressure and collected oxygen. It is then possible to envisage – not to see – other layers above these that are sensed and defined by scientific apparatuses.

The scenario drawn on the causes and consequences of climate change is dramatic and haunting. Worse still, it identifies causes on the very roots of the Western culture, science, and its technological applications that, seen from another point of view, have been so successful in, e.g., extending drastically human life expectancy all over the planet. This may seem contradictory – so good, so bad? – and not making sense to the laypeople. It may be a reason for the success of denialist and conspiracy theories, at the same time raising a challenge for the scientific endeavor of reaching wider audiences with understandability, which could be of help to subsidize mutual support.

There is a lot to be done with respect to scientific literacy at large, on how to guarantee ethical utilization of scientific knowledge. Scientific education should be more than distributing technology to all kinds of entrepreneurs and let them go non-scrutinized to make easy profit and to acquire political power, to the point of causing harm to the population of citizens that should not be reduced to consumers. Much of the fight of consumer protection groups that should be directed against the unethical components of big industry, big pharma, etc., end up spilling over the science that gives them technical support and does not participate in its misutilization by them.

Scientific disciplines should also adopt adequate training for mediation of conversations with religious citizens and institutions in order to avoid unlearned and undue rejection based on misinterpretations. We may hear a lot of religious learned people, e.g. physicians, saying of the "tyranny of science." It might be a better approach to say of complementariness in attempts at mutual help than by taking sides and possible conflicts or rejection.

Ethics, compassion, and empathy might help, for instance, in the proposition that science and religion should be kept each restricted to its proper realm, not allowing for disputes that are typical of the invasion of each other's arenas. In these conflictive spaces, we should ask for the help of ethicists, psychologists, and philosophers for mediation. The objective is to avoid rejection and stalled discussions for the sake of obtaining support for action in the direction of correcting the course of – besides uncontrolled population increase – environmental devastation, for society to become aware of and friendly toward the environment. People should learn that such attitudes will be life-saving and profitable, beyond the emotional or esthetical. Science is much more than utilitarianism. It has much to say about reality, "weltanschauung," well-being, and needs to be enriched with participation on ethical debates.

5 To Live Is to Metabolize

Metabolism is driven by consumption, but there are earthly limits to the number of metabolic units (organisms, ecosystems), to their avidity and modes of processing. It is disturbing to realize that the challenge posed by the crisis requires an almost complete revision of "the system" that prevails in our society. The utilization of energy derived from the burning of diverse kinds of fuels – the Pyrocene aspect of the Anthropocene – shall be stopped, as well as the application of the fast and bulky industrial scales and dimensions to all aspects, from mere extraction to production, which may result in nearly unrecoverable devastation. We are immersed in an over-explored, saturated, overstuffed world, clogged in diverse senses.

Society is embedded in an ecosystem and will be largely digital, but built on the basis of metabolism. It should privilege solidarity, which has the collectivity together with individuals, and sustainability, which has the environment together with humans and all other aspects of the ecosystem. Accordingly, health shall refer to all components – the "One Health" concept – physical, biotic, and human (Destoumieux-Garzón et al., 2018).

Humans are no different from bacteria with respect to metabolism and its environment degradation aspect. Activities and behaviors may be considered extensions of metabolism: niche-building is part of exchange behaviors. Some human adaptations may differ from the generality of species due to involvement of intentionality and complex sociocultural interactions. However, the complexity of evolutionary populations is not reducible to individualisms. The discovery of evolution and of the role of environments require humility with respect to (non)predictability, and the capacity of association of bacteria to overcome the obstacles to the continuity of life!

We dare to propose a biochemically possibly oversimplified but functionally evident definition: *living beings are protein synthesis systems* (Guimarães, 2019). The stance is utilized that the metabolic system is defined by its final product, its reason for being. The product defines all previous steps. To reach such unity, multiple feedback loops are necessary in the network. It happens as if the development of the system goes through loosely defined sub-closure steps and goes on until it can find its form in a structurally finished state. Although not necessarily functionally finished, it may be considered cognitive, in view of its specificities.

The descriptive approach is organized with respect to the partial functions. DNA and RNA correspond, respectively, to (1) the cellular memories, which define the sequences of proteins, that are the protein-coding genes. (2) Other genes may code for RNAs that are vehicles of information for the expression of the genetic memories: the generic protein synthesis machinery (ribosomal RNAs, transfer RNAs, etc.); the chromosome structures (most important, e.g., for reproductive processes is the correct partition of DNA among descendants); and the regulation of gene activities. All these activities of DNA and RNA maintenance and expression of memories are, in fact, realized in association with proteins, that is, in the form of DNPs and RNPs, which are organized aggregates, deoxyribonucleoproteins and ribonucleoproteins. It can be said that proteins are "taking close care" of the mechanisms of construction of their memories, by checking the quality of the system all along its constructive steps.

A third component is the most actively "living," a term which denotes the dynamics of being alive – metabolism. This may, in fact, designate all aspects of the living, including its relations with the environment. The classical minimal living entity (a living body) is the cell, from bacteria and archaea to protists, fungi, animals, and plants. All living beings have unicellular stages. Some phyla are exclusively so, others have uni- and multicellular stages or body forms.

The biochemical description of metabolism says of uptake of matter (mass and energy) from the environment, its transformation into derivatives that are proper to the cellular materials and activities, and extrusion of non-serviceable remains, excreta, and secretions. In larger cells, bodies, or organisms, there are other descriptive characters, such as behaviors of hunting for preys, escaping from predators, looking for sexual partners, spreading appendages or gametes, and socializing, which may all be included in the category of extensions of the three just cited: uptake, transformation, and extrusion. All these functions, from the interactive receptors to the active motors, are basically jobs of proteins, aside with the nucleic acids – memory components, plus the structural and functional helpers – lipids and carbohydrates.

Viruses and prions are not living or metabolizing entities. They are segments of cells that are replicated and transmitted between other cells. They evolve along the journey and may transport materials between cells, accomplishing the role of increasing the plasticity and the evolutionary dynamics of the systems. Their variety is enormous and may get close to challenging the limits and frontiers of standard concepts.

6 Network Plasticity and Self-Reference

This description of components is integrated in bodies through their organization as networks of molecular activities. Network structures are integrated combinations of linear chains and cycles. Cycles are typically multiconnected in the network, which provides for their nearly uninterrupted activities, therewith accounting also for the function of memories. Network memories are of the epigenetic kind, that is, systemic, corresponding to the conjoint activity of various genes together, as in the long-term maintenance of phenotypic states. These are sometimes inheritable, in the category of "epigenetic transgenerational transmission".

It is then possible to summarize the necessary concepts.

1. *Living beings* are protein synthesis systems (Guimarães, 2019).
2. To live is to metabolize, to maintain active the network of uptake-transformation-extrusion.
3. The system of evolutionarily organized polymer sequences (nucleic acids and proteins) is continuously regenerated and reproduced by the metabolic activities that are instantiated by the polymers themselves, in self-reinforcing – self-referential – cycles.
4. *Life* is the process instantiated by living beings, which is adaptive and evolutionary.
5. The adaptive system is based on the *plasticity* of the components, lower in DNA, intermediate in RNA, higher in proteins, and still higher in the network.
6. Stability of the system is obtained mainly through the memory components, (a) genes for the constitution of the polymers, and (b) epigenetic cycles, for the network functions, such as the differentiated and the adapted phenotypic states, including the homeostasis and resilience processes.

7 Metabolism Is Contradictory; Evolution Is Mandatory

With respect to the interactions with environments, organisms are essentially degradative, which requires continuous (re)adaptations and evolution. Otherwise, with respect to internal "economy" (Holmes, 1986), the self-referential character is applicable in its full generality: genes code for proteins that help replication (of genes) and reproduction (of the system, the cell); proteins instantiate the metabolic networks, whose products are materials for the synthesis of proteins. The communication between components is mainly through direct contacts (e.g., protein binding) or at distances (enzyme products or secretions). The wider notion of self-reference or self-reinforcement is preferable over other similar but more restrict to special cases, such as autocatalysis.

The self-referential processes are constructive, in the organismal side, but at the same time destructive, in the relational side, with respect to the environment. We arrive then, at the great contradiction: living beings degrade the environments on which they depend. Left alone in this direction, they will exhaust resources and have to stop reproducing, entering stationary states, in a path that leads to possible death. There are many species, including the simpler microbial, that persist up to the present in this unregulated regime. Their survival depends on short and fast reproductive cycles, where the plain mutation rate is sufficient to generate variations that facilitate survival across time and the adequate range of environments.

Regulatory developments arose in some species with formation of resistance states such as cysts and spores, which require complexity of controls. These are dormant states with quiet metabolisms that could survive under restrictive and stressful conditions, as if just waiting for reestablishment of environments favorable to germination. Higher levels of regulation come in with interspecific interactions developing mutualist regimes. One species enriches the environment of another and the conjoined set may form a higher-order network of mutual help. The plasticity of the network is synergistically higher than the mere sum of the plasticity of the components, in confronting the physicochemical environments, which is a reason-for-being of ecosystems. Some simpler cases are, for example, of syntrophy, more complex the symbiosis, among a large diversity of arrangements.

8 History

It is apparent that humans may have developed some kinds of communities and cultures where such widespread friendliness, among the cultures and in relation with the other biologic and physicochemical components prevailed, forming stable, long-lasting ecosystems. Most probably, peaceful stability would have fluctuated around regimes of the saltatory equilibrium types, punctuated with crises, either internal or relative to the environments, and even war-like, with respect to neighboring cultures. To the contrary, the historical period of humanity seems more war-filled-like than of the long-lasting friendship and peacefulness kinds. Our history differs from that of nonhuman ecosystems, in general, mostly in view of the dominant role of complex cultural habits, filled with intentionality and planning.

The metabolic essence is a generator of crises. This dampens somewhat the blame that is being thrown in the back of the humankind, by spreading it to the whole of the life process and of the living beings. We may behave more destructively than other species, but we are not the inventors of the habit. The problem with humans in causing the environmental crises is therefore, not in the *original sin* of being metabolic but, in having the intelligence tool, of misdirecting it. The original contradiction is amplified in all directions and with low discriminatory power to choose the direction toward long-term sustainability. The situation is described with similarity to the case of mutations, paraphrasing and extending François Monod's *Chance and Necessity* (1971): there are many pathways, in all directions, while just a few of them result in the sustainability, or the fitness, or the living success criteria. Would the consequence tend to favor the direction of catastrophe?

The living contradiction is much obvious, to the point of participating in literary observations, among which the Brazilian João Guimarães Rosa: "living is very dangerous" (Grande Sertão Veredas 1956, The Devil to Pay in the Backlands, 1963; Rocha, 2023). We add that it is as much dangerous to the living cell or organism as to its environment. The environment that sustains us is placed in danger of disruption by its own creatures, analogs of "parasites."

The time course of humans, along phylogeny, and in the ontogenesis of individuals is short in relation to the biological tradition of ecosystem maturation based in Darwinian processes. We may circumvent the difficulties with the help of cultural learning and education, which anthropologists, ecologists, and many other domains are fighting for. The road is bumpy and tortuous, but we'll get there successfully at least in some lucky territories and paradigmatic islands, seeds for the necessary spread. We have first to survive the crisis, then learn from it, participating fruitfully in the trajectory of our Gaia, Earth, and Tellus geochemical cradle and home.

If to live is to metabolize, what to say of a plant seed found in a Pharaoh's tomb 5000 years of age? Is it dead or alive? – a biological version of *Schrödinger's cat* paradox (1935). Metabolism is tested stopped and morphology OK, but we cannot tell until it passes the test of germination, by placing it under conditions adequate for living or rebirth. If metabolic dynamics is reassumed, it was not dead; if not, it was dead. If it were not dead but not manifesting the dynamics, we'd say it was in a state of suspended life, defined as being capable of "resurrection," that is, resume or restart. Suspended life belongs to the "spectrum of variation" categories, which include dormancy, numbness, and hibernation. Techniques for manipulation of such states are desiccation, freezing in the liquid state (without allowing for the growth of ice crystals), lyophilization, and freeze-drying. The biological answer to Schrödinger's quest is in the hands of the relations between organisms and the contexts, if they are productive or not.

The organism with cardiorespiratory or cerebral inactivity that is legally diagnosed as dead may still have parts that are not dead. The legal definition is utilized for the sake of allowing extraction of parts or organs for transplantation, for therapy of other members of society. The concepts of mutuality and symmetry are followed that the individuals serve society, mirroring the previous state where society served the individuals along their lifetime.

The derived lesson refers to the importance of the environment in its relations with organisms. Most interesting would be the radical stance of saying that it is upmost, full, integral 100%. If this is the case, we reach the situation where the environment is not there, outside, but it is also inside, an integral part of the living system. A partial conclusion is that the living system is the cell and its immediate supporting environment, a system that cannot be separated into independent parts. Each individual is a composite, a mini-ecosystem (Ribeiro et al., 2011). In this sense, the individual and its immediate environmental cut co-evolve. Images of such an ensemble would be similar to the astronaut and its indispensable bubble, the passenger in an airplane.

9 Mother Earth

The analogies may be weak but the message is strong: we have to take very good care of the environment on which we strictly depend. A provocative comparison can be made with our mothers: the environment would be more extensively essential even than them. Mothers can be temporary: they and their wombs are essential for the gestational and early ontogenetic process. After parturition and separation of the umbilical cord, the newborn is loose in the world and can be supported by other caretakers and helpers. Contrarily, our environment and our ecosystem have to take care of us for the whole of our lives, as well as the lives of our mothers. Biological mothers are temporary, the ecosystem of long duration. This just reminds us of and enforces the concept that "living systems are differentiated states of the life-giving, proto-living or life-precursor geochemical systems (grand-mothers of the biological mothers)." Such new states remain being fed, sustained, and supported by geochemistry. Ilya Prigogine's physicochemical and thermodynamic statements are well-suited in saying of "dissipative systems," models for some aspects of the living (Prigogine & Lefever, 1968).

The environment and the ecosystem in our (and of our mother's) vicinity have to be healthy uninterruptedly and along our entire lives: there are no umbilical cords that could be cut. In this realm, other tests can easily be conducted to demonstrate the premise of the essentiality of environments through blockades. If you stop breathing, you die in a few minutes; stop drinking water, in a few days; and stop eating, in a few weeks. If you ingest toxics, it is all the same. Mother Earth is "forever," according to the duration of the planetary conditions. Biological lineages and collectivities have originated at ~4Ga, with open-ended evolutionary futures.

10 Ahead from the Generalized Neglect

Other cultures before or aside with ours might have felt the importance of the environment or of the surrounding natural world, but, at least in the Western tradition, those were treated as inexhaustible resources. They might have gone through crises of unsustainability, but they did not leave us efficient and timely messages. Instances of punctual acquisition of ecological awareness are described, for instance, in cases of infectious diseases. They were thought to have been caused by the so-called miasmas and fetid vapors and gases evaporated from rotten materials in swamps. The same is with various cases of toxicity. There would be something from the surroundings affecting us. Without an encompassing notion of the environment as a bona fide entity or identification of the bad and good components, generic neglect was the dominant behavior. The same behavior happened with respect to the waste disposal, considering the environment an infinite sink.

Much worse was the notion that nature existed just to serve us and should be subdued. This comes from the religious foundations of the west.

The Sixth Day …*27* So God created man in His own image; in the image of God He created him; male and female He created them. *28* God blessed them and said to them, "Be fruitful and multiply, and fill the earth and subdue it; rule over the fish of the sea and the birds of the air and every creature that crawls upon the earth." *29* Then God said, "Behold, I have given you every seed-bearing plant on the face of all the earth, and every tree whose fruit contains seed. They will be yours for food". Berean Standard Bible (https://biblehub.com/genesis/1-28.htm 04Oct2022 16:25 Brasilia Std time)

A surprise came in from the unsuspected notion of evolution. Together with the industrial revolution, there entered the need for knowledge on the building and maintenance of machinery, which gave a big push to thermodynamics. Its second law became a generic law of evolution, together with, for example, the direction given the growth of entropy and by the obedience to descending energy/mass gradients. Paleontology was accumulating evidence for the existence of extinct species and their chronology according to the geological layers.

Finally comes in Darwin, introducing himself and Wallace conjointly to the Linnean Society, to present a theory on how would biological species evolve, in concert and interacting with the environment. This side of the story marks the birth of the environment, as a bona fide player in the evolutionary records, the birth of a new concept about the partners and forces that shape the evolutionary scenarios, and the new "weltanshauung" of the dynamic world.

Darwin was one among the naturalist voyagers inheriting and developing the influences of the German "Naturphilosophie." His iconic image is the hand-drawn small first phylogenetic tree with the title on top of the handbook page "I think" (ca 1837). Gathering data from his trip recollections and the information detailed by the museologists back home, the small tree organizes the more than 20 species or specimens into a continental source and a few derived groups, spread in concert with the geographical distribution of the islands of the Galapagos archipelago. Then we have the island environments in correlation with the species morphology, which surprisingly made a lot of sense.

In the incoming theory, the environments participate with "forces" as big as the biological in shaping the species, through interactions. The aging *Scala Naturae* mode of organizing the known entities, including the biological species, is substituted by the scientific phylogenetic trees. The *Scala Naturae* mixes together all kinds of entities, natural, religious, and other mental artifacts. The Great Chain of Being starts with God and descends in angels, demons, stars, Moon, kings, nobles, plebeians, wild animals, domestic animals, trees, other plants, precious stones, precious metals, and other minerals (Lovejoy, 1964).

11 Concluding Remarks: The Self-Inflicted Crisis

We are repeating what primitive living beings always did – unregulated growth, unresponsive to environmental clues of saturation – radical, extreme, and far-self-reference. We consider ourselves intelligent and we may be so, but we privileged some directions that are now proving to reach exhaustion of resources and saturation of space with our activities and accumulated waste. The rationale for the causal diagnosis is the plain metabolic.

If it is considered that biochemistry is a differentiated state and a chapter of geochemistry, then the Mother Earth meme is correct and the path being followed is matricidal. If it is considered that we are fully and continually dependent on respiration and metabolic transformation of material and energetic nutrients taken up from the environment, then the path being followed is also suicidal. It may be wise to utilize such dramatic terms in order to hit loudly the bells of alert.

How come, can we find roots of the crisis on characters of the Western culture? Yes, we can, but I don't have details on how similar or different ours is from other cultures. I don't know them enough to risk non-justifiable opinions. We can say that living beings, in general, share the same metabolic trend of manifesting the self-referential behavior and also that such manifestations may acquire cultural idiosyncrasies and tonalities.

Self-reference means that we involve ourselves in the explanations we are giving. It occurs in natural or formal languages when a sentence, idea, or formula refers to itself. We chose this term in view of its generality and plasticity in being applicable to diverse contexts of the life process and their local expressions. An example is the first among the Great Extinction Events in the paleontological record, which was caused by the rise of cyanobacteria and resulting in the Great Oxidation Event. The evolutionary event of photonic water breakage would have been fixed due to being driven by energetic advantages to the cyanobacteria. This went on irrespective of the ecological consequences, which reside in the exterior and in the obviously unknown future scenarios. Then the self-referential behavior of the cyanobacteria is "blind" to collaterals that belong to contexts other than those included in the source object and "sees" only the internal context, which is "itself."

The reality of self-reference in the life process adds another meaning to the famous (in the Brazilian context) statement of Riobaldo in João Guimarães Rosa (1956, *The Devil to Pay in the Backlands*): "Living is very dangerous." Yes (and I ask for excuses on the repetition), dangerous with respect to the living beings in general and to their environments. Everything a living being does in the world refers to that living being, irrespective of consequences to others in the same species, to other species and to the physicochemical components of the environment.

This is the primary type of behavior, which may be modulated, regulated or even reversed in the course of adaptations and evolution of behavioral traits to ecosystem and social contexts. When we feel in love with someone, the feeling is not only directed to that someone; it is also directed to ourselves. A more radical proposition with bases on the self-referential principles is autopoiesis, which we will not detail. It seems it needs to reflect more largely on the environments.

There are some parallels between the building of social systems and of ecosystems. In both there is some taming and regulation of self-reference, which depend mostly on the network structures that are developed. The basis of networks is the interdependence and mutuality among the components where some degrees of freedom of a component are sacrificed in favor of the distributed benefits that are spread through the whole system. This is the network version of the series of common maxims: "the whole is greater than the sum of the parts," "everything is interconnected." and "the top-down effects," where "the top" is the whole system that acts upon each of its components.

Network structures are dynamic and plastic, with wide adaptive capabilities. The workings of the internal substructures are rarely deterministic, demonstrating local plasticity via fluctuations in the communication channels among components. Notable exceptions are the eusocial insects and the naked-mole-rats, where the effects of developmental chemicals are quasi-deterministic. Such kinds of effects are retained in the more complex vertebrates mostly in the realms of sexual mechanisms, via the pheromones.

There is still a lot to be developed with respect to the utopic global stable and sustainable network. In the sense of Eduardo Galeano, the sustainability utopia remains a guiding attractor (1994): "Utopia resides there in the horizon. I get two steps closer she gets two steps backward. I walk ten steps the horizon runs also ten. The more I can walk, I'll never make it. What good does the utopia serve? It's just for this: so that I don't stop walking."

The *evolutionary imperative* applies, running together with the *metabolic imperative*. The latter incorporates a contradiction that provokes the former, as an attempted solution, but it only recreates the condition of the previous type. Therefore, no linearity but circular configurations of the uroboric and lemniscate types. Some degree of rigidity may be necessary for stability of the system, but it should not be impeditive to the generation of novelties and inventions, which are the sources of ingredients for the adaptations and evolution. The latter are essential for the survival under variable environments and, most important, under the condition where the constituents of the system are themselves fragile, mutable, and variable. Therefore, rigidity is both impossible and disadvantageous. A reason for the evolutionary choice of proteins, for composing the main stuff of living is their intrinsic plasticity.

A similar internal conflict applies to our social condition. A maxim type of statement would say the following: (1) *Individuals* are necessary for the adaptive and evolutionary process but are not sufficient. (2) *Collectivities* are the fundamentals of the living systems, individuals are components. (3) Living inside communities gives rise to structured interaction networks where the component individuals acquire identities through the interactions and effects of *mutuality*. This would be a biologic background for the analytical psychologies' concepts of the relevance of "the others" for the development of any component. Populations are full-fledged entities as much as the individuals. The former subsumes the latter as higher-order entities, but they remain fully dependent on the latter. Therefore, (4) the system dynamics is what matters, not the proposition of the conflictive situation of parts x wholes.

Acknowledgments To the long period of learning I spent in the self-organization group of UNICAMP. To the LBEM that hosts me more recently. To my professors Luigi Bogliolo, Giorgio Schreiber, Willy Beçak, Hertha Mayer, David Bloch, Winston Gutteridge, Volker Erdmann, and Edward Trifonov. To the many friends who make our lives happier.

References

Brentari, C. (2015). Jakob von Uexküll: The discovery of the umwelt between biosemiotics and theoretical biology. ISBN 978-94-017-9687-3. https://doi.org/10.1007/978-94-017-9688-0

Concílio, V., & Koudela, I. D. (2019). (Portuguese) Protocols and the theater pedagogy – From the translation of student's protocols about 'The yes sayer' to the protocols of 'joyous work'. *Urdimento (Florianópolis), 1*(34), 246–255. https://doi.org/10.5965/1414573101342019246

Darwin, C. (1859). *On the origin of species by means of natural selection, or the preservation of favoured races in the struggle for life*. John Murray.

Destoumieux-Garzón, D., Mavingui, P., Boetsch, G., Boissier, J., Darriet, F., Duboz, P., Fritsch, C., Giraudoux, P., Roux, F. L., Morand, S., Paillard, C., Pontier, D., Sueur, C., & Voituron, Y. (2018). The One Health concept: 10 years old and a long road ahead. *Frontiers in Veterinary Science, 5*, 14. https://doi.org/10.3389/fvets.2018.00014

Galeano, E. (1994). *Las Palabras Andantes, El Derecho al Delirio, Siglo 21, Buenos Aires.* https://www.revistaprosaversoearte.com/para-que-serve-a-utopia-eduardo-galeano/

Guimarães, R. C. (2019). Roots of complexity in the self-referential genetic code. Ch. 6, pg 117–143. In L. H. Wegner & U. Lüttge (Eds.), *Emergence and modularity in life sciences. Part II from modules to emergent holistic properties in living organisms.* Springer. ISBN 978-3-030-06127-2. https://doi.org/10.1007/978-3-030-06128-9

Holmes, F. L. (1986). Claude Bernard, the Milieu Intérieur, and regulatory physiology. *History and Philosophy of Life Sciences, 8*, 3–25.

Horgan, J. (2016). *The mind–body problem, scientific regress and "Woo"*. https://blogs.scientificamerican.com/cross-check/the-mind-body-problem-scientific-regress-and-woo/Thescience of consciousness

Huxley, A. (1932). *Brave new world.* Chatto and Windus.

Lovejoy, A. O. (1964). *The great chain of being: A study of the history of an idea.* Harvard University Press. ISBN 0-674-36153-9 Wikipedia 09Aug2022, 20:39 Brasilia Std Time.

Meadows, D. H., Meadows, D. L., Randers, J., & Behrens, W. W., III. (1972). *The limits to growth.* Potomac Associates – Universe Books. ISBN 0-87663-165-0. ClubOfRome.org

Monod, J. (1971). Hasard et la Nécessité: An essay on the natural philosophy of modern biology. Knopf. ISBN 10 0394466152.

Prigogine, I., & Lefever, R. (1968). Symmetry breaking instabilities in dissipative systems II. *The Journal of Chemical Physics, 48*, 1695–1701. https://doi.org/10.1063/1.1668896

Ribeiro, J. A. G., Cavassan, O., & Brando, F. R. (2011). (Portuguese) Constructing a model of the concept of environment through the scientific models of ecological units: Contributions to ecology teaching. In Teaching Sciences and Mathematics V, *History and philosophy of science* (pp. 169–189). Org. AMA Caldeira. Cultura Acadêmica/UNESP. ISBN 978-85-7983-214-7.

Rocha, L. O. S. (2023). Kindly offered me five citations of the author João Guimarães Rosa to the theme, in the book (Portuguese) 'Grande Sertão: Veredas' 1956 José Olympio, Rio de Janeiro, Brazil. Literal English 'Great Backlands: Oases'. English Publication: The Devil to Pay in the Backlands 1963.

Saramago, J. S. (1995). Ensaio sobre a cegueira; Blindness 1997. Ed. Caminho – The field of the word. ISBN 0-15-100251-7.

Schrödinger, E. (1935). The present situation in quantum mechanics. *Naturwissenschaften, 23*, 807–812. https://doi.org/10.1007/BF01491891

Romeu Cardoso Guimarães is Medical Doctor (1965) and PhD in Pathology (1970). He has post-doctorals in Molecular Biology, Genetics, and Evolution. He developed the systemic concept of the gene in 1992, described the reduction in genetic polymorphisms in animal evolution in 2008, and developed the self-referential model for the genetic code in 2008. He was a full professor retired from Universidade Estadual Paulista (UNESP) and senior researcher at the Laboratory of Biodiversity and Molecular Evolution, Department of Genetics, Ecology, and Evolution, Institute of Biological Sciences, Federal University of Minas Gerais (UFMG), 31270.901 Belo Horizonte MG Brazil.

13 The Rise of Consciousness in Social Practices: A Pathway to Peace and Benevolence

Joanna Cameron

Abstract

A mental framework is essential for our approach to governance issues concerning intentions in decision-making and political conduct. A bioethical approach should take into account the scientific, philosophical, and spiritual aspects of human personal and social consciousness. In the reality of social practices, consciousness is the awareness of the Self and the Other. This awareness is an important component of human decision-making and can be improved by means of spiritual practices. A peaceful mentality, essential to a constructive governance in times of conflict, can be conceived as the harmony between oneself and the world, leading to a disposition toward benevolence in social practices. This chapter discusses some practical aspects and offers ideas of how we can produce good will and benevolence in our relationships. The author encourages a perspective to promote compassionate outcomes. As human consciousness rises, it brings opportunities to find new possibilities in life. Meditation and similar practices in religious contexts are the keys to unlocking these opportunities. We suggest that the rise of consciousness should be directed to peace and benevolence, instead of current trends in misdirected (pseudo)religious behaviors, which have led to the propagation of hate in social networks, the iatrogenic conduct of governments in health crises, and massive death of innocent people in war.

1 Introduction

Let me take you to a time in the not-too-distant future. Feeling confident, a patient checks into a treatment facility called "The Cancer Healing Center." After lying comfortably in a theater, he notices a picture of the body tumor on the surrounding screen above. There are no drugs or sedatives in his treatment. His excitement increases as a new reality is introduced. It is an immersive-augmented relaxing environment customized for his unique preferences by artificial intelligence (AI). He sees his therapist is with him in this new reality and assists him in setting the intention in his body to heal. After his nod of agreement, the auditorium comes more alive. It is like a multidimensional dream as he watches the tumor shrink. He relaxes further as tears of joy stream down his face. He feels his heart energy and the incredible unconditional love for himself

J. Cameron (✉)
Trance Lady Productions, Inc., McLean, VA, USA

A. Pereira Jr, F. Sousa (eds.), *Principles for Governance*, Sustainable Development Goals Series,
https://doi.org/10.1007/978-3-031-40025-4_13

and all humanity—a coherence of Oneness. He sees his heart rate variability, HRV, on the screen above and realizes he is in the healing zone. After the session, scientific tests confirm no cancer in the patient's body.

This essay discusses some practical aspects and offers ideas of how we can produce good will and benevolence in our relationships to prevent escalation of differences and promote compassionate outcomes. As human consciousness rises, it brings opportunities to find new possibilities for quality of life. How do we, as humans, live in a peaceful world where basic needs of food, shelter, and healthcare are taken care of? Collaboration and cooperation are the keys. That tells me that we must focus on what we have in common. Inventions continually make our lives easier as we learn new ways to grow healthy foods, heal our bodies, heat and cool our homes, and cooperate with each other through the miracles of modern communication.

Human nature has come a long way. Remember it was not long ago that on weekends we watched our friends and neighbors getting beheaded. You may have even packed a picnic! I make the argument that consciousness is rising and people are becoming more aware and self-observant. Eighty-five percent of the people believe we have one Soul. The founder of Buddhism, Gautama Buddha, tells us there is not a transcendent God to be worshipped, but an inward journey! (Muhammed 19). The four noble truths are the truth of suffering, the truth of the cause of suffering, the truth of the end of suffering, and the truth of the path that leads to the end of suffering. Happiness cannot be found in our belongings and accomplishments but rather by forsaking the ego for the path of enlightenment. Meditation, kindness, compassion, balance, simplicity of life, right action, and honesty are all part of the path. Inviting compassion for ourselves, compassion for others, and compassion for the planet, and a plan to increase Oneness will bring more Awakening.

2 Summary of Hypotheses

This chapter will make the following points:

1. Meditation is a feeling of benevolence (compassion) for self and others, Oneness, unconditional love, coherence, and connection to your Soul. Over 85% of people in the world believe in the Soul.
2. We can measure the physical electromagnetic effects of meditation by using an electroencephalograph (EEG). Different brain waves signal different levels of consciousness. Also, magnetic resonance imaging (MRI or brain scans) shows an increase in the activity of brain areas correlated with the meditation states. Finally, the "flow-FISH" test measures the telomere length in each cell within a patient's blood sample. Telomere length is related to the cell's health and subsequent reproduction abilityn
3. Our biological structure (DNA and its machinery of reading, transcription, etc.) is multidimensional. (When you sleep, you are experiencing multidimensionality.)
4. There is a shift in consciousness (awareness) happening on the planet. We are possibly moving to a higher dimension of consciousness, where honesty, collaboration, and cooperation are the norms.
5. We can use the tools learned from studying the future of AI in meditation to design public policies, which promote integrity, peace, and happiness for all humanity.

As human consciousness rises, it brings opportunities to find new possibilities in life. Meditation and similar practices in religious contexts are the key to unlocking these opportunities. Therefore, we suggest that consciousness-raising be directed toward peace and benevolence, rather than the current trends of misdirected (pseudo)religious behaviors, which have led to the spread of hate on social media, the iatrogenic conduct of governments in health crises, and mass death of innocent people in war.

3 Meditation

Meditation is a vital tool. In the early 1980s at the height of the Middle East war, the Maharishi of Lebanon hypothesized that if enough people collectively used transcendental meditation (TM) and radiated peace at the same time, the progress toward peace would be facilitated. What happened during that time possibly has proved his theory. It was recorded that the more people meditated in Jerusalem daily, the more desperate acts of terrorism measured by deaths and other cultural behavior dynamics such as burglaries and violent crimes decreased (Johnson et al., 1984).

Meditation is crucial for world peace. Studies such as the example in Jerusalem represent a way to measure the effectiveness of meditation with data, empowering people to have integrity and a better quality of life. Subsequent experiments have continued to statistically confirm, for some authors that 900 people radiating peace would mitigate the negative energy of 1 million. This has become known as "The 1% Effect" (World Peace Group, 1987).

These studies suggest that one's consciousness is directly connected to an underlying, universal field of consciousness. By collectively enlivening that universal field through meditation, such a group can increase integrity, benevolence, and kindness in society. Meditators with loving intentions transmute negative energy into positive energy.

In 2014, an MRI brain-scan study of participants in a Mindful Awareness group at Harvard, who meditated for 27 min daily, had increased brain gray matter density in the hippocampus. This is essential for learning and memory structures associated with self-awareness, compassion, and introspection. Meditators also correlated stress reduction data with decreased gray matter density in the amygdala, which plays a role in anxiety and stress (Johnson et al., 1977).

Our brains are ever-changing and neuroplastic. With intention and meditation, we can increase self-awareness, reduce stress, and feel a sense of well-being and superb quality of life. On the subject of happiness, it is widely understood that the main factor necessary to increase our joy and satisfaction is the quality of our relationships. To have healthy, loving relationships with others, we must first have self-compassion and belief in ourselves. So how do our neuroplastic brains change this energy into another form so we can feel it?

The first law of thermodynamics states that energy cannot be created or destroyed but can be transferred from place to place or transformed into different forms (Watt, 1774). This transmutation happens in your body's energy chakras (energy centers). Have you ever had the chills, and you know that your intuition, or maybe Spirit energy, is talking to you? Transmutation is a switching of energy out of the physical, which in this case is a synaptic disturbance through nerve cells, into a higher feeling dimension. You can sense this in your body as a bolt of pleasure, warmth, or tingling.

So let us talk about your Soul in connection with your DNA. How is it that, according to several religions, a loving God can understand millions of prayers/intentions at the same time? The answer is in your DNA system, which is increasingly understood as multidimensional. In 1962, James Watson, Frances Crick, and Maurice Wilkins jointly received the Nobel Prize in Physiology or Medicine for their 1953 determination of the structure of deoxyribonucleic acid, DNA (Crick et al., 1962). Using X-ray crystallography, they proposed a double helix (like a ladder) structure of four proteins and phosphates held together by hydrogen bonds. The sequence of the four proteins determines each gene. Then we move fast forward to 2003 when the human genome project was completed, generating the first sequence of the human genome, thereby opening the door for fantastic collaboration with medicine and subsequent improvement of human life (NGHRI, 2003).

Interestingly only a small percentage of the DNA is used to produce genes. The term "junk DNA" was used for a while as some scientists argued that this other DNA is functionless (Ohno,

1972). It is not junk DNA. It simply means we cannot see the quantum field of DNA with today's tools.

Human DNA has 23 chromosomes (Tijo & Levan, 1956). The number is a mystery, as all the workings of physics are a patterning at a 4 and 12 level. Did that resonate with you? We can only study DNA from our dimension—commonly called 3D or 4D (if you add the dimension of time.) A mystic of our time, Kryon, has indicated that we have 24 chromosomes (Kryon, 2022). The 24th is multidimensional. It is impossible to see your multidimensional chromosome with the tools we have now.

Quanta are packets/particles of a general type of energy that includes physical, informational, and psychic or spiritual submodalities (Pereira Jr et al., 2018). In 1905, Albert Einstein proposed the wave-particle theory of electromagnetic radiation. This theory states that electromagnetic energy is released in discrete packets of energy—called photons—acting like waves. It is interesting how the wave appears, according to the perspective of the observer. Light can be both a particle and a wave, according to how it is measured (Young, 1801). This is a perfect example of the multidimensional nature of DNA.

Recent research on DNA has shown that meditation increases the lengths of the telomeres on your chromosomes (Blackburn et al., 2009). It is as if your DNA has shoelaces with caps. The caps are protective, like the plastic ones on your shoelaces. They prevent the shoelaces or the DNA from unraveling and being damaged. Damaged cells cannot reproduce. Telomere length is related to aging in human beings. Lifestyle factors, including smoking, lack of physical activity, obesity, stress, and exposure to pollution, can potentially increase the rate of telomere shortening, cancer risk, and pace of aging. It is ideal to have long telomeres.

Let us talk about Oneness and compassion. The Hindus are associated with the term "Namaste." It is a greeting of respect, often used in yoga. It simply means "the God inside me recognizes the God inside you!" Buddhism has been described as not so much a religion but a way of life. That is pertinent in discussing how different societies can become honest, fair, and compassionate. Everybody can have the opportunity of life, liberty, and the pursuit of happiness. Every human being is born with free choice. Sadly, many do not know how grand and expansive they are. Their beliefs have conditioned them that they are less than magnificent. The ego mind magnifies this drama.

Buddhism is one of the largest and oldest cultural ways of embracing God in life. The Hindus and the indigenous peoples also hold the concept of a God inside us and accept the principle of Oneness. There is one God. As I mentioned earlier, 85% of the world believes in an eternal Soul. People all over the planet understand the magic of going within and connecting with their Souls. Inviting compassion for ourselves, compassion for others, compassion for the Earth, and a plan to increase Oneness will bring more Awakening.

The Mayans had no calendar or predictions after the precession of the equinox in 2012. Many other indigenous tribes looked and studied the stars and predicted that the world would end. Let me explain. The Earth was believed and measured to be on a 26,000-year wobble as it hurtled through space (Ptolemy, 2023). The wobble ended on December 12, 2012. Contrary to doomsday predictions, we are here in a new frontier, and many of us know we are Awakening to a higher, more compassionate human consciousness. It is a shift to a higher dimension of energy. It is like going from black and white to color as we revolve around the sun in a new space in orbit. Your consciousness is affected by its environment including the Earth's magnetic field, the heliosphere of the sun, gravity, and possibly other invisible forces.

Another popular belief is that the shift to a higher energy dimension is related to the shift in the Earth's magnetic grid(s). The position of Earth's magnetic north pole was first precisely located in 1831 by the Australian explorer, George Hubert Wilkins. Since then, it is gradually drifted north-northwest by more than 600 miles (1100 km). The brain also has an electromagnetic field, and many view consciousness as an electromagnetic phenomenon and a quantum field. The electromagnetic field may be the dominant energy in bodily functions through the synaptic system, whereas the quantum field is possibly more prevalent in brain understandings (cognitions).

Can mind turn thought into matter? A study was done in which a group of healers and prayers did three different experiments on three vials of human DNA:

1. The healers were asked to meditate and focus on winding and unwinding the double helix of their DNA.
2. In this experiment, the healers focused simply on feeling grateful.
3. In the last experiment, the healers focused on winding and unwinding the test tube DNA in their head. Simultaneously, the healers were imagining the successful outcome and feeling extreme gratitude. Presuming and imagining that the DNA was already unwound was a clear instruction. There was no change in the DNA for either experiment 1 or 2. It did not wind or unwind. However, the DNA in experiment 3 was 25% unwound. Individuals capable of generating high ratios of heart coherence, measured by heart variability rate (HRV), could alter DNA structure according to their intention. The healers had used the power of intention, the focused thought of the desired outcome, and the extreme feeling of gratitude to unwind the DNA (McCraty et al., 2003).

The evolution of meditation has been closely correlated with the evolution of communication. Artificial intelligence allows religious leaders, gurus, swamis, and yoga instructors to reach their flock in a new way. The Virtual Ashram is here. Meditators will be provided with every convenience to make their inner journey a time of connection and healing. Apps, the MUSE are just the beginning.

There are efforts to integrate meditation with other public health and social policies, such as workplace wellness programs, health equity initiatives, and climate change mitigation strategies. Identifying and tracking indicators for the adoption level of meditation in public policy can help to inform policy decisions and resource allocation, as well as monitor and evaluate the impact of different interventions over time. For example, policymakers can use data on meditation adoption levels to identify gaps and areas of need in public health and social services and to allocate funding and resources accordingly. They can also use indicators to assess the effectiveness and scalability of different interventions and to make evidence-based decisions on whether to expand, modify, or discontinue certain programs.

4 Exercises

Here are a few meditation exercises for you, dear reader, to practice as you increase your health and manifestation powers.

Exercise #1 to open up your imagination

- Sit comfortably in your chair. Breathe in and count 1, 2, 3, 4, 5. Notice how your body expands.
- Breathe out 5, 4, 3, 2, 1. Feel your body soften. Gently close your eyes. Continue breathing.
- Allow your breath to lead your mind, rather than your mind leading your breath. Notice the sensation of your breath and where you feel it in the body. Follow that rising and falling sensation, allowing thoughts to come and go. When you are ready, gently open your eyes.
- Stay inside. Notice how you feel. With your body softened, you are relaxed and ready for the next step.
- Remember this feeling. Lean into it. You want to teach your body to lean into this relaxation. If a disturbance arises in your thoughts, lean out to teach your body that disturbances are not wanted, and then lean in as you set intention and relax the body.
- Repeat this exercise and use it when you have a disturbing thought.

Exercise #2 to relax your mind

- Breathe. Focus on one word on this page, and one word only. Stare at it. As you stare at it, do your eyelids have a tendency to want to blink? Continue staring, and now relax your eyes.
- Breathe. Every time you read the word "blink." Blink your eyes: BLINK, BLINK, BLINK, BLINK, BLINK, faster, BLINK, BLINK, BLINK. Now relax your eyes. At some level, which you may be beginning to be aware of, you can totally relax.

Exercise #3 to "Empty Out" your thoughts

Breathe. Become aware of your surroundings. Look at the walls and notice the pictures and colors. Also notice the enormous amount of space. Most rooms are predominantly space. Fixate your eyes on my words. Notice the space between the words, and stay inside your body. As you do this, can you imagine that your mind is like a container? Empty out that container, your thoughts, your personhood, your to-do list, your story, your roles, any judgment, or reaction. Repeat to yourself—"Empty Out. Empty Out."

Feel the alert stillness as awareness rises. Feel it. When you empty the container, your body, what you are left with is yourself. It is an IS-NESS—pure awareness, pure consciousness. Acknowledge it. This is your essence. Acknowledge any thoughts that might arise, but do not engage. Do not log into them. You are the sky, and the clouds are passing thoughts. Hold this space. Ask yourself the following six questions as you observe this moment:

Does this space have a beginning or an end?
Can this space be manipulated?
Does this space have a north, south, east, or west?
Does this space have motion?
Does this space ever judge you?
Is this space yourself?

Breathe in again. Open your heart doors. Feel an upward rush of pleasure fill your brain. Revel in it. It is the connection with your Source. Some people call this God Within.

Personally, the ability to Empty Out thoughts has brought great peace and is pivotal in my satisfaction with meditation and the joy of spiritual growth. If you meditate long enough, you get to a place of no thought. Empty Out speeds this process. You will be able to hold yourself in a meditative state in the awake state of mind. It is a backdrop on the stage of life. It is a constant reminder that you are loved, never alone, and choose to make every moment better in your life.

5 From Personal to Social Consciousness

In the reality of social practices, consciousness is the awareness of the Self and the Other. The word "Other" stands here for the reality that is not contained in the Self, encompassing not only other persons but the world the Self inhabits.

Considering the mirroring of the inner and outer worlds (Velmans, 2021), when a person knows herself, she also knows about the reality of which her Self is part of. This awareness is an important component of human decision-making and can be improved by means of spiritual and holistic practices. A governor who conceives herself as in a position that is in conflict with the rest of the world has a tendency of not being benevolent with others, but, on the contrary, has dispositions toward triggering mortal wars and destructive drives in the population.

A peaceful mentality, essential to constructive governance in times of conflict, can be conceived as the harmony between oneself and the world, leading to a disposition toward benevolence in social practices. This type of disposition can be seen in historical figures as Mahatma Gandhi displaying social abilities to actively work for the construction of peace, within and among countries.

6 Concluding Remarks

As human consciousness rises, it brings opportunities to find new possibilities in life. Meditation and similar practices in both religious and holistic contexts are the key to unlocking these opportunities. Therefore, we suggest that the rise of consciousness should be directed to peace and benevolence, instead of current trends in misdirected (pseudo)religious behaviors, which have led to the propagation of hate in social networks, the iatrogenic conduct of governments in health crises, and massive death of innocent people in war.

References

Blackburn, E. H., et al. (2009). *How chromosomes are protected by telomeres and the enzyme telomerase*. Available at https://www.nobelprize.org/prizes/medicine/2009/press-release/

Crick, F., et al. (1962). Available at: https://www.science-history.org/historical-profile/james-watson-francis-crick-maurice-wilkins-and-rosalind-franklin

Johnson, O., et al's. (1977). *Scientific research on the Transcendental Meditation program: Collected papers* (Vol. 1). MERU Press. Available at: http://www.feelguide.com/2014/11/19/harvard-unveils-mri-study-proving-meditation-literally-rebuilds-the-brains-gray-matter-in-8-weeks/

Johnson, O., et al's. (1984). *The collected papers of Scientific Research on Maharishi's Transcendental Meditation and TM-Sidhi Programme Volume 4 Paper 337*. Available at: https://www.worldpeacegroup.org/lebanon_peace_project_research.html

Kryon. (2022). Available at https://www.bitchute.com/video/3rOHIs3USzHw/

McCraty, R., et al. (2003). *Modulation of DNA conformation by heart-focused intention* (Publication No. 03-008). HeartMath Research Center, Institute of HeartMath. Available at https://www.aipro.info/drive/File/224.pdf

Muhammad 19., The pillars of La Illaha Illa Allah (*There is no God to be worshipped*). Available at https://knowingallah.com/en/articles/the-pillars-of-la-illaha-illa-allah-there-is-no-god-to-be-worshipped-but-allah/

NGHRI. (2003). *National Human Genome Research Institute generate the first gapless sequence of the human genome*. Available at https://www.genome.gov/news/news-release/researchers-generate-the-first-complete-gapless-sequence-of-a-human-genome

Ohno, S. (1972). So much "junk" DNA in our genome. *Brookhaven Symposia in Biology, 23*, 366–370.

Pereira, A., Jr, et al. (2018). Consciousness and cosmos building an ontological framework. *Journal of Consciousness Studies, 25*(3–4), 181–205. Imprint Academic. Available at: http://hdl.handle.net/11449/160217

Ptolemy, C. (2023, January 13). *Encyclopedia Britannica*. https://www.britannica.com/biography/Ptolemy. Accessed 9 Apr 2023.

Tijo, J., & Levan. (1956). The chromosome number of man. *Hereditas, 42*(1–2), 1–6.

Velmans, M. (2021). *Is the universe conscious?* Reflexive monism and the ground of being. In E. Kelly & P. Marshall (Eds.), *Consciousness unbound* (pp. 175–228). Rowman and Littlefield. Available at: https://www.researchgate.net/publication/351308047_Is_the_Universe_Conscious_Reflexive_Monism_and_the_Ground_of_Being

Watt, J. (1774). *Patent*. Available at https://www.livescience.com/50881-first-law-thermodynamics.html

World Peace Group. (1987). *The Basinka study*. Available at https://www.worldpeacegroup.org/baskinta_peace_project.html

Young, T. (1801). Double-slit science: *How light can be both a particle and a wave*. Available at https://www.scientificamerican.com/article/bring-science-home-light-wave-particle/. 2013.

Joanna Cameron, British-born, graduated in Zoology (St. Andrews University) and came to the USA in the 1970s to do scientific research on mitochondrial DNA at the University of Utah. She has operated for 44 years with history and cultural tours in China and the USA. As a hypnotherapist and a stage hypnotist, she has taught hypnosis worldwide and participated on both America's and Britain's Got Talent. She produced a documentary called DrainThatPain (2017), a holistic approach to chronic pain elimination. She has published 16 books on hypnosis, meditation, and other healing modalities. She is an audiobook producer for Amazon and produced and narrated an additional 14 books. She resides with her husband in McLean, VA. Site: www.joannacameron.com

Part V

Education and Social Service for Human Development

Well-Being and Well-Living: New Directions in Social Assistance Based in Psychosocial Esteem

14

Maria Candida Soares Del-Masso and Marta Bartira Meirelles-Santos

Abstract

The purpose of this chapter is to analyze the conceptions of well-being and well-living and good living in the current social context and their actions in public policies, especially in the field of national policy of social assistance with the System Social Assistance System (SUAS), carried out in Brazil. Faced with the current crisis, we place special emphasis on initiatives aimed at possible solutions to minimize social inequalities exacerbated by the current environmental, health, economic, and social crisis resulting from the COVID-19 pandemic. From this perspective, we present a case study associated with the protection of women's rights. This is the Special Coordination of Policies for Women linked to the Social Assistance Secretariat of the Municipality of Valinhos, State of São Paulo. This service's mission is to articulate municipal public policies for women to face the various types of violence in the family, as well as guaranteed women's rights by reducing vulnerability and social and economic inequalities. For this emancipatory study, we discussed the category of psychosocial esteem – taking care of oneself and others in the individual, group, institutional, and community spheres – in the practice of this service. In all these areas of care, there are conflicting and resolving links between people and institutions. This was demonstrated in three examples: (a) first diagnosis on violence against women and the proposal for a combat plan, (b) women's income generation group, and (c) the case of Ceci. The results of the analysis of these actions indicate that the expansion of personal, group, institutional, and community awareness of the construction of communication processes (problematic, welcoming, and democratic) can generate various links both for confronting domestic violence and for reducing economic inequality and social gender even partially.

1 Introduction

The purpose of this text is to analyze the conceptions of well-being and well-living in the current social context and their actions in public policies in the face of the current crisis, with special emphasis on initiatives that aim at possible solutions to minimize inequalities potentialized by the current environmental, health, economic, and

M. C. Soares Del-Masso (✉)
São Paulo State University (UNESP), São Paulo, Brazil
e-mail: del.masso@unesp.br

M. B. Meirelles-Santos
Psychiatry Department, Graduation - University of Campinas (Unicamp), Campinas, SP, Brazil
e-mail: mb@martabartira.com.br

A. Pereira Jr, F. Sousa (eds.), *Principles for Governance*, Sustainable Development Goals Series,
https://doi.org/10.1007/978-3-031-40025-4_14

social crisis resulting from the SARS-CoV-2 virus pandemic, known as COVID-19, decreed by the World Health Organization (WHO) in the early 2000s. The current world cenario unfolded, even in rich countries conditions never seen since the Spanish influenza.

The Spanish flu, also known as the 1918 flu, was a vast and deadly influenza virus pandemic. From January 1918 to December 1920, it infected an estimated 500 million people, about a quarter of the world's population at that time. As cited by Neto and collaborators (2022):

> 102 years ago, Brazil was hit by the Spanish flu pandemic in the context after World War I. At the time, it is known that the scourge entered the country on the Demerara ship, departing from Liverpool to Recife. Due to many days of travel, several deaths occurred. With the arrival of the transport to the port, a flag that signaled the quarantine was hoisted, but the Brazilian medical authorities thought that the conduct was premature, as it was a simple flu. From then on, the disease spread out and devastated the rest of the country. In this perspective, the context was disorganized in the political, social, economic and health field with the closure/paralysis of ports, transports and various other services. (Neto et al., 2022, p. 2)

Suggesting mere coincidence, Brazil did not follow the lessons experienced by the devastating and deadly Spanish flu and reproduced, in the midst of the twenty-first century errors, misconceptions, and wrong news about the current COVID-19 pandemic. Reinforcing our point of view, Neto et al. (2022) state:

> This pandemic scenario in the old days reminds the present time, with the new disease caused by a new type of coronavirus – COVID-19, which leads to the question: what are the lessons left by the Spanish flu for the current COVID-19 pandemic? […] it may even seem redundant, but it is necessary to recall the historiography of the great epidemics and remember that they examine the epidemiological ruptures, which shook the social, economic, cultural, political, and demographic structures. (Neto et al., 2022, p. 2)

> The Spanish flu, which dates to 1918, originated in the United States of America in the post-war period, but received this name because Spain was the first country to report that several soldiers had to leave the front after showing viral symptoms, in addition to registration of deaths due to clinical pneumonia. Thus, there was the false impression that Spain was the most affected country, or that the flu would have originated there. (Neto et al., 2022, p. 3)

The current scenario demonstrates the ills we still face in a growing scenario of the COVID-19 pandemic and its variants that every moment brings heartbreaking news of contamination, deaths, and the cult of nonscience and denialism, which is more worrying with such development sociocultural and informational. What leads individuals to break with history? To denialism of science? What do we have on the scene that the eyes don't see? Are they just economic, social, political issues? These are extremely complex topics that we would not be able to delve into each one of them. Thus, in this text, we will address issues involving social well-being and possible necessary public policies so that society can overcome this challenge of inequality with women in situations of social vulnerability. We know that there is much to do, to change, but we intend to reflect on the extent to which social assistance, at first and later social work and social psychology at the level of human promotion, could subsidize and guide public policies aimed at minimizing social, economic, health, and survival inequalities aiming the protection of women's rights.

2 Well-Being and Well-Living in the Current Social Context

The terms well-being and quality of life are understood differently in different areas of knowledge. Some areas, however, consider the terms as synonymous expressions. The terms are also associated with the conception of inequality, which in our view is the opposite of well-being and which generates social exclusion.

According to Neri (2015, p. 268), inequality is:

> a relational concept that does not consider the individual taken in isolation, but rather a property of the relationship between individuals. It differs from the concept of well-being, a dimension that can initially be observed individually, or through the aggregation of a group of people, when talking

> about social well-being. Bearing this in mind, it is possible to define inequality as a property of the social welfare function, insofar as, to measure it, we move from measuring the levels of well-being of each person to calculating the welfare of the whole of people who make up a society.
> [...] the different dimensions of analysis at play in considering well-being and inequality. First, objective measures of well-being are distinct from subjective measures. In the first case, we deal with tangible variables such as income, consumption and level of education. In the second case, we have measures based on people's perceptions of these or other topics. As aspirations and value judgments vary between individuals, and even for the same individual over time, in adapting to new life situations, subjective measures tend to be imprecise, even though they can be important.

Conceptions of well-being and inequality are dynamic, procedural, and often self-organized through new variables that are imposed at every moment in different social contexts. From this perspective, objective individual and/or collective actions are measurable and can receive or suffer actions – positive or negative, from individuals and/or political-social groups. However, we observe that many subjective actions end up impacting and generalizing objective actions, which affect people and groups, leading them to the difficulty of surviving with dignity. Thus, we return to the example cited by Neto and collaborators in the comparison between the Spanish flu and the current COVID-19 epidemic, when they refer that:

> The epidemic moment, imposed by the Spanish flu, was one of abrupt ruptures, imposing feelings and emotions permeated with an effect of sadness. This reinforced the arguments of dismantling the situation, that is, a reflection of the whole by the part in people's lives. To give you an idea, we know that many deaths occurred in the 20–45 age group, especially among those with heart, kidney, and tuberculosis diseases, also including those who lived in conditions of social and biological deprivation. (Neto et al., 2022, p. 4)

In the current sociocultural scenario impacted by the devastating effect of COVID-19, we also identified that the most vulnerable suffer the most. This fact is aggravated by the effect of governmental actions without scientific bases being imposed and generalized as absolute truths. Both the severity of contamination by the contemporary pandemic virus and official denialist actions affected and disrupted individuals, families, and social groups. These people, many times, fight only for survival, and most of the times do not have the support of public policies that aim at well-being and well-living.

How to deal with these impacts? What does the current scenario show us? What else is needed for effective government actions to take place and minimize the installed social crisis? One of the lessons left by the Spanish flu pandemic and which can be observed in times of COVID-19 is the possibility of thinking about changes at the Federal, State, and Municipal levels.

The current scenario of disinformation persists, often based on fake news, added to the anti-vaccination movement that ends up making it difficult to extinguish the disease in the world context. What else is needed in the twenty-first century with full access to information, culture, and education? Socioeconomic and cultural inequality grows visibly, generating a complex situation of social exclusion with increasing unemployment, domestic, and social violence generated by the close coexistence of people and family groups in tiny spaces. As de Souza et al. (2023) argue, the situation resulting from COVID-19 affected different contexts, municipal, state, national, and global, which underwent changes and needed to be reorganized. Faced with this new scenario, social facilities had to reorganize themselves in their social, professional, educational, and health dimensions. Thus, as the authors mention, they were as follows:

> [...] different impacts making us look for viable alternatives so that we could, as far as possible, maintain and move forward with our actions, both professional and personal.
> In the professional aspect, there were significant changes that forced us to use the different technologies available at the time to align our actions and achieve our goals in maintaining a reasonably geared professional life together with our students, professors, co-workers, and family members. The school had to adapt to the new reality to continue to serve the school community and respect health protocols. The use of digital technologies to support the teaching and learning process has become essential in the new scenario. The teacher, in a short period of time, needed to readjust his profes-

> sional work to the new reality to continue the pedagogical activities to minimize the losses that occurred in this context. (de Souza et al., 2023, p. 83)

As well as schools, other social facilities such as social assistance, health, etc., must reorganize their actions, using different technologies and tools to maintain, at a minimum, good living in society, even if synchronously and remotely. Among the most used tools is WhatsApp. This application allows people to use it to perform a multitude of tasks, among which "we highlight SMS messages, voice messages, sharing videos and documents, making video calls, among others" (de Souza et al., 2023, p. 88). Despite the intention of the application to be at the service of society in a positive way, it has been used to spread disinformation and fake news, which makes it difficult to act and have real and true knowledge of what is happening in society at a national and international level. For these actions to have a reliable basis, the information policy must be changed, which suggests overlapping with current public policies. Considering this perspective, how are public policies elaborated and possible based on well-being and good living?

The good living, cited by Acosta (2016, p. 27):

> [...] bets on a different future, which will not be conquered with radical speeches lacking proposals. It is necessary to build relations of production, exchange and cooperation that provide sufficiency – more than just efficiency – sustained by solidarity.
>
> People must organize to recover and take control of their lives. However, it is no longer just about defending the workforce and recovering free time for workers – that is, it is not just about opposing the exploitation of labor. Also at stake is the defense of life against anthropocentric schemes of productive organization, which cause the destruction of the planet.

The author completed that (Acosta, 2016, p. 27):

> Good Living – as a philosophy of life – is a liberating and tolerant project, without prejudice or dogmas. A project that, by having added countless stories of struggle, resistance, and proposals for change, and by nourishing itself with existing experiences in many parts of the planet, stands as a starting point for democratically building democratic societies.

In that, we bring to reflection in the Brazilian context anchored in the public policy of social assistance and how it enables and acts in the perspective of well-being and good living.

3 Possible and Necessary Public Policies: A Municipal Case Study on Social Protection for Women

The social assistance is a citizen's right and a duty of the State, established by the Federal Constitution of 1988 (Brasil, 1988). From 1993, with the publication of the Organic Law of Social Assistance[1] (Brasil, 1993), the Social Security Policy is created, composing the Social Security tripod, that is, Social Assistance along with Health and Social Security. Social Assistance, unlike Social Security, is not contributory, that is, it must serve all citizens who need it. It is carried out based on integrated actions between the public, private, and civil society sectors. Its objective is to guarantee social protection for the family, childhood, adolescence, old age, and support for underprivileged children and adolescents. Added to these actions is the promotion of integration into the labor market, rehabilitation, and promotion of integration into the community for people with disabilities, including the payment of benefits to the elderly and people with disabilities.

In 2005, the Unified Social Assistance System[2] – SUAS (Brasil, 2005) – is established, decentralized, and participatory, whose function is to manage the specific content of Social Assistance in the field of Brazilian social protection. In this way, it consolidates the shared management method, co-financing, and technical cooperation between the three federative entities that, in an articulated and complementary way,

[1] In Portuguese language, LOAS means Lei Orgânica da Assistência Social.

[2] In Portuguese language, SUAS means Sistema Único de Assistência Social.

operate the noncontributory social protection of social security in the field of social assistance. On July 6, 2011, Law 12,435 (Brasil, 2011) was enacted, guaranteeing the continuity of the SUAS. The system organizes social assistance actions into two types of social protection. The first is Basic Social Protection, carried out at the Social Assistance Reference Center[3] – CRAS, aimed at preventing social and personal risks, by offering programs, projects, services, and benefits to individuals and families in situations of social vulnerability. The second is Special Social Protection, for example, CREAS Specialized Reference Center for Social Assistance, aimed at families and individuals who are already at risk and who have had their rights violated due to abandonment, mistreatment, sexual abuse, and drug use, among other aspects.

The SUAS also offer the provision of Assistance Benefits, provided to specific audiences in conjunction with other public services such as, for example, the Basic Health Unit[4] – SUS, schools, etc., contributing to overcoming situations of vulnerability. It also manages the linkage of social assistance entities and organizations to the system, keeping the National Register of Social Assistance Entities and Organizations up to date and granting certification to charitable entities, when applicable. The management of SUAS actions and the application of resources are negotiated and agreed upon in the Bipartite Intermanagement Commissions[5] (CIBs) and in the Tripartite Intermanagement Commission[6] (CIT). These procedures are monitored and approved by the National Council for Social Assistance[7] (CNAS) and their local peers (State and Municipal Councils), who carry out social control.

About the protection of women's rights, we bring a case study of the Special Coordination of Policies for Women[8] (CEPM) linked to the Social Assistance Secretariat of the Municipality of Valinhos, State of São Paulo. This service, formalized by Municipal Law No. 3861, of 12/29/2004 (Valinhos, 2004), has as one of its objectives to articulate the Women's Protection Network in that city. That is, this institution seeks to integrate the various public policies (social assistance, health, education, public safety, economic development, culture, sports, etc.) to mitigate, as a priority, the number of citizens who deal with different types of family violence, in addition to curb the various gender inequalities.

After years of inactivity, this institution was restructured in 2019. That year, the team consisted of the coordination, usually a politically appointed person, an administrative professional, and a technician. We called this trio a core group whose mission was to carry out the integration of public policies to face domestic violence at first. Since 2019, the technician, the second author of this chapter, was a civil servant with a background in social psychology and dance. His method of working with groups, critical and complex, in an institution is called Operative Play Group[9] (GOL) (Laban, 1990; Meirelles-Santos, 2018, 2021a, b; Valinhos, 2019b). Its role was to encourage coordination in carrying out CEPM's mission.

The representation of the various municipal departments was also part of the CEPM. This group, which we call extended, met once a month. Its task was also to promote the institution's mission, to resolve economic and social inequalities in the municipality, especially for women. To understand this unjust social phenomenon of inequality, as previously presented, we also bring up the critical proposal to the capitalist society – in which we live – by the sociologist Boaventura de Souza Santos (Santos, 2020).

[3]In Portuguese language, CREAS means Centro de Referência Especializado em Assistência Social.

[4]In Portuguese language, SUS means Sistema Único de Saúde.

[5]In Portuguese language, CIB means Comissões Intergestores Bipartite.

[6]In Portuguese language, CIT means Comissões Intergestores Tripartite.

[7]In Portuguese language, CNAS means Conselho Nacional de Assistência Social.

[8]In Portuguese language, CEPM means Coordenadoria Especial de Políticas para Mulheres.

[9]In Portuguese language, GOL means Grupo Operativo Lúdico.

The author points out that we built a predominantly capitalist, colonialist, and patriarchal society. It is important to highlight that these three characteristics occur simultaneously and structure an unequal economy and a violent culture among people. In a recent e-book called "The Cruel Pedagogy of Virus" (Santos, 2020) about COVID-19, the sociologist comments that for 40 years, since the 1980s, the world has been dealing with recurrent crises. This occurs due to a dominant mode of the capitalist system called neoliberalism, characterized more and more by subjection to the logic of the financial sector. Added to this, the permanent crises are not to be resolved because the crisis in the dominant capitalist logic both legitimizes the enormous and degrading concentration of wealth and boycotts measures to prevent the imminent ecological catastrophe, in addition to this crisis being used to justify cuts in social policies (health, education, social security) or in the degradation of wages.

In addition to capitalism, regulated by the dominance of the logic of capital, we also deal with colonialism in society. Even though it may seem to have finished with the independence of Latin America and Africa colonies, as an example. But, as a matter of fact, "neocolonialism, imperialism, dependency, racism, etc" continued to happen (Santos, 2020, p. 12). Colonialism can have both a material and a subjective dimension. The first can be characterized by the aspect of Latin American extractive economies. The second refers to a low valuation of our Latin American identity, that is, a low psychosocial esteem, a theme that will be deepened in this chapter. Still on colonialism, a form of relationship submissive to the dominant Eurocentric and North American logic, we highlight the text on good living, already presented in this work, by the former Minister of Economy of Ecuador, Acosta (2016). He, like other authors, comments that the understanding of the meaning of good living:

> [...] it cannot simply be associated with 'Western well-being', it is necessary to start by recovering the cosmovision of indigenous peoples and nationalities. This recognition does not fully mean denying society's own modernization, incorporating many valuable technological advances into the logic of 'Buen Vivir'. Nor are important contributions of human thought being marginalized, which are in tune with the construction of a harmonious world, as derived from the philosophy of 'Buen Vivir'. For this reason, one of the fundamental tasks lies in the permanent and constructive dialogue of ancestral knowledge and knowledge with the most advanced part of universal thought, in a process of continuous decolonization of society. (Acosta, 2016, p.201)

In addition to criticizing colonialism and neoliberalism, Santos (2020) also describes our society as patriarchal. In his words, the author says that "[...] patriarchy induces the idea of being dying or weakened due to the significant victories of feminist movements in recent decades, but, in fact, domestic violence, sexist discrimination and feminicide they do not cease to increase" (Santos, 2020, p.12). This type of violence, with great influence from the dominant patriarchal culture, is defined in Art. 5 of Law Maria da Penha n° 11.340 (Brasil, 2006), of August 7, 2006, pointing out that "Domestic and family violence against women is defined as any action or omission based on gender that causes death, injury, physical, sexual or psychological and moral or patrimonial damage."

This problem, which affects not only women but all components of many families, has been answered by CEPM, even though this institution has faced many obstacles. In particular, the coordination had to deal with behaviors associated with recurrent cultural and institutional patriarchy over the 18 years of its existence. That is, in this process of building its identity, we identified that this institution either fulfilled its functionality (2009, 2012, 2019) or was dampened. Sometimes it had structure; sometimes it was unstructured. For example, we identified that, in the social environment, institutions such as the legislative branch, a civil society institution and the Public Prosecutor's Office pressured the Municipality of Valinhos so that CEPM was recognized and carried out its emancipatory mission. We can say that there was a dialectic between the institutions, that is, CEPM was included and excluded. In other words, there were social institutions – such as those mentioned – that ensured that this institution for confronting gender inequalities interfered in the present patriarchy, both in public institutions and in the Valinhos community in general.

This fact, among others learned in 20 years of work with GOL groups in some municipal public services (Valinhos, 2019a, b; Meirelles-Santos, 2021a, b; Del-Masso & Meirelles-Santos, 2022), encouraged us to conceive the notion of psychosocial esteem. In other words, based on critical social psychology (Lane & Codo, 2001), Antonio da Costa Ciampa's theory of identity, PUC/SP (Ciampa, 2009; Amleida, 2017; Furlan, 2020), and our praxis complex and interdisciplinary work with groups, Operative Play Group (GOL), in an institution, we present the characterization of the notion of psychosocial esteem. This socio-emotional skill, complementary to the themes of self-knowledge and autonomy, is one of the four teaching-learned skills in this model. It is worth noting that in this way of working with groups, welcoming and problematizing dialogues, as taught by Freire (2005), are always encouraged and developed.

Esteem from the Latin comes from *aestimar*, that is, to estimate, to value. According to Pichon-Rivière (1907–1977) (Pichon, 2000), an Argentinean psychoanalyst and psychiatrist, it is not possible to think of psychology other than as a social psychology. That is, the internal world (desires, fantasies, thoughts, feelings, beliefs, etc.) of the subject is in a dialectical relationship with the external world, socioeconomic structure. Given this, we can understand that the term psychosocial can be understood as the interrelationship, inseparable, between subjectivity and social relations. Intuitively, psychosocial esteem is the ability people must value each other. It is the process of awareness in which "I value myself and I value you" and vice versa. And for this activity of giving value to materialize, care is necessary. Leonardo Boff, a Brazilian theologian with an immense body of work, teaches us that:

> The essence of the human being resides in care rather than in intelligence and freedom. This is because care is the prerequisite for these attributes to erupt in evolution. Without care, neither the human being nor any living being has its life guaranteed. Everything needs care to continue to exist. Care represents a loving relationship with reality. Where care for one another prevails, fear disappears, the secret origin of all aggression and all violence. (Boff, 2006, p.82)

It is not only necessary to develop intelligence, self-awareness, and awareness of others. It is essential that this awareness is materialized in a concrete activity of care. Taking care of needs (subjective, bodily, social, and material) to be met in everyday life. In this sense, when conceptualizing psychosocial esteem as an aspect of the identity process, we can propose that caring can be directed toward two dimensions. The first is associated with aspects of the body, mind, and the external world or three areas of human conduct as proposed by Pichon-Rivière (Pichon, 2000). Let us remember that the field of human behavior, delimited in time and space, must be understood in its totality and breadth.

The second dimension of care can be directed to the four areas of psychology (psychosocial, sociodynamic, institutional, and community). The psychosocial scope refers to the individual. The other areas are characterized by the dimensions "sociodynamics, which analyzes the group as a structure; and the institutional one that takes an entire group, an institution or an entire country as an object of investigation" (Pichon-Riviére, 2000, p.2). The community scope, evidently, associated with the community, it can be defined "as a group of people who live together, in the same place, and between which there are established certain links, certain functions in common or certain organization" (Bleger, 1984, p.83).

When we integrate these two dimensions into the notion of psychosocial esteem, we see that it is a socio-emotional skill, learned through group teaching-learning processes, to favor awareness of some care. Care is associated with the following aspects: (1) of the mind, (2) of the body, (3) of bonds, (4) of material resources, (5) of financial resources, and (6) of physical nature, and bonds can be cared for in the areas psychosocial, sociodynamic, institutional, and community. In this chapter, with a focus on reducing inequalities through actions of the Special Coordination of

Policies for Women, we will delve into the care of the bonds in these areas presented above. This is because we understand that if we increased awareness of these variables, associated with the notion of psychosocial esteem (taking care of oneself and others), we can favor the mission of CEPM. In all these areas of care (individual, group, institution, or community) – which are not mutually exclusive – there are links between people. We know that in these relationships' conflicts, impediments, violence, and possibilities of overcoming with different degrees of valuing oneself and the other are evident.

Below, we present how the category of psychosocial esteem, an integral part of the OPG group praxis, will dialogue with the description and analysis of fragments of three actions carried out in the period from 2020 to 2022 at CEPM. The actions were as follows: (1) "First Diagnosis, territorial and intersectoral, on Violence against Women and the Proposed Plan to Combat" (Valinhos, 2020), (2) group of women from the community (2022), and (3) the Case of Ceci (2022).

3.1 First Diagnosis, Territorial and Intersectoral, on Violence Against Women and the Proposal for a Combat Plan

The diagnosis and proposal to combat domestic violence was an old goal of CEPM. Unfortunately, combating this worrying phenomenon is a priority in the field of women's rights. This task was performed under the influence of concomitant factors. First, there was an increase in domestic violence due to COVID-19. On May 15, 2020, when the Ministry of Women, Family and Human Rights launched the campaign against the violation of women's rights, there was an increase of 35% of complaints, via telephone call to the number 180, in relation to the month of April (Brasil, 2020).

As for global data, the increase in gender violence, because of social isolation, has been proven in Spain, Italy, and France. In China, police records of this human rights violation have tripled (Vieira et al., 2020). The other factors that stimulated this work, to guide public policies for women in the city of Valinhos/SP, was both the political decision to carry out this survey – something unusual – and the technical competence of a group of municipal servants who were at the ahead of this action. These three variables are as follows: (1) increase in domestic violence due to COVID-19, (2) political decision, and (3) engagement of the group of servants demonstrate the three scopes of bonds: community, institutional, and sociodynamic. That is, for the making of this diagnosis, there was an awareness, in different areas, to take care of women.

Data were collected from five municipal public services, four from public health and one from social assistance, from January to June 2020. It was evident that (a) of the 90 women in the study, 71% suffered physical violence and/or psychological. These two types of violence were the most frequent. These data are in line with the survey by the Municipal Council for the Rights of Women in the city of Valinhos, and the results of the research showed that:

> The general data relating to the classification of violence have the highest percentage rate of injury/defamation and threat with 38.81%, followed by physical violence with 34.23% and psychological violence with 15.09%. These data demonstrate the cycle of violence, which begins with name calling and ends with physical aggression, which can lead to death. (Valinhos, 2019a, p.15)

The other result shows that (b) of the 47 women notified by the National System for Notification of Diseases (SINAN), data collected in the 3 municipal emergency rooms in Valinhos/SP, 64% made suicide attempts by performing self-inflicted violence (use of medication in their majority). These two main serious results show that the phenomenon of violence is also a public health problem, as mentioned by de Minayo and Souza (1993) and Reichenheim and collaborators (2011), as well as a public safety problem. In view of this information, it was proposed that the teams of the main public services, which assist women in situations of violence, hold meetings with some community leaders in the city to build an efficient FLOWCHART, as a group, for better intersectoral care for these women/families. The urgency of welcoming and treating women who attempted suicide was also reinforced. However, perhaps due to the phenomenon of the naturalization of violence, the management did not prioritize caring for the mental and physical health of these women. Finally, in this document, a request was made for municipal servants to carry out scientific research applied to women's rights without burdening the municipality, which was partially denied. It is important to emphasize that these proposals are in line with the proposal of this book insofar as it was also valued that the people participate constructively with the government in the definition of strategic objectives and their implementation, as well as the participation of public universities so that they support and redirect their efforts to work in synergy with local communities.

We emphasize that in carrying out this diagnostic task, essential for improving public policies, there was a harmony of interests and ties in different spheres (community, institutional, and sociodynamic/group) as already mentioned. But a scope has not yet been made explicit. It needed a servant with technical knowledge and political will to lead this action. That is, in the psychosocial field, a professional was needed to technically articulate all the other areas for the delivery of the final report. It was even identified as a hypothesis that this professional, due to the great effort undertaken, seemed to take good care of her mind, in relation to her body and the bonds she established. Or could it be that after leading this endeavor she had psychological or physical symptoms? Finally, it was evident, in this first example, how the notion of psychosocial esteem was applied, especially about the integration of ties in different areas to take care of coping, even if partial, with domestic violence.

3.2 Women's Income Generation Group in the Community

From our point of view, the "great gem" of the 2022 CEPM management, coordinated by a leader with a great community work in the city, was the implementation of the women's group, in the territory covered by the Social Assistance Reference Center (CRAS), located in Jardim das Figueiras. CRAS is the basic social protection service for families in psychosocial vulnerability linked to SUAS. The objective of this group, proposed in partnership by CEPM and CRAS, is to generate income through handicrafts and strengthen psychosocial esteem, which includes group interaction, knowledge of existing public services, and awareness of social rights. We would like to highlight in this second example that, to take care of women in the community, CEPM and CRAS servants need to take care of the bonds between themselves and between these two institutions. In other words, women servants take care of women in the community. For this, the servants of these two institutions need to take care of their bonds. We know that building partnerships is not easy and there are often communication difficulties. It seemed to us that the main difficulty experienced between these two institutions was associated with language differences.

Since the CEPM coordination presented a commonsense language associated with the work experience – volunteer – with women from the city, the CRAS coordination, on the other hand, presented a technical language in the field of psychology and social assistance applied to this service basic protection. This difference was made explicit, for example, in some conflicts when communicating with the women in the group, what to say, what not to say, how to say it, and how to link. Another situation that disturbed the communication between the institutions was the invitation that CRAS made to the supervisor of Guardian Maria da Penha to hold a conversation with the group of women without previously notifying the CEPM coordination, which was surprised by this initiative. Finally, we could see that more careful communication between these institutions would require more meetings for planning and evaluating both the progress of the women's group and the inter-institutional partnership.

In other words, we learned that innovating in the field of women's rights, which in this case presupposed the partnership of public services, it is necessary to take care both bonds between the servants of these institutions and with the community of women that receives this service. Incidentally, the participation of the women in the group in the planning and evaluation of their activities was not carried out. In this first moment of structuring this group, the women did not participate, along with the servants, in the democratic creation of this service. Finally, in this example, we highlight the importance of deepening the understanding of communications, critical and complex, conscious, and unconscious, that permeate the bonds in the various spheres, both inter-institutional and with the community of women who live in the territory covered by CRAS Figueiras.

3.3 The Ceci Case

Ceci is 37 years old, Caucasian, and a female head of the family with two children, one aged 8 and the other aged 6. She did not inform his schooling. She was a cleaning professional at CEPM.

One morning when the CEPM core group was talking about the violence that women suffer, Ceci, who was sitting listening to the conversation, reported on the violence that she suffered on a psychological, physical, and patrimonial level. In her report, Ceci highlighted that at the beginning of March 2022, her former partner, named Peri, broke into her home and stole her cell phone. He said he would reset the phone. He also said that even if she called the police, she would have no way of proving that she owned the device. According to Ceci, in that episode, the citizen pushed her and threatened to kill her and her 8-year-old son. This child was also pushed and witnessed both the physical and psychological aggressions and the theft of the mother's cell phone. Ceci called the Military Police (PM) to report the incident. As Peri lives close to Ceci's house, the PM went to find out if the cell phone was in that man's possession. Peri denied the theft to the officer. As Ceci did not have the invoice for the cell phone, the PM ended the investigation. It is important to highlight that Ceci has a work contract which includes a discount on the value of the stolen cell phone. However, that document is with Peri.

From this report of multiple forms of violence (psychological, physical, and patrimonial), CEPM carried out the following actions with Ceci: (1) preparation of a report on Ceci's psychosocial reception for referral to the Women's Defense Police Station (DDM), where the Bulletin is made of Occurrence; (2) monitoring, indirectly, the implementation of the Maria da Penha Law by notifying the aggressor to go to the DDM; and (3) dialogue with the Guardian Maria da Penha (GMP). This service, linked to the municipal public security policy of the city of Valinhos, aims to visit women regularly so that the aggressor does not disturb their daily lives, as determined by justice and (4) referral of Ceci to the Social Assistance Reference Center – CRAS/SUAS for applicable actions.

In the perspective of psychosocial esteem, in this third example, we demonstrate the psychosocial scope of a woman who suffered domestic violence. This violence affected her mind to the extent that she suffered psychological violence. It affected her body because she suffered physical violence. And her bond with her son was disturbed as he too was threatened with death. In addition, she suffered because her material asset (cell phone) was stolen. If, on the one hand, it was left unprotected in these complex dimensions, on the other hand, it was cared for by the institutions CEPM (sociodynamic core group), DDM, and GMP at first. In a second moment, the CRAS can protect the family in a situation of psychosocial vulnerability. Finally, Ceci had her cell phone returned.

4 Where Are We Heading?

The lessons left by the Spanish flu epidemic in times of coronavirus have been diluted in political, social and care aspects. The possibility of changes/adjustments in the field of policies, especially in health, considering the knowledge of science, and the direct relationship between the economy and health, go against the feelings of fear, dread and horror, which are destabilizing in a health emergency. Public health. (Neto et al., 2022, p. 7)

Despite the great difficulties encountered in distressing times of the COVID-19 pandemic, we can observe that there is vitality in some responses to people's needs, especially in the field of women's rights as presented in the case of CEPM. We seek to demonstrate in the three examples that the notion of psychosocial esteem, as a goal of human development, can partially favor the minimization of family violence. This notion is a socio-emotional skill, learned through group teaching-learning processes, to encourage awareness of some care. Care is associated with the following aspects: (1) mind, (2) body, (3) bonds, (4) material resources, (5) financial resources, and (6) physical nature, and bonds can be cared for in areas psychosocial, sociodynamic, institutional, and community.

Regarding the action of the first diagnosis/proposal to face domestic violence, we identified, on the one hand, that this process and result, carried out by a group of servants and managers, was innovative. On the other hand, we were disappointed with the low resolution of the information collected, which is not uncommon in the municipal public service in our city. Until today, we hope that the integration of services that assist these women will be effective. Is it the women (victims, servants, and managers) who naturalize violence, and that's why they don't unite? Is there not only an institutionalized patriarchy but also internalized in all of us? Is it not possible to sensitize the Municipal Mayor of the city so that she prioritizes tackling domestic violence that mainly affects women, but also all family members (children, teenagers, the elderly, the disabled, etc.)?

These questions, if answered, could open new possible paths. Perhaps personal, group, institutional and community awareness will be expanded to build communication, problematizing, and welcoming processes, which can generate links (e.g., in different institutional spheres) to reduce the number of abused women in the city. In example two, a group of women in the community, we identified the challenge of building good understanding between people and institutions (CEPM and CRAS), when there are different languages, visions of the world and of human beings. It seems to us that if this communication is well aligned, it can be strategic both for the development of the psychosocial esteem of the servants and of the women in the community. The awareness that women care for women can encourage attentive and loving care in intrapersonal, interpersonal, and institutional relationships, for example. In this type of psychosocial conduct, all women win! In example three of the woman Ceci, we witness different communicational processes, personal and institutional, very resolute, which generated full protection for a victim of domestic violence. That is, this report was emblematic because it showed a flowchart of adequate public services for overcoming the situation of individual violence. In addition, the perpetrator was blamed. He had to attend the DDM to answer for the property violence committed and returned the mobile phone.

In this chapter, we walk in tune with both the concept of well-being and good living, as their perspectives seek to build democratic and, evidently, non-colonized societies. It is public services, strategic resources of this humanist social vision, which have the mission of reducing inequalities in general. Hence, we present a case study of an institution that aims to articulate municipal public policies to reduce inequalities of gender. The notion of psychosocial esteem, in dynamic and procedural construction, based on the praxis of the GOL group model, can be a resource that considers the uncertainties and surprises of human relationships, in various areas, to favor new emancipatory and therefore more egalitarian paths.

References

Acosta, A. (2016). *O bem viver: uma oportunidade para imaginar outros mundos*. Autonomia Literária, Elefante.

Almeida, J. (2017). Identidade e Emancipação. Psicol. Soc., Belo Horizonte, v. 29, e170998, Available from: http://www.scielo.br/scielo.php?script=sci_arttext&pid=S0102

Bleger, J. (1984). *Psico-Higiene e Psicologia Institucional*, Ed. Artmed, 2ª ed, Porto Alegre.

Boff, L. (2006). *A força da ternura-pensamentos para um mundo igualitário, solidário, pleno e amoroso*. Ed. Sextante.

Brasil. (1988). Presidência da República. Casa Civil. Subchefia para Assuntos Jurídicos. Constituição da República Federativa do Brasil. Disponível em https://www.planalto.gov.br/ccivil_03/constituicao/constituicao.htm. Acesso em 07 abr 2023.

Brasil. (1993). Presidência da República. Casa Civil. Subchefia para Assuntos Jurídicos. Lei N° 8.742, de 07 de dezembro de 1993. Dispõe sobre a organização da Assistência Social e dá outras providências. Disponível em http://www.planalto.gov.br/ccivil_03/leis/l8742.htm. Acesso em 07 abr 2023.

Brasil. (2005). Presidência da República. Sistema Único de Assistência Social – SUAS. Disponível em https://www.gov.br/cidadania/pt-br/acoes-e-programas/assistencia-social/gestao-do-suas. Acesso em 07 abr 2023.

Brasil. (2006). Presidência da República. Casa Civil. Subchefia para Assuntos Jurídicos. Lei N° 11.340, de 7 de agosto de 2006. Cria mecanismos para coibir a violência doméstica e familiar contra a mulher, nos termos do § 8° do art. 226 da Constituição Federal, da Convenção sobre a Eliminação de Todas as Formas de Discriminação contra as Mulheres e da Convenção Interamericana para Prevenir, Punir e Erradicar a Violência contra a Mulher; dispõe sobre a criação dos Juizados de Violência Doméstica e Familiar contra a Mulher; altera o Código de Processo Penal, o Código Penal e a Lei de Execução Penal; e dá outras providências. Disponível em http://www.planalto.gov.br/ccivil_03/_ato2004-2006/2006/lei/l11340.htm. Acesso em 07 abr 2023.

Brasil. (2011). Presidência da República. Casa Civil. Subchefia para Assuntos Jurídicos. Lei N° 12.435, de 06 de julho de 2011. Altera a Lei N° 8.742, de 07 de dezembro de 1993, que dispõe sobre a organização da Assistência Social. Disponível em http://www.planalto.gov.br/ccivil_03/_ato2011-2014/2011/lei/l12435.htm. Acesso em 07 abr 2023.

Brasil. (2020). Ministério da Mulher, da Família e dos Direitos Humanos, 2020. Governo Federal lança campanha contra violência doméstica. Disponível em https://www.gov.br/mdh/pt-br/assuntos/noticias/2020-2/maio/governo-federal-lanca-campanha-contra-violencia-domestica. acesso em: 18 mai. 2020.

Ciampa, A. (2009). A estória do Severino e a História da Severina: um ensaio de psicologia social, 11ª reimpressão, Ed. Brasiliense.

de Minayo, M. C. S., & Souza, E. R. (1993). Violência para todos. Cad. Saúde Pública, Rio de Janeiro, *9*(1), 65–78. Available from http://www.scielo.br/scielo.php?script=sci_arttext&pid=S0102-311X1993000100007&lng=en&nrm=iso. Access on 02 Sept. 2020. https://doi.org/10.1590/S0102-311X1993000100007

de Souza, R. R., De-Paula, E. T., Del-Masso, M. C. S. (2023). Whatsapp Como Ferramenta de Mediação da Aprendizagem: estudo de caso no período da pandemia. In: L.R.R. Carreiro, M.V. de Araújo, E.F.A. de Prado, E.B. Bissoli, *Vivências na pandemia da covid-19 e aprendizados sobre família e escola*. CRV, p. 83–95.

Del-Masso, M. C. S., & Meirelles-Santos, M. B.(2022). *Wellbeing, well-living: New directions in social assistance*. Trabalho apresentando no Seminário do Grupo de Pesquisa Desigualdade Social e Econômica/CLE - UNICAMP.

Freire, P. (2005). *Pedagogia do oprimido*. 44 ed. Paz e Terra.

Furlan, V. (2020). Biopolítica, Reconhecimento e Identidade. Ed CRV. Curitiba.

Laban, R. (1990). *Dança educativa moderna*. Ícone.

Lane, S., & Codo, W. (Eds.). (2001). *Psicologia social: o homem em movimento* (3a reimpressão). Editora Brasiliense.

Meirilles-Santos, M. B. (2018). Sistema GOL (Grupo Operativo Lúdico) aplicado em ludoterapia de grupo com crianças: uma construção interdisciplinar com resultados, Trabalho apresentando no Seminário do Grupo Interdisciplinar- CLE Auto-Organização- UNICAMP.

Meirilles-Santos, M. B. (2021a). Estima psicossocial minimiza violências domésticas? Trabalho apresentando no Seminário do Grupo de Pesquisa Desigualdade Econômica e Social /CLE – UNICAMP.

Meirilles-Santos, M. B. (2021b). Lutas e (mu) danças com mulheres na pandemia de COVID-19: uma experiência no município de Valinhos/SP. In C. P. Alves, S. C. Miranda, D. Portugueis, & C. R. S. Santos (Eds.), *Identidade, Metamorfose e Emancipação diante da COVID-19*. Ed. Amavisse.

Neri, M. (2015). Desigualdade. In G. Di Giovanni & M. A. Nogueira (Eds.), *Dicionário de Políticas Públicas* (2nd ed., pp. 268–270). Editora Unesp; Fundap.

Neto, M., et al.(2022). Lessons from the past in the present: News from the Spanish flu pandemic to COVID-19. *Revista Brasileira de Enfermagem* [online], *75*(01) [Accessed 16 January 2022], e20201161. Available from: https://doi.org/10.1590/0034-7167-2020-1161. Epub 06 Sept 2021. ISSN 1984-0446. https://doi.org/10.1590/0034-7167-2020-1161

Pichon-Rivière, E. (2000). *O processo grupal*. 6 ed. Martins Fontes.

Reichenheim, M. E., Souza, E. R., Moraes, C. L., Mello Jorge, M. H., Silva, C. M., & Minayo, M. C. S. (2011). Violence and injuries in Brazil: The effect, progress made, and challenges ahead. *Lancet, 377*(9781):1962–1975. p. 4.

Santos, B. S. S. (2020). A Cruel Pedagogia do Vírus. Coimbra, PT; Edições Almedina S.A., 2020.

Valinhos. (2004). Lei N° 3861, de 29 de dezembro de 2004. Cria a Coordenadoria Especial das Políticas para as mulheres – CEPM e dá outras providências. Disponível em https://leismunicipais.com.br/a/sp/v/valinhos/lei-ordinaria/2004/386/3861/lei-ordinaria-n-3861-2004-cria-a-coordenadoria-especial-das-politicas-para-as-mulheres-cepm-e-da-outras-providencias. Acesso em 07 abr 2023.

Valinhos. (2019a). Prefeitura Municipal de Valinhos. Ata da 008ª Reunião Ordinária do Grupo de Trabalho da Coordenadoria Especial de Políticas para Mulheres – CEPM: Ordem do Dia – Continuidade do Programa "Autoestima Sistêmica com Mulheres" e Histórico de Prestação de Serviços Públicos, em psicologia social, via Abordagem Grupal (GOL), em instituição. Boletins, n. 1892, Página 2, 06 nov. 2019. Disponível em: http://www.valinhos.sp.gov.br/sites/valinhos.sp.gov.br/files/boletins/2019/1892.pdf. Acesso em: 18 dez. 2020.

Valinhos. (2019b). Conselho Municipal dos Direitos das Mulheres. Estudo dos dados coletados na Delegacia da Mulher –Valinhos. http://www.valinhos.sp.gov.br/sites/valinhos.sp.gov.br/files/comunicacao/estudo_-_ddm_-_2019_-_corrigido.pdf. acesso em 07 de outubro de 2020.

Valinhos. (2020). Prefeitura Municipal de Valinhos. Primeiro Diagnóstico, territorial e intersetorial, de violências contra Mulheres e Proposta de Plano de Enfrentamento –CEPM. Disponível em http://www.valinhos.sp.gov.br/sites/valinhos.sp.gov.br/files/boletins/2020/2051.pdf. acesso em 06 de maio 2021.

Vieira, P. R., Garcia, L. P., & Maciel, E. L. N. (2020). Isolamento social e o aumento da violência doméstica: o que isso nos revela?. *Rev. bras. epidemiol.*, Rio de Janeiro, *23*, e200033. Available from http://www.scielo.br/scielo.php?script=sci_arttext&pid=S1415-790X2020000100201&lng=en&nrm=iso. Access on 18 May 2020. Epub Apr 22, 2020. https://doi.org/10.1590/1980-549720200033

Maria Candida Soares Del-Masso is a PhD in Special Education, Professor at the Universidade Estadual Paulista Julio de Mesquita Filho (UNESP), Brazil, retired, with degrees in Education and Special Education. She is working in the UNESP Graduate Professional Program in Physical Education and Graduate Professional Program in Special Education. Coordinator of GEPIS – Study and Research Group on Social Inclusion (UNESP).

Marta Bartira Meirelles-Santos is a social psychologist and dancer. After working with children who lived in the streets of São Paulo city, she systematized this experience in her social psychology master's degree in the Pontifical Catholic University of São Paulo and led her to develop a group approach called Operative Play Group (OPG). This method was applied in different public services (training for public health workers, group play therapy, women rights and vulnerable families protection). She is finishing her doctoral degree in the Psychiatry Department of University of Campinas (Unicamp). She also develops workshops about psychosocial esteem through dance. Site: www.mb@martabartira.com.br. and www.martabartira.com.br

A Plea for Rationality: Education and Bioethic Governance

15

Ralph Ings Bannell

Abstract

In this chapter, I discuss the relationship between epistemic rationality, education, and bioethic governance. After briefly outlining a definition of bioethic governance, I locate the importance of normative standards for judging belief and the need to educate decision-makers to be guided by those standards. This leads to a discussion of the rationality of belief and, therefore, to aspects of epistemic rationality. I focus on the role of affectivity in the production of meaning, especially our affective and aesthetic sensibility to the world, as an essential aspect of epistemology. During the discussion, I raise some questions in the epistemology of education, especially the aims and curriculum content necessary to educate rational decision-makers in the promotion of good governance.

1 Introduction

Education has as perhaps its principal objective the learning of how to live in the world that exists. But it is more than this. It also aims to help people of all ages to live better lives than they now do. Assuming that we do not live in the best of all possible worlds, how can we improve the world and improve ourselves? It is generally assumed that a specific form of governance is necessary for a transition toward a better society. It is also assumed that individuals and societies need to use their rationality – both theoretical and practical – in order to promote well-being. This, at any rate, has been the guiding idea since Ancient times, through the Enlightenment, and to the present day.[1]

Bioethic governance is one of the proposals on the table nowadays for a form of governance that leads to development and well-being. The principal question I wish to try and answer in this text is the following: what is epistemic rationality, how can it help in bioethic governance, and what kind of education is necessary to promote it? Part of trying to live better is to have standards by which to judge our present lives, knowledge, and action. These involve standards of belief and

[1]This assumption is not shared by all but I do not have space to enter into these debates, especially those that come from a "post-structuralist" or "post-modern" perspective. I believe their suspicion or even (in some cases) rejection of the concept of reason to be a profound mistake that has helped create a cultural environment hostile towards truth and rationality, with profound and disturbing consequences. In this chapter, I will deal only with the rationality of belief for lack of space. The rationality of action is equally important but will have to be left aside.

R. I. Bannell (✉)
Pontifícia Universidade Católica do Rio de Janeiro, Rio de Janeiro, Brazil

A. Pereira Jr, F. Sousa (eds.), *Principles for Governance*, Sustainable Development Goals Series,
https://doi.org/10.1007/978-3-031-40025-4_15

standards of action and good governance. Korsgaard (1996) notes that insisting on such standards is as old at least as Plato and Aristotle. Discussions of epistemological normativity have traditionally been the subject matter of theories of knowledge, while discussions of practical normativity have traditionally been the subject matter of practical philosophy. Since the historical period we call modernity, epistemology has been one of the central branches of philosophy but so has practical philosophy. In this chapter, I will deal only with epistemic rationality, for lack of space to deal with both.

I start this chapter by assuming that no account of the transition toward a better form of governance can ignore normativity, be it the normativity of belief or action. Therefore, as a contribution to this book, I would like to discuss the question of the rationality of belief. I will comment on the education necessary to promote the rationality of belief, and therefore of governance, as I proceed.

2 What Is Governance?

What kind of society do we want? What kind of economy is conducive to that kind of society? What form of polity and governance is necessary to promote the society we want? These are big questions and I will not try to answer them here. They involve thinking about questions of equality, justice, democracy, freedom, and the like, but also about questions related to what to believe and how to act.

"Governance" is a relatively recent term to refer to "rules of the political system to solve conflicts between actors and adopt decision making (legality)." It has also been used to describe the "proper functioning of institutions and their acceptance by the public (legitimacy)."[2] Now, my contention is that to talk in terms of the proper functioning of political institutions and their legitimacy, we need to look at how beliefs are justified, that is, the normative basis that underpins the beliefs that feed into the decisions made by those whose job it is to develop and implement government policies, resolve conflicts between actors and legislate on behalf of the majority. It is also necessary to look at the normative basis of the beliefs of ordinary citizens about politics, economics etc., because the legitimicy (or not) of political institutions and their functioning depend on these beliefs.

If we wish to talk of *good governance* and the kind of reasoning and decision-making necessary for such governance, we also need to look, even if briefly, at a definition of such governance. One common definition is as follows: "good governance relates to the political and institutional processes and outcomes that are necessary to achieve the goals of development. The true test of 'good' governance is the degree to which it delivers on the promise of human rights: civil, cultural, economic, political and social rights." The notion of rights is fairly recent in historical terms, dating from the Modern period, but has been sedimented into current moral and political discourse. The concept of human rights has become normative, that is, a value that should oblige us to act in certain ways and not others. What political and institutional decisions and processes will realize the promise of human rights? How can we ground human rights in a way that makes them universal, necessary norms and not simply the social ethic of a specific group or geographical region and, therefore, ignorable by those who rule and/or make decisions in other countries or regions?

Of course, not all good governance requires moral rules or action in accordance with human rights. The authority of rulers and the obligations of others toward them, as well as the obligations they have toward the population in general, are not always moral obligations. Good governance could mean acting for the good of a community, where the "good" is here defined in terms of the specific cultures and needs of that community. For example, the Yanomami community in Brazil has its norms of governance even if they are different to those of the wider political society in which the community is embedded. However,

[2]These definitions come from Wikipedia and are only meant to show how the term is commonly understood today. I am using them here as a working definition for the purposes of developing my discussion.

they and the wider society try to preserve the rights of the Yanomami community and its development. Thus, if these definitions are anything to go on, promoting human rights, which are moral, universal rights, seems to be the principal objective of good governance, even by those who accept a distinction between the right and the good.[3]

Conceptually, then, governance refers to the rule of the rulers. Another definition of governance could be "all processes of governing, the institutions, processes and practices through which issues of common concern are decided upon and regulated."[4] This definition refers more to the processes of decision-making that take place when individuals and institutions govern. It also places emphasis on the beliefs and decision-making procedures of those who rule. It is this dimension I would like to discuss in this chapter. Whichever definition of governance is adopted, all imply the rational formation of belief and rational action. Processes of belief formation and action are central to any decision-making process, whoever the decision-makers are and whatever the institutions through which government and other State policies are formed and implemented.

2.1 Bioethic Governance

The discipline of bioethics, as Farrarello (2021) describes it, was developed by Potter (1970) in order to think of ways to improve the quality of life or "living systems – meaning not limited to humans (…) – through an interdisciplinary integration of life sciences, philosophical reflection, and moral choices" (Farrerello, 2021, p.8). It has had a great deal of influence in medical ethics and procedure and has been extended to other areas of human practice, including government. According to Farrerello (2021, p.9), it was influenced by Kantian thought, especially the notion of the categorical imperative "according to which all living beings have the right to be treated not as means to an end but as ends in themselves – that is, as unique individual agents that possess the sanctity of life." But it also comes from the Ancient Greek word *bios*, which "indicates the conditions of possibility for our life not just to be but also to thrive if we so choose." In other words, it is "the way we choose to live; that is, the decisions we make to express the living biological principle that animates our being (…) to improve the quality of our unreflective (*zoé*) life and reflective (*bíos*) biological life" (Ferrarello, 2021, p. 9). The Ancient term "ethos" referred to an "accustomed place" and "habits" that someone or a society develops through time (Farrarello, 2021, p. 9). For this reason, Potter thought that bioethics results from "an understanding of how our thinking brain can combine biological knowledge with social and philosophical consciousness" (Potter, apud Ferrarello, 2021, p.9).[5]

These roots of the term are interesting because they show the centrality of reflection for the improvement of life, both in Ancient Greek thought and in the Enlightenment thought that we have inherited in our scientific and philosophical culture. It also shows that the question of governance involves both belief and (moral) action.

[3]Here we can look at documents produced by UNESCO for sustainable development. See UNESCO: https://www.sightsavers.org/policy-and-advocacy/global-goals/?gad=1&gclid=CjwKCAjw6IiiBhAOEiwALNqncXfyWu5h68YXGYM-E5YnT3PbCTqEenqIu6Ss65EMpQziRrH2qG0zWhoCyDQQAvD_BwE.

It is common in contemporary political theory to distinguish between the good, which is relative to a specific community or group, and the right, which is universal. I do not have space to elaborate more on this.

[4]https://www.ohchr.org/en/good-governance/about-good-governance. This definition assumes a representative democratic regime. I personally favor a more direct democracy in which social control – including economic control – is in the hands of the many and not the few, but I cannot go into this here. For the sake of this article, I will assume a representative democracy and a capitalist economic order.

[5]I leave aside here the assumption that it is the brain that does all the thinking. I do not agree with this, favoring a conception of cognition that includes the brain, the body, and the environment (physical and cultural). I will touch on this briefly below.

When Farrerello says that bioethics "should globally refer to all the sciences of well-being, namely, environmental, political, psychological, and social well-being" (Ferrarello, 2021, p. 10), she implies that belief formation in these sciences is rational and this implies looking at the human being from a specific point of view. This is not exclusively the third-person scientific point of view but also includes a first-person perspective in order to theorize the point of view of the agent who is acting. In doing this, she focusses on human emotions as a phenomenon central to what it is to be human. Human emotions are central for the kind of practice necessary to "contribute to cultural and biological evolution"; they are central to the practices that need to be adopted by practitioners in medicine, education, law, economics, and politics, to mention only a few. I will return to this question of emotions later.

According to Ferrarello (2021, p. 28) once again, "the job of the bioethicist is to facilitate complex decisions for the well-being of the society from a social, environmental, medical, and psychological perspective. The decision-making activity changes, of course, its feature according to the political and geographical context in which it is practiced." Such decision-making requires knowledge, of course, to feed into the correct decisions, and such knowledge needs to be formed rationally. But it also requires practical reason in the sense of acting rationally, be this in the doctor's consultancy, the classroom, the law courts or government, and legislative bodies. And this, ultimately, concerns the norms of rational belief formation and action that need to guide decision-makers.

Now, this means that the bioethicist needs to be educated, if she is a politician, a government employee or the person on the street who is to legitimize the decisions taken in the name of well-being. How can we ensure, if we can, that decision-makers have the theoretical and practical reasoning skills necessary to make correct decisions? An example is economic decisions. Political representatives and "experts" make most of the important decisions about economic structures and practices, be they members of the government or the State apparatus, members of private institutions within the economy or academics. Are these decisions based on true and reliable beliefs? Are these decisions being taken rationally? In short, what kind of rationality does a decision-maker require? In what follows, I will focus on rational beliefs, but it should be noticed that rational action is also necessary for decision-making in the interests of well-being.

3 Rationality

3.1 Epistemic Rationality

Knowledge is a form of belief. Traditionally, since Plato, it has been defined as true, justified belief. Now, clearly to arrive at a true, justified belief, one needs the capacity for perception, conceptualization, and reflection.[6] Conceptions of perception, conceptualization, and reflection have changed over the centuries, but all have been attempts to characterize humans as rational creatures, subject to the normative dictates of theoretical reason. This means the capacity to produce knowledge rather than opinions about the world.

I do not want to discuss the traditional definition of knowledge and the various debates that it has spawned (Gettier, 1967). I will concentrate on the question of rationality and how we can rationally justify our beliefs. Pritchard (2018) has noted that epistemic rationality involves judgment. We criticize or praise someone's judgment, be they a judge, a politician, an economist, or someone forming a common sense belief about the world. We praise a rationally formed judgment and criticize one based on chance or some arbitrary process of belief formation. But what is the distinction being made here? This is the epistemological question. Pritchard also remarks that we usually only accept rational beliefs as candidates for knowledge but how can we define rational belief? The question here is one of belief

[6] Some would argue that a priori knowledge does not require perception, but my main concern in this chapter is empirical knowledge of the world.

justification and not truth, although truth remains a goal of belief formation. Even if a belief is true, we criticize the person who justifies this belief arbitrarily rather than through a rational procedure. For example, it is true that the world is spherical in shape and not flat. But we intuitively criticize someone who believes this solely on hearsay just as much as he who believes the world is flat because he read this on a WhatsApp group message. Rational beliefs are those that are arrived at in a specific way that eliminates chance, prejudice, and other irrational means. So, a justified belief is a rational one, and justification is at least necessary for knowledge, if not sufficient.

The rationality involved here is epistemic rationality, a form of rationality aimed at justified belief. This is contrasted with practical rationality, which is aimed at the best course of action. In fact, as Pritchard notes, rational belief can, at times, undermine rational action. Take the following example from Pritchard. If I am running away from someone intent on harming me and I approach a ravine, I have to rationally decide to jump or not. Norms of rational action may well dictate that I jump, because that it is the only way to survive, but the norms of rational belief would probably force you not to jump, because you doubt your physical or psychological capacity to do so. It does not matter here if it is true or not that you have the strength to jump the ravine; what matters is how you justify or not that belief to yourself. In this case, focusing on epistemic justification is likely to get you killed.

But there are many cases in which epistemic rationality is central to how we live our lives. A judge is expected to weigh evidence before judgment and to form his belief as to the guilt or innocence of the accused on that basis. Of course, he might not have all the relevant evidence and convict someone for whom it is false that they committed the crime. Miscarriages of justice can happen. But, in general, we accept the decision of the judge or the jury if it was arrived at rationally. The same applies to those who make decisions related to governance. We want their decisions to be based on true beliefs but, since it is not possible to collect all the evidence necessary to confirm the truth or not of a belief, we want, at the very least, that their decisions are based on justified beliefs. We want them to be epistemically rational. What does this mean?

Although the relationship between epistemic rationality and true belief is a complex one, as mentioned above, we do aim at truth. However, in most contexts, we relax the goal of truth and aim more for justified belief.[7] As Pritchard argues, if we maintain the goal of truth, then one obvious way of being epistemically rational is to maximize beliefs, aiming for the most true beliefs possible, even if most of them are trivial. But this is clearly impractical as well as unnecessary if our goal is rational justification rather than truth. For example, the most rational strategy here might be to believe just about anything because, that way, one would ensure a better chance of believing the truth. But that is counter intuitive, because this would lead to accepting a great many false beliefs as well as true ones just to increase our odds.[8] It might seem strange to some that our epistemic goal is not necessarily to aim at *true* belief but only *justified* belief, but, given the complexities the goal of truth implies, I am inclined to suggest that truth is too strict a condition on rational belief formation.

We have to take the risk that some of our beliefs will be false even though we arrived at them rationally. This is a dilemma faced by anyone who has to decide if something is to be believed or not. Is it irrational to believe something that eventually turns out to be false? I do not think so, given that we weigh the best available evidence and arguments at the moment of deciding. That is, our goals at the time of forming

[7]Pritchard, for example, does not relax the goal of truth, which leads him to discuss the complex problems of epistemic rationality if this goal is maintained. My own approach is more in line with pragmatic approaches to knowledge, where truth is best understood as coextensive with justified belief. See Habermas (2003) for a well-known statement of this view.

[8]Pritchard (2018, p. 45). Pritchard suggests that one way to deal with this problem might be to maximize falsehood rather than truth (something like Popper's falsifiability criterion instead of a verifiability criterion for scientific knowledge) but points out that this could lead to not believing anything, which is also counter intuitive as well as impractical for those who need to make decisions.

a belief need to be those other than simply gaining true belief for its own sake. Knowledge has to serve nontrivial goals, related to well-being and human flourishing (Pritchard, 2018). A life worth living is not one simply oriented toward gaining true beliefs, whatever they happen to be. We need to acquire important beliefs and risk the fact that often we will not even acquire them because of our limits at gathering evidence or assessing arguments.

What are the goals of epistemic rationality, then? One clear goal, it seems to me, is to arrive at beliefs responsibly and to employ belief-justifying procedures that are equally responsible. But this is the same as being as rational as possible in one's judgments. The fact that it might not secure truth does not eliminate it as a good way to secure our claim to justified beliefs. Scientists accept, for example, that their theories and explanations might turn out to be false. They are fallible and provisional but that does not mean we give up on the search for truth or on the methods consecrated in a specific field of inquiry. But this does not relieve us of our responsibility, not only in the application of methods of inquiry but also in our questioning of those very methods when they produce problematic results. If we can be forgiven about the truth of our beliefs, it seems we must be responsible for their justification and for the justification of the methods used to arrive at them. We need to be responsible for our beliefs and thought. That is, presumably, one of the epistemic goals of education.

But can we always be held responsible for forming our beliefs? Are we responsible for perceptual beliefs, for example, many of which could be seen as involuntary? What about beliefs formed on the basis of optical illusions? What about other beliefs we have no control over? Would beliefs based on emotional reactions be included here? Pritchard claims that if we are taught a wrong epistemic norm, where this is a supposed rule to follow in order to arrive at true beliefs, then we cannot be held responsible for this. He claims that such a person could be seen as epistemically rational, since she is forming her beliefs responsibly and on the basis of testimony she has no reason to doubt. However, I think we have to qualify this argument because we have to arrive at the epistemic norms in the right way, too. More than this, I think that, if we cannot blame someone who has not acquired the right epistemic norms, we can, at the very least, blame the educational institutions and practices that did not teach them. Here I am assuming that belief formation is external as well as internal. This distinction is part of epistemology and can be thought of in the following manner. An internalist thinks that everything which makes a belief justified must be internal to the agent. Moreover, it must be cognitively or introspectively available. Externalists, on the other hand, will include among the features that make a belief justified the process by which the belief was formed, which sometimes is not available to the agent (Kornblith, 2001).

3.2 The Epistemic and Cognitive Role of Affectivity

Another important aspect of belief formation concerns the emotional aspect of our lives. We are embodied creatures and the study of mind cannot ignore lived experience, and this involves the emotions and affectivity in general (Colombetti, 2014). What is the role of affectivity in lived experience and does this experience sometimes help us form justified beliefs? This might sound odd, given that feelings and emotions are commonly understood as in opposition to rational belief, but they can also be understood as contributing toward belief. We often think emotions give information about what matters to us, what has value, and that this can or even should guide our belief. Brady (2013), for example, argues that emotions help serve our epistemic needs by capturing our attention, and by facilitating a reassessment or reappraisal of the evaluative information that emotions *themselves* provide. As a result, emotions can promote understanding of and insight into ourselves and our evaluative landscape. As an example, he uses the film *12 Angry Men* (1957), in which jurors use their emotional insight to arrive at a decision. I want to argue that all thought requires an affective dimension. I do not mean that we should let our emotions take control of our thought, something that could lead to distortions in belief formation. However, from the idea that emotions can *sometimes* distort belief, we cannot conclude that they *always* do so.

Colombetti (2014) goes further, arguing that we cannot ignore the bodily and affective dimensions of cognition, when we try to make sense and engage with the world. Let me explain this a little. Here we need to accept a naturalized epistemology that addresses empirical questions about how we arrive at beliefs and not only conceptual or linguistic analysis of the definition of knowledge and the logical relations between beliefs. Naturalized accounts include normative models of how we ought to arrive at our beliefs as well as empirical accounts of how we do, in fact, arrive at them. So, a naturalized epistemology does not reject normative benchmarks for belief formation. Once we include empirical questions about how we form beliefs, we are in the terrain of cognition, but the question we are trying to answer is the same: how do we come about the beliefs we have of the world and which are more rational, if any?

For a time, especially in Analytical Philosophy but also in Phenomenology, under the influence of anti-psychologism, epistemology was understood as a purely conceptual and logical analysis of knowledge, thus ignoring the psychology of belief formation. Quine (1969) was one of the first Analytical philosophers to suggest a naturalized epistemology (using the behaviorist psychology of the time). The second half of the twentieth century saw an approximation between the philosophical and the psychological approaches to knowledge, in developmental psychology and genetic epistemology (the work of Piaget and his followers) and the work of philosopher such as Goldman (1986) and others. This led to a redefinition of epistemology, which came to include scientific research on cognition. This included the so-called cognitive sciences, whose objective, in their early stages, was to explain cognitive mechanisms, including those of belief formation, in terms of the computational model of cognition, on the presupposition that it is the brain that is responsible for cognition. More recently, the cognitive sciences have included many other approaches, including those that see cognition as a combination of the body, brain, and environment acting together.[9] This moves our attention to how we make sense of the world rather than the logical or linguistic analysis of beliefs.

Colombetti combines an enactive approach to cognition[10] with affective science, the result of which is an explanation of how we make sense of and engage with the world through our emotions and affectivity in general. This approach presupposes a deep continuity between life and mind (Thompson, 2007), and one of the main theses of the Enactive approach to cognition is that the organizational properties of life are sufficient for and continuous with mind. One of these properties is affectivity. Emotions in particular are necessary for processes of self-regulation and regulation with others in our exchange with the environment (physical and social), making it impossible to separate emotion and cognition. As Colombetti (2014, pp. 2–3) says:

> Affectivity as I understand it (…) goes beyond (…) fleeting events: it is a broader phenomenon that permeates the mind, necessarily and not merely contingently. The mind, as embodied, is intrinsically or constitutively affective; you cannot take affectivity away from it and still have mind. Affectivity (…) refers broadly to a *lack of indifference*, and rather a *sensibility* or *interest* for one's existence (…) It refers to the capacity to be personally affected, to be 'touched' in a meaningful way by what is affecting one. (…) It is not necessary to be in a specific emotion or mood to be in an affective state; one is affected when something merely strikes one as meaningful, relevant, or salient.

This means that if cognition is sense-making then it is necessarily affective. But if affectivity

[9]Colombetti belongs to one of these new approaches called enactivism. For a good overview of the 4E approach to cognition, see Newen et al. (2018).

[10]For an explanation of this approach, see Colombetti (2014), Varela et al. (1992), Thompson (2007), and Di Paulo et al. (2017, 2018), among others.

has this cognitive role, can we also say it has an epistemic role, in that it can provide a benchmark for justifying belief?

Colombetti speaks of a primordial affectivity that is basic for human existence. She traces the philosophical lineage of the notion of affectivity, the theories in which make explicit reference to the body, leading her to the view that "primordial affectivity is a dimension of embodiment" (Colombetti, 2014, p. 3).[11] The upshot of this analysis is that the human being is characterized by caring for its activities, as in Heidegger's concept of *Dasein*. This means, among other things, that we cannot escape our moods. Because moods are primordial, a purely contemplative stance toward the world is impossible for human beings. And we feel moods in our body, because our body is the experiential counterpart to sense-making. This gives us a sensibility that is a kind of "empathetic understanding" (Colombetti, 2014, p. 14).[12]

Can this idea be used to help us understand epistemic rationality? If such rationality is a question of the manner in which we arrive at our beliefs about the world, I think we need to include empathy and affectivity in general in our practical and professional dealings with the world, even in belief formation. But we have to see it as part of thought and reflection and not as separate from them, as suggested even by some of the theorists discussed by Colombetti. To see this, we need to understand how affectivity can shape our thought and reflection.

If this is true, what does it tell us about how we can rationally justify beliefs? In the first place, it obliges us to include affectivity as a dimension of belief formation. We are embedded in and create environments that have a special significance and value. This idea is sometimes captured with the concept of *Umwelt* (von Uexküll, 1957). All organisms regulate their transactions with the world, and, in the case of humans, this involves regulating transactions with other humans and other nonhuman animals, as well as other life forms, inanimate nature and cultural artifacts. It is in this process of regulation that we transform the world into "a place of salience, meaning, and value – into an environment (*Umwelt*) in the proper biological sense of the term. This transformation of the world into an environment happens through the organism's sense-making activity" (Thompson, 2007, p. 25).

"Cognition" here has the meaning of "behaviour or conduct in relation to meaning and norms that the system itself enacts or brings forth on the basis of its autonomy" (Thompson, 2007, p. 126). This means that the meaning and the norms are generated in interaction with the world. It is in the activity of sense-making that norms are generated, not by some supposedly internal epistemic process separated from our active engagement with the world. This concept of cognition does not necessarily require content in the form of propositions that represent the world, at least at the basic level of mind, although obviously most cognitive activity of human beings requires representations in the form of propositional beliefs.[13] However, it is still a cognitive relation to the world that "implies a discerning perspective on the part of the organism" (Colombetti, 2014, p. 18). This is why "cognition is simultaneously also *affective*" (Colombetti, 2014, p. 18). In short, living systems care about their existence, and human beings are living systems in this sense.[14]

However, if we take this approach to cognition, we must recognize that human beings are, unfortunately, failing in their ability to discern what helps them survive and what does not. The climate crisis and the increase in inequality are just two examples of how things have gone

[11]This lineage includes Aristotle and, in the Modern period, Spinoza, Maine de Biran, and Michel Henry, as well as Heidegger and Patocka in contemporary thought. A more recent discussion has been in the context of embodied theories of the mind and cognition. See, for example, Johnson (1987, 2007, 2017, 2018).

[12]For a very recent discussion of this in relation to online interaction, see Torbjørnsen and Hipólito (in press).

[13]For a radical view on this, see Hutto and Myin (2013, 2017).

[14]It also relates theoretical and practical reason in a new way that shows their common roots, but I do not have space to go into this here.

wrong. Have we lost our "cognitive-discriminative capacity," which is at the same time an "affective-evaluative one that requires the living system to be affected or struck by the suitability of an event for its own purposes"? (Colombetti, 2014, p.19). I do not think we have lost this capacity because we are still sense-making organisms of a complex kind. Moreover, if affectivity is the source of meaning that grounds our sense-making activities, then we should look toward the so-called higher emotions in humans to help guide us out of the mess we have created. Maybe guilt, shame, and other such emotions are in order here but, as the saying goes, do not hold your breath. Later I will reflect on some possible impediments to rational sense-making and affective sense-making.

This discussion emphasizes what could be called an aesthetic sensibility to the world, using this term in its original sense. I do not have space to discuss this aesthetic dimension in greater detail, but I believe it to be essential, in theory and in practice, if we are to overcome the grave problems facing humanity and other life forms on the planet. For a slightly different but equally important discussion of this dimension in perception, the formation of concepts and thought, see the work of Mark Johnson (1987, 2007, 2017, 2018).

If sensitivity to the world is a primordial capacity of human beings, how can we harness it in order to address current crises and to help in rational belief formation? I think this requires reflecting on the norms of epistemic rationality.

3.3 The Norms of Epistemic Rationality

What are the epistemic norms necessary to develop epistemic rationality? How can we ensure, if we can, that our epistemic norms are the right ones? In this section, I look briefly at two questions: (1) What kind of knowledge is necessary for bioethic governance? (2) What personal characteristics are necessary in order to produce that knowledge?

3.3.1 What Kind of Knowledge?

One of the central questions in epistemology is, as I have emphasized, how can we justify our knowledge and, therefore, render it reliable? There is a huge literature on this, but our contemporary culture assumes that scientific knowledge is the most reliable, although not the only type of knowledge. This is because it employs specific methods and has a specific logical structure, even though the latter has been contested within the philosophy of science. Methods can be contested, of course, in the name of more reliability. For this reason, each science has its own specific methods that are tried, tested, and contested. In the area of the natural and life sciences, these methods can be contested, but, in general, there is a consensus around which work within the scientific community. When it comes to natural science, I think the most reliable methods are those customarily used by researchers in these sciences. If someone comes along with a rejection or negation of science, we should be at the minimum suspicious that their belief will not be reliable. Of course, scientific revolutions are promoted by radically new beliefs and methods of investigation, but they are rare (Kuhn, 1970). But even revolutions are based on rational procedures.[15]

In relation to the social and human sciences, the debate has been largely around the nature of the reality being investigated – that it is essentially meaningful – and that, therefore, the best method is interpretative. This approach is based on phenomenological and hermeneutical approaches to knowledge. In addition, there are the so-called critical approaches to the social and human sciences that defend a science in the interests of emancipating certain groups in society. For this reason, critical theories of science will try to relate scientific methods and conclusions to

---

[15] Kuhn's theory of scientific revolutions is not an exception to this because, although he points to economic, religious, and political determinations for scientific norms and theories, he does not ignore the fact that the paradigms that are questioned also have to be resistant to anomalies and internal incoherence. If there were not anomalies and problems in argumentation, it is unlikely that the external determinations would be enough for a scientific revolution.

specific social classes or, more recently, cultural groups, as well as widening the conception of reality beyond the beliefs and meanings of social actors. Marxist methods are examples of the former critical approach, where science is understood as related to the interests of capital or labor. Here the question of ideologies plays a central role. I cannot analyze these approaches here but suffice it to say that all aim at the truth and justified belief – that is knowledge – and all are, or should be, nondogmatic in that they are willing to revise both the scientific norms and the explanations produced by applying them.

When it comes to social and human science, given the ideological underpinnings of the various methods, I favor critical perspectives and methods because they are more likely to create knowledge in the interests of those who are excluded in contemporary societies, be they social classes or racial, cultural, ethnic, and gender groups. Ideology is not an epistemological category and needs to be taken into consideration when evaluating all scientific approaches.

Take economics as an example here. Orthodox neoclassical economic theory tries to explain how the economy works, and it has consecrated concepts and methods for doing so. However, I think it safe to say that most are in the interests of capital and its reproduction. But recent years have seen the rise of alternative economic thought that is gaining visibility not only in academic circles but in society as a whole. The chapters collected in this book are an example of this. To the extent that there is deep disagreement between specialists in each camp, we have to look carefully at the scientific norm in question before we assent to it. However, I think it can be said that orthodox economics is an ideology in favor of capital, while critical economics is an ideology in the interest of the working class and other excluded groups.

My suggestion is that here the affective dimension of cognition should enter. If the "science" does not respond to what matters to us, what we care about, then we should be suspicious of its analysis and conclusions. We all care about having enough food to eat, and shelter, as well as satisfying work and care in our old age. If an economic system is not supplying these goods to the majority in a country, it needs to be substituted by another that does. Referring back to my discussion of sense-making in the previous section, I think we can appeal to the affective and aesthetic dimension of thought in order to lead specialists in economics to consider alternatives. Of course, such theorists (and the practitioners who also follow their guidelines) will often argue that such economic thought *is* in the interests of the well-being of humanity and that they *are* guided by affective considerations. Here we come across deep disagreements, which are often extremely difficult to resolve. However, we need to consider to what degree such thinkers are deceived or deceiving themselves or suffering from some other kind of cognitive problem. If an economist is shown that the policies he defends are creating more poverty and hardship, and continues to defend them, then I think we need to ask what pathologies are operating, either individual or social in nature.

One more point I would like to make regarding epistemic norms is in relation to norms of inference. Much of our knowledge is arrived at by inference. There are correct and incorrect ways of inferring. The classic types of inference are deduction, induction, and abduction, but I think we can include rhetoric as a means of securing a conclusion from premises. Rhetoric has had a bad press for a very long time but has been revived recently in the so-called new rhetoric (Perelman & Olbrechts-Tyteca, 1969), which tries to classify and analyze argumentative forms that are not included in the classical canon. Theories of argumentation try to spell out what makes a good argument, and, although there is disagreement in this field, as there is in all fields of inquiry, we should not let that fact lead us to reject the idea that there are norms for argument.[16] Moreover, and in line with the above argument, we cannot ignore the affective dimension of arguments and inference. Dewey (1984) spoke of "qualitative thought," which is thought

[16] Other advocates of a theory of argumentation rather than formal theories of inference are Toulmin (2003) and Habermas (1998, 2003).

guided by our bodily feelings and emotions. He even went as far as to suggest that:

> Even 'the greatest philosopher' exercises an animal-like preference to guide his thinking to its conclusions. He selects and puts aside as his imaginative sentiments move. 'Reason' at its height cannot attain complete grasp and a self-contained assurance. It must fall back upon imagination – upon the embodiment of ideas in emotionally charged sense. (Dewey, 1934 p. 34)

We also need to remember something defended by Aristotle a long time ago, namely, that "episteme" or correct belief can be divided into different categories, each with its sphere of influence. According to the standard interpretation, Aristotle defended three distinct types of knowledge: episteme (knowledge of reality, usually translated as scientific knowledge these days), techne (knowledge necessary for technical skill in crafts and other practical endeavors), and phronesis (often translated as practical wisdom, to be applied in the social, legal, and political spheres). This is worth remembering because our culture overvalues episteme and techne to the detriment of phronesis in the economic sphere. We are told that economic decision-making should be guided by technical considerations and that economics is a science. However, as many point out, economics is inseparable from politics, as the original name of the discipline attests (political economy). Technical hypotheses and theories should always be at the service of practical wisdom in the social sphere, which includes economics. And practical wisdom requires an ethical and aesthetic sensibility, as Aristotle knew well.

The last point worth mentioning here is the importance of testimony for acquiring knowledge. We acquire a lot of knowledge because we believe in the testimony of others. Which sources are more reliable is an important question to ask. The most reliable sources are those who have looked at as much of the evidence as possible and reached their conclusion through correct inference, including nonformal arguments or through a form of practical reasoning. The testimony of teachers in schools and universities is, perhaps, one of the strongest sources of knowledge, but we need to be extra careful here. We depend upon testimony for our knowledge, but we also have a responsibility to make sure that the testimony we accept comes from a reliable source. This is not always easy, and we can be led astray by a number of factors but accepting that something is true just because someone you know has posted it on WhatsApp is not a responsible attitude if the objective is to secure the reliability of one's beliefs. This leads me to the personal characteristics necessary for knowing. I will say more about social epistemology below.

3.3.2 What Personal Characteristics?

Even if we cannot always blame agents for their ignorance, we can point to character traits that can favor the acquisition of reliable belief. What personal character traits are necessary in order to produce or legitimize the knowledge outlined in the previous section?

This topic is often discussed under the rubric of virtue epistemology, but it can also involve other areas of research, such as agnotology. What are the principal intellectual virtues? They include care, attention, curiosity, thoroughness, open mindedness, courage, tenacity, humility, and autonomy (Baehr, 2011), to which I would add responsibility, for the reasons stated above. Intellectual vices include pride, negligence, idleness, cowardice, conformity, carelessness, rigidity, prejudice, wishful thinking, closed-mindedness, insensitivity to detail, obtuseness, and lack of thoroughness (Zagzebski, 1996; Baehr, 2011; Cassam, 2019). There has been something of an explosion of literature recently on epistemic virtues and vices. Looking at epistemic vices can help us understand how humans often *actually do* think and inquire, since they are so ubiquitous, while analyzing epistemic virtues helps us understand how they *ought to* think and inquire. Epistemic vices obstruct knowledge; they are "intellectual defects that get in the way of knowledge, and calling them vices is to suggest that they are blameworthy or in some other sense reprehensible" (Cassam, 2019, p. ix). The epistemology of education has analyzed intellectual virtues (Baehr, 2016), with some very interesting results. I think we should include these virtues in school and university curricula in order to improve belief formation.[17]

But even if someone has all the virtues necessary to form justified beliefs, we still encounter the psychological problems that might interfere with her cognitive capacities. Bad beliefs are not confined to bad people. This is a vast topic in cognitive psychology, and I do not have space to deal with it satisfactorily here. But its importance is highlighted nowadays with the contemporary problems of fake news and science denial, which is not always a result of the fact that the norms of scientific discovery are not well taught or that the agent has more epistemic vices than virtues, although vices do obviously get in the way here.

Recent research shows that bad beliefs are prevalent, even among those who have knowledge of scientific method. Bad beliefs are also held by good people (Levy, 2022), and bias is a problem for everyone (Kelly, 2022). The science of agnotology reminds us that we are all prone to ignorance (Proctor & Schiebninger, 2008). Ignorance is as much a subject of epistemology as is knowledge. Sometimes ignorance is deliberately manufactured, or knowledge is suppressed, but not always. If it is, then the problem is not belief formation but the ethical character of the person spreading the disinformation and a question of ethics or practical reason.[18] Sometimes it is the consequence of the social or gender position one occupies in society, ideological influences, or the socio-material conditions in which people live, which I will discuss below.

And, of course, there is the problem of what to do when experts do not agree over something. Deep disagreements, as mentioned before, are common in philosophy and science. The case of economics is again an important example here. Although orthodox neoclassical economic theory is prevalent, alternative theories that explain not only what *actually happens* in the economy but what *ought to happen* are gaining more ground (once again, see the chapters in this book, for example). This illustrates the importance of justifying the norm before we justify knowledge based on that norm.[19]

But there is little investigation into the affective dimension necessary to form good, justified beliefs. If my discussion above of the affective dimension of cognition is on the right lines, this cannot be ignored. Here we come to the aesthetic dimension of understanding. Here I can do no more than suggest more research on this and suggest the reading of the texts already referred to in this chapter.

3.4 Socio-material Conditions

However, it is important to realize that the rationality of belief (and action) is not all down to the individual agent and her internal cognitive processes. Apart from the bias and ignorance we are all prone to, we live in a world that is socially, materially, and ideologically structured in a way that does not always permit rational thought and action. These are the socio-material conditions in which we form beliefs and react to others' actions. Also, much belief justification is not internal to the knower, external factors being important for belief formation. This brings me back to social epistemology.

Social epistemology is the sub-area of the theory of knowledge that usually deals with these issues. This has its "analytical" branch and its "continental" branch. In the former, Goldman's book *Knowledge in a Social World* (1999) was perhaps one of the first to develop a social epistemology (although work in the philosophy of science also broached relevant issues, especially that of Thomas Kuhn). In the latter tradition, the idea that knowledge is (partly) determined by social, political, and economic factors has a much longer history, going back at least to Hegel if not

[17] See Baehr (2016) and the site: https://intellectualvirtues.org/. See also The Jubilee Centre for Character and Virtues, The University of Birmingham: https://www.jubileecentre.ac.uk/527/character-education/framework, and Kristjánsson, 2020.

[18] Zagzebski (1996) argues that intellectual virtues are grounded in ethics, but I do not have space to analyze her arguments here. I also believe they are grounded in aesthetic and ethical sensibilities.

[19] See also rethinking economics: https://www.rethinkeconomics.org/.

before, especially in Marxist thought and so-called critical theory but also in the tradition of hermeneutics.

I mentioned testimony above and this is one of the central topics of social epistemology. We often accept the opinion of others as knowledge, without trying to justify it ourselves. Are we being rational in doing so? Another social practice that is central to belief formation is argumentation, which, by definition, involves at least two people. This has been the center of analysis in the work of Toulmin and Habermas, for example. Habermas (2018) goes as far as to elaborate ideal social conditions for argumentation that he believes will secure a rational outcome. He also argues that the political public sphere should be designed in such a way as to allow such argumentative practices to shape public policy and public opinion. Here the conditions for rational inquiry are coextensive with those for a democratic public sphere.[20]

Here the question of ideology has to be mentioned. Although some theorists have tried to persuade us that we live in the era of the end of ideology, I think this is false. As Mészáros (1989) says quite rightly, ideology is everywhere, even in science and supposedly rational processes of belief justification. The question is not one of trying to avoid ideology by some supposedly neutral method of justification but to acknowledge that your ideas and arguments will support one or other social class or group. We should not try to get rid of ideology under the false belief that this is possible, which only masks and promotes one ideology in detriment to others.[21]

Goldman (1999) also analyzed these social conditions on the production and spread of knowledge. He includes the technology and economics of communication and how it influences the knowledge that is produced and spread. This involves the computer-mediated communication of the 1990s and the economics of communication between academics and other specialists, all of which have intensified during this century. Nowadays, both are of even more importance since the advent of artificial intelligence (AI) systems, especially ChatGPT.[22] Also, fraud in scientific publication is a problem on the rise, not to mention the political economy of funding for research. We also have to consider what Goldman called the "marketplace of ideas," including the media as one of the central agents in this marketplace. This requires State regulation and a heightened awareness among the public that consumes what is produced in this marketplace.[23] Here, as Goldman urges, we must look at social interaction and knowledge-enhancing practices. This brings us back to the contribution new theories of cognition can make, especially enactive theory, in the analysis of social interaction and the role of language as mediation (Di Paulo et al., 2018). Unfortunately, I do not have space to go into these questions here. However, I have no doubt that we are facing a crisis of knowledge in the world today. This is a crisis of ignorance as well, including political ignorance, biases and self-deception, scientific ignorance, fake news and conspiracy theories, collective ignorance and ignorance and big data.[24] One way (but only one way) of addressing this problem is to look at the epistemic goals of education.

[20] See Habermas (1996, 1998). He has recently analyzed the role of information technology in the public sphere. See, for example, *Constellations*, vol.30. 2023. Personally, I do not think these conditions sufficient for avoiding ideological influence in the interests of capital. However, to be fair to Habermas, he has no intention of questioning either the capitalist economy or the democratic constitutional state.

[21] I think Habermas can be accused of this, but I cannot go into this here.

[22] Zuboff (2019) describes in detail the rise of surveillance capitalism since 2000, when Google changed their business plan to include advertising revenue. The problems of fake news and conspiracy theories are also of considerable importance today. See Cassam (2019).

[23] Some advance in regulation has taken place in Europe (see https://wayback.archive-it.org/12090/*/https://ec.europa.eu/digital-single-market/en/policies/open-internet), but it is still almost nonexistent in the USA and in Brazil, for example.

[24] This list comes from a conference announcement.

4 The Epistemic Goals of Education for Bioethic Governance

What kind of education is required for bioethicists to develop their epistemic rationality? The category of bioethicists includes, in my view, not only the so-called decision-makers and policy-makers but all citizens in that governance has to be legitimized by all. Which goals, methods, and interests should educational institutions have, use, and defend? What curricular content should be privileged, if any? Can we talk of the epistemic authority of such institutions, their curricula, and their teachers? These are the questions Goldman (1999) raises at the end of his book and deserve our attention.[25]

Goldman defends what he calls a veritistic conception of education, aligned with traditional Enlightenment thought, that is, one that emphasizes that the principal aim of education is to help students acquire knowledge understood as justified true belief. I do not disagree with him and would defend the view that the knowledge taught should be the best available scientific knowledge of the world but always with the proviso that it is fallible and provisional and that, especially at university level, it can be contested. But contestations should never be dogmatic but always based on a rational (if ideological) approach to research. Of course, not all knowledge is scientific, and there is room in the curriculum for other forms of knowledge. Goldman includes aesthetic judgment but doubts whether this is knowledge. Given what I have said above about the aesthetic dimension of cognition and sense-making, I would include the development of aesthetic and ethical sensibilities as a central function of education.[26] So, if we do not want to include artistic skill and judgment as knowledge, because it does not admit of truth and falsity, it should still be included, especially if it can help develop the aesthetic and ethical sensibilities necessary to have the world in view, so to speak.[27]

What about testimony? Testimony should be trusted at school level but probably less at university level. Or, perhaps it would be best to say that university students should always apply their critical thinking skills to all content taught on their courses. But here, as well as at school level, the education of the educators has a responsibility in that care must be taken to educate teachers and researchers in ways that teach beliefs and theories nondogmatically, both in their own education and the education of their students.

But here someone could object by saying that the veritism advocated by Goldman, based as it is on traditional epistemology, which emphasizes the truth of what is to be taught, may be too strict an aim for education. I argued for something like this at the beginning of this chapter. Would not it be better to simply aim at justified belief, that is, to aim at developing the skill of having good reasons for believing something rather than true belief. That is, perhaps warranted belief is a better and more reachable aim for education. This is a criticism Goldman examines in his book in the context of the claim that the most important educational aim is to teach critical thinking skills. Let me look at this a little more closely.

There is a plethora of books, articles, and courses urging the teaching of critical thinking in schools and universities.[28] Goldman argues that critical thinking is not an end in itself but a means to the epistemic end of truth. He also analyses the question of testimony, suggesting that students do not always need independent reasons for believing something the teacher tells them (the vast majority of teaching is still in the form of exposition) and that we can endorse a principle of unsupported trust. His argument is that, assuming there is no ultimate foundational belief on which the belief in testimony is founded, and assuming we want to avoid skepticism, we cannot avoid the principle of unsupported trust, at least on some

[25] He also asks what should we think about the current "post-modern" approaches to education, with their skepticism, if not disdain, in relation to truth and rationality? But, as I said at the beginning, I do not want to enter this debate here.

[26] For an excellent analysis of practical reason and ethics that emphasizes this aspect, see Blackburn, 1998.

[27] Of course, some might contest the claim that aesthetic judgements, for example, do not admit of truth and falsity.

[28] They are far too numerous to mention here. For a good example, see Siegel, 1988, quoted by Goldman (apud, 1999, p. 363).

occasions. Either we very rarely have any justified beliefs or we have to trust others sometimes.

I agree that students do not always need independent reasons for trusting what teachers say to them, on pain of an infinite regress as well as making teaching and learning an almost impossible task. If the student has negative reasons for trusting a teacher, then her mistrust would be justified, according to Goldman (following Burge and Fricker). However, there is no need for positive reasons for trust. Are students obliged to look for defeating evidence? I think this depends on the age of the student. Small children would not be so obliged, but university students should always be taught to look for falsifying evidence as a part of scientific method (or counterfactual evidence in the case of philosophy and the other humanities, where scientific method is not applicable). Also, there is the question of the maturity of the student and the fact that, at times, it is necessary to give a slightly false statement when students are young, to be corrected when the student gets older.

Critical thinking is to be endorsed, therefore, especially in relation to the interpretation of evidence for and against statements of fact. Here we come back to the conditions for rational argumentation necessary for rational belief formation. I have mentioned the conditions advanced by Habermas, but there are others, including ones advocated by Goldman. This involves public debate and participation in rules that promote the cause of truth. However, there is no consensus over these rules. Apart from this, some rules are contested in the name of emancipation from Western epistemological ideas, including the idea of truth itself.[29] I think we have to reject strong relativist versions of these approaches.

How about the teacher as knower? Let us go back to the question of responsibility. Can we hold teachers responsible for the norms they follow and the beliefs they teach? As Pritchard (2018) says, and Goldman agrees, sometimes the best way to get at the truth is not discernible to the agent. If epistemic rationality is a question of how we justify beliefs, then sometimes the agent does not know how to arrive at justified beliefs or can only do so in the context of institutions and processes that are not of her making but, nonetheless, she still arrives at them. Is she being rational in accepting the authority of the institution? Here I think there is a responsibility on the part of teachers, and their teachers, to look for negative evidence and counter arguments to all they teach. They might not be successful, but it is their responsibility nonetheless, as it is for researchers in all fields of inquiry. This also applies to journalists and policymakers and for the same reasons.

Maybe one way of avoiding this is to place the blame not on the individual agent but on the institutions and their practices that lead someone to believe things by external means, when those means can create false or unjustified beliefs. Here the responsibility is on the institution and those who work within in. The external mechanisms leading to the belief can be made responsible even if it is more difficult to praise or blame the agent herself for the belief in question. Schools are such institutions, but so are the media and other institutions, including State institutions, that are supposedly to be trusted in the information they use and transmit.

Teachers, researchers, reporters, and editors cannot be relieved of this responsibility, even though they often implement methods, curricula, and practices not of their choosing. It is true that in this case their practical reason might be undermining their epistemic rationality. For example, they may decide that keeping their job is a rational decision even if they do not believe what they are doing or telling others. But they are still responsible for their research, teaching, and reporting, just as their training courses and teachers are responsible for teaching them the correct norms, as well as the government responsible for including such norms in the school and university curricula, as well as the evaluation of their policies. They can be blamed or praised for their epistemic virtues and vices, as well. If epistemic rationality and justification are some of the central

[29] I have already said I cannot go into this "post-modern," "post-structuralist," and "post-colonial" arguments against traditional epistemology. My own view is that much of it, when analyzed closely, reduces to something like the traditional search for truth and reliability or involves a performative contradiction (Apel, 1980; Goldman, 1999).

aims of a curriculum, then educational institutions and practices need to take them seriously. This also goes for cultivating the aesthetic and ethical sensibility necessary to make sense of the world and those in it.

The last point I want to raise is that of epistemic authority. Here we touch on curricula content but also on the information circulating within the information age. As Goldman 1999, p. 367) says, "a truth seeker in the information age must negotiate all sorts of treacherous communicational terrain, full of hazards of every kind." With relation to curricular materials, should they include religious dogma or appeals to patriotism? What about history, where alternative narratives are central to the current struggles for the curriculum, in all countries I can think of? Is it possible to measure which truths are more important to teach? What about the claim that alternative truths exist, especially from the point of view of first nations or ethnic groups?

Certainly, the fact that all knowledge is revisable and provisional should not be a reason for not teaching it as true knowledge, as long as it is the product of rigorous research and always open to negative evidence. But it should never be taught as absolute truth, which does not exist. Previously unexamined aspects of any question should also be given space in the curriculum because we never know what new evidence might be important. However, this requires the proviso that small children should not be confused with contradictory evidence for a proposition when they are still acquiring the basic skills of thinking and the basic aesthetic and ethical values of their community. At university level, the answer should be "teach the conflicts," together, and as part of, the development of the critical thinking skills of students. This also applies to journalism, where the norm of presenting both (or more) sides to a question is becoming rare, if it ever existed.[30] Finally, these considerations also apply to decision-making bodies, especially those that decide on government policies.

[30] Habermas (1996) lists the norms of good journalism, most of which are not followed in today's corporate media.

Perhaps one possible solution to this problem is to say that truths in certain areas of knowledge can be accepted, while in others the search for negative evidence will have to be increased. Mathematics and Physics would be examples of the former, while biology and the life sciences might be examples of the latter. When it comes to cultural values and aesthetic and ethical sensibility, we need to take into consideration the cultural context of those affected by the teaching, reporting, and decision-making. Here there are no universal truths or, if there are, they are very few. Perhaps human rights are examples of such truths, but even here there are contradictions.

Clearly, if bioethic governance is to be promoted, we need to educate people with the epistemic rationality necessary to form rational beliefs. My argument in this chapter has been that doing this involves developing and employing the correct norms of epistemic rationality in order to think and act in ways conducive to the well-being of society, and that this, in turn, means understanding human beings as affective beings, who have ethical and aesthetic sensibilities. Understanding affectivity as part of cognition will help us strengthen the ethical and aesthetic sensibilities necessary to rationally form beliefs and, thus, to make decisions in the interests of human beings and other life forms. And this, in turn, will strengthen the actions adopted by those whose job it is to promote well-being, as well as the ability necessary for ordinary citizens to legitimize (or not) those decisions. Experts and authorities are fallible, which is why a good dose of skepticism is necessary in all fields. But we also need literacy in all fields, including science, philosophy, and ethics, in order for the general public to have the necessary means to evaluate beliefs and values and to legitimize or not the decisions taken in their name.

5 Conclusion

Some might think that I have simply repeated some timeworn ideas about knowledge and have added nothing new, even in educational epistemology. If this is true, it is partly because this chapter was written not for specialists in epistemology but for those who might like to know how epistemological questions bear on the central concern of this book, which is good governance. Some might comment that when things started to get interesting, I only suggested more research or gave a promissory note. That is true but I hope one day to make good on that promissory note. However, if something is worth saying, it does not lose any of its force in the retelling, especially in a sociocultural – and philosophical – environment that often denies these basic ideas. This chapter is a plea for rationality in an irrational world.

References

Apel, K-O. (1980). *Towards a transformation of philosophy* (G. Adey & D. Frisby, Trans.). Routledge and Kegan Paul.

Baehr, J. (2011). *The inquiring mind. On intellectual virtues and virtue epistemology*. Oxford University Press.

Baehr, J. (Ed.). (2016). *Intellectual virtues and education. Essays in applied virtue epistemology*. Routledge.

Blackburn, S. (1998). *Ruling passions. A theory of practical reasoning*. Clarendon Press.

Brady, M. S. (2013). *Emotional insight. The epistemic role of emotional experience*. Oxford University Press.

Cassam, Q. (2019). *Vices of the mind. From the intellectual to the political*. Oxford University Press.

Colombetti, G. (2014). *The feeling body*. MIT Press.

Dewey, J. (1934). *Art as experience*. Penguin.

Dewey, J. (1984). Qualitative thought. In J. A. Boydston (Ed.), *The later works of John Dewey. Volume 5: 1929–1930*. Southern Illinois University Press.

Di Paulo, E. A., Buhrmann, T., & Barandiaran, X. E. (2017). *Sensorimetor life. An enactive approach*. Oxford: Oxford University Press.

Di Paulo, E. A., Cuffari, E. C., & De Jaegher, H. (2018). *Linguistic bodies. The continuity between life and language*. MIT Press.

Ferrarello, S. (2021). *Human emotions and the origins of bioethics*. Routledge.

Gettier, E. L. (1967). Is justified true belief knowledge? In A. Phillips Griffiths (Ed.), *Knowledge and belief*. Oxford University Press.

Goldman, A. I. (1986). *Epistemology and cognition*. Harvard University Press.

Goldman, A. I. (1999). *Knowledge in a social world*. Clarendon.

Habermas, J. (1996). *Between Facts and Norms. Contributions to a discourse theory of law and democracy* (W. Rehg, Trans).. Polity Press.

Habermas, J. (1998). *On the pragmatics of communication* (M. Cooke, Ed.). MIT Press.

Habermas, J. (2003). *Truth and justification* (B. Fultner, Ed., and Trans.). MIT Press.

Habermas, J. (2018). *Philosophical introductions. Five approaches to communicative reason*. Polity.

Hutto, D. D., & Myin, E. (2013). *Radicalizing enactivism*. MIT Press.

Hutto, D. D., & Myin, E. (2017). *Evolving Enactivism. Basic minds meet content*. MIT Press.

Johnson, M. (1987). *The body in the mind. The bodily basis of meaning, imagination, and reason*. The University of Chicago Press.

Johnson, M. (2007). *The meaning of the body. Aesthetics of human understanding*. University of Chicago Press.

Johnson, M. (2017). *Embodied mind, meaning, and reason. How our bodies give rise to understanding*. University of Chicago Press.

Johnson, M. (2018). *The aesthetics of meaning and thought. The bodily roots of philosophy, science, morality, and art*. University of Chicago Press.

Kelly, T. (2022). *Bias. A philosophical study*. Oxford University Press.

Kornblith, H. (Ed.). (2001). *Epistemology: Internalism and externalism*. Blackwell.

Korsgaard, C. (1996). *The sources of normativity*. Cambridge University Press.

Kristjánsson, K. (2020). *Flourishing as the aim of education*. Routledge.

Kuhn, T. S. (1970). *The structure of scientific revolutions* (2nd ed., enlarged). University of Chicago Press.

Levy, N. (2022). *Bad Beliefs. Why they happen to good people*. Oxford University Press.

Mészáros, I. (1989). *The power of ideology*. Harvester Wheatsheaf.

Newen, A., De Bruin, L., & Gallagher, S. (2018). *The Oxford Handbook of 4E Cognition*. Oxford: Oxford University Press.

Perelman, C., & Olbrechts-Tyteca, L. (1969). *The new rhetoric. A treatise on argumentation*. University of Notre Dame Press.

Pritchard, D. (2018). *What is this thing called knowledge* (4th ed.). Routledge.

Proctor, R. N., & Schiebninger, L. (Eds.). (2008). *Agnotology. The making and unmaking of ignorance*. Stanford University Press.

Quine, W. V. (1969). Epistemology naturalized. In W. V. Quine (Ed.), *Ontological relativity and other essays*. Columbia University Press.

Thompson, E. (2007). *Mind in life. Biology, phenomenology, and the sciences of mind*. Harvard University Press.

Torbjornsen, R. R., & Hipólito, I. (in press). *Widening the screen: Embodied cognition and audiovisual online social interaction in the digital age*.

Toulmin, S. E. (2003). *The uses of argument* (Updated edition). Cambridge University Press.

Varela, F., Thompson, E., & Rosch, E. (1992). *The embodied mind: Cognitive science and human experience.* MIT Press.

Von Uexküll, J. (1957). A stroll through the worlds of animals and men. In K. S. Lashley (Ed.), *Instinctive behaviour: The development of a modern concept.* International Universities Press.

Zagzebski, L. T. (1996). *Virtues of the mind. An inquiry into the nature of virtue and the ethical foundations of knowledge.* Cambridge University Press.

Zuboff, S. (2019). *The age of surveillance capitalism. The fight for a human future at the new frontier of power.* Hachette Book Group.

Ralph Ings Bannell is an associate professor of the Philosophy of Education, Pontifical Catholic University of Rio de Janeiro (PUC-Rio). He has BA in philosophy (Stirling) and MA and DPhil Social and Political Thought (Sussex). He is a visiting fellow at the Institute of Education (IoE), University College London (UCL) 2010–2011. He is the founding member of the Brazilian Society for the Philosophy of Education (SOFIE) and past coordinator of the Philosophy of Education Working Group of the National Association for Research in Education, Brazil (ANPEd). He is the author of various books, book chapters and articles in the philosophy of education.

Economic Complexity in the Information Age: Challenges for Professional Education

16

Luiz Gonzaga Chiavegato Filho

Abstract

One of the current circumstances of the world of work refers to advances in the new digital age, which transforms information and knowledge into the main assets of our economy. Taking this reality as a starting point to its quest, the theory of economic complexity manages to demonstrate how the distribution of information and knowledge is unequal in our society, reinforcing the importance of policies that not only improve the distribution of these resources but additionally favor their development and accumulation. Within this scenario, the professional education strategies that stand out are more focused on favoring collaborative and reflective knowledge construction processes. Preceded by work analysis, such strategies intend to expand the capacity of individual and collective action of workers on their own activity, in order to achieve better performance, efficiency, and preservation of health. Therefore, while considering data issued by the economic complexity theory, one of the challenges of professional education is the development of empowering work environments that lead to sustainable economic growth in the information age. Some practical experiences strengthen these reflections, as is the case of Orkestra – Basque Institute for Competitiveness and the programs of incentives toward the digitization of social life in Estonia.

1 Introduction

In today's world, digital technologies are irreversibly redefining our way of life. This profound transforming power places education as one of the primary areas for investment, as it is necessary to supply the extensive demand for human capital[1] that technological advances require. In this regard, in the labor world, work analysis and

[1]Human capital refers to the work of Bourdieu (2001) who expands the concept of capital to fully explain the relationship between socioeconomic background and educational outcomes. In this way, in addition to the economic dimension of capital, it develops social and cultural dimensions. The two elements that constitute social capital are networks of social relationships, which allow individuals to access the resources of group or network members; and the quantity and quality of the group's resources. The notion of cultural capital arises from the need to understand the inequalities in school performance of individuals from different social groups.

L. G. Chiavegato Filho (✉)
Federal University of São João Del Rei, São João Del Rei, Minas Gerais, Brazil
e-mail: lgcfilho@ufsj.edu.br

A. Pereira Jr, F. Sousa (eds.), *Principles for Governance*, Sustainable Development Goals Series,
https://doi.org/10.1007/978-3-031-40025-4_16

professional education assume a central role, given the importance of knowledge and information in gaining competitiveness, in the development of qualified jobs and in sustainable economic growth.

The theory of economic complexity delivers important reflections on the role of information and knowledge in the economy and society in general (Hidalgo et al., 2007; Hidalgo & Hausmann, 2009; Hausmann et al., 2014; Hidalgo, 2017). Hidalgo (2017) claims that information is the main economic resource in the twenty-first century; it is the raw material of the modern economy, an increasingly important asset in the creation of value and innovation. However, information and knowledge are unevenly distributed around the world, both among countries and within countries. This condition stems from the fact that economies are embedded in social and professional networks, shaped by historical, cultural, and institutional factors, which, despite the fact they can restrict economic activity, on the other hand, they are fundamental networks since they are part of the only set of structures available to store knowledge and know-how in a large scale.

The above-mentioned theory reinforces the importance of investing in public policies that not only promote equitable access to education and technology but also encourage the creation of broader, more collaborative, and inclusive human networks that favor the creation and accumulation of knowledge, information, and innovation. Guile (2008) sustains that education, in the knowledge economy, should focus on building critical and reflective thinking skills, as well as on developing skills related to the creation and use of knowledge, reinforcing the role of work analysis and professional education in this process.

2 Professional Education

Historically, from the Psychology point of view, it is possible to understand the presence of men at work based on the assumption that they are part of a sociotechnical system. This system involves complex interactions between technologies and workers, whose multiple dimensions (social, psychological, cultural, etc.) are interrelated and interdependent, not only the ones linked to the task itself (technical side) but also those linked to who performs it (social side). However, throughout the previous century, trying to answer the question of how this system presents itself and can be analyzed, different theoretical perspectives of psychology at work have been developed, emphasizing each of the dimension of this system depending on the occasion. Thus, different perspectives of analysis of productive action and the respective forms of psychology at work were defined, as well as the implications of workers and the way in which their behavior is affected by work situations (Leplat & Cuny, 1977).

The conceptual bases of the vision of the man and the world that supported the initial practices of psychology, at the beginning of last century, arise from the notion of homo economicus, developed by Adam Smith, one of the fathers of political economy in the eighteenth century. This notion presupposed a man whose actions would be predominantly rational, mechanical, and focused on his own interests. The psychology practices originated from this conception, such as the selection of people, education and professional education, and performance evaluation, among others, had, and still have, the human being as an object, understood as an adjustment variable of the productive system.

For Spink (1996, p.98), this reductionist tradition represented the separation between an experimental field (theory) and an applied field (practice), which, linked to the determinism of technical efficiency and the vision of organization as a machine, "were combined to produce a fertile field of an unproblematic expansion," as if the work environment were not full of antagonistic and contradictory interests. The logic underlying this approach was historically connected to the Taylorist principle of division between the conception and execution of tasks in the work environment and constituted the applied version of work psychology at the time, almost autonomous from the rest of psychology.

Despite numerous criticisms, this conception was developed over the century and survives up to now carrying the commitment of building scientific strategies (under the positivist paradigm) for the prevention, measurement, and control of human behavior at work in order to improve the workers' performance (Montmollin, 1974; Baumgartl & Primi, 2006; Baumgartl et al. 2010; Barbosa & Cintra, 2012; Ryan & Ployhart, 2014).

Around the 1950s, influenced by developments in Ergonomics of Activity and Psychopathology at Work, other approaches to psychology developed critically, each in its own way, in relation to the positivist view mentioned above. Such conceptions intended to present and demonstrate that, before looking for specific skills in certain workers to perform certain tasks, it was necessary to return to the work situation, showing all its complexity and the multiplicity of factors that compose it. This would be the only way to find the foundations to develop the skills, operating modes and competences required for productive environments. Ergonomics, of French origin, prioritized the analysis of activity in a real work situation, without disregarding the psychophysiological characteristics of the worker, now understood as an actor in the work process, not only a subject who knows, but a subject affected by his or her social condition (Abrahão & Pinho, 1999; Guérin et al., 2001; Clot, 2010).

This vision of ergonomics takes up the central distinction for the analysis of work, which concerns the difference between "task" and "activity," which are now called "prescribed work" and "real work." Since concrete work, historically and socially, is the result of a division between conception and execution, it is understood that the "prescribed work" is linked to the conception and the "real work" to the execution. Work analysis, then proposed by ergonomics, focuses precisely on the differences between what is prescribed and what is real, and on the corresponding consequences of this difference for the worker, for the company and for society in general. Such positions will define and renew the objects of action and intervention of various disciplines, including work psychology (Daniellou et al., 1989; Clot, 2010).

In methodological terms, the Ergonomics of the Activity favored the workers' view of their own work, in other words, of the execution conditions, difficulties, complaints, and verbalized problems. In addition to it, there was the observation of work in real situations and the survey of the most important indicators on the production process. Then, they promoted the confrontation and analysis of these data with the questions raised by the specialist and with the existing findings in the literature. The main purpose was (and still is) to overcome the Taylorist dictates, which reign in work organizations, and of "scientific positivism" among the fields of knowledge, which have human work as an object of study. Professional education processes play an important role, given the fact they target to analyze work, develop skills, preserve health, and contribute to transforming work situations (Wisner, 1987; Abrahão & Pinho, 2002; Daniellou, 2004).

In the ergonomics of the activity, the professional education of workers must be motivated and outlined in terms of the job analysis. This tradition is consolidated with the works of Pastré (2017, p. 635). According to his work, "It is necessary to go through the study of the situation to understand the activity. Being that analysis the one responsible for the identification of organizing conceptual elements that its actors retain from the situation." Work analysis occupies a central place in professional education, as it enables the definition of the contents to be worked on and then works as a support for the development of the required skills. It is worth emphasizing that the learning process is intrinsic to the activity, given its complexity and variability. In this way, actions that provide workers with moments of reflection on their own activity, contribute to developing these professionals' ability to act on them.

This reflection on one's own work is fundamental for the whole formative process, since, as described by Durrive (2011, p.58):

> the work situation is organized in advance by the prescription, the operating mode is thought by others and a whole instrumentation accompanies the man at work. For the latter, they are all precious resources. However, the knowledge inscribed in

the activity of the operator is important, which will lead and guide the action towards its result.

Santos and Lacomblez (2016, p.15) also point out that a professional education action should provide moments to analyze oneself and others in the activity to provide "the knowledge and reflection capable of promoting the personal and professional development of workers as well as the development of work situations." Such moments should be based, according to Lacomblez and Teiger (2007, p.593), on the following principles: "the recognition of the specificity of each one's experience; the need to create conditions that guarantee the communication and confrontation of knowledge; and the importance of the activity of reflection in the process that leads to action."

For Engeström (1987, 2013), people and organizations are learning something all the time, in a dynamic, indefinite way that cannot be understood a priori. Thus, it is necessary to develop professional education practices that allow a given collective or group to radically question the sense and meaning of the context, aiming to build a broader alternative professional education context. The idea is to make a group of people work collaboratively to tackle a contradiction and to develop the activity in a certain direction, producing new forms and culturally new patterns of work activity (Engeström, 1987, 2013; Engeström & Sannino, 2010; Virkkunen & Newnham, 2013).

In short, there are countless different interventional methodological tools, such as simple and crossed self-confrontation interviews (Fernández & Clot, 2007), the three-pole dynamic device (Trinquet, 2010), and the change laboratory (Virkkunen & Newnham, 2013). Each one, in its own way, plans to develop strategies in favor of healthier and more stimulating work environments. In the same way, they all have an urge for contributing to the development of collective spaces for reflection and evaluation of the meanings of the actions produced, which consider the human variability present in work situations, with a purpose of expanding the power of work collectives so they can act on the activities.

The undeniable and inevitable advances of digital technologies have been responsible for changing the universe of work. Artificial intelligence, robotics, autonomous vehicles, 3D printing, nanotechnology, and quantum computing are innovations that have already been present, even if only occasionally, since the so-called Third Industrial Revolution, in the middle of the last century. However, for many researchers, three characteristics, added to these innovations, inaugurated a new process called by some the Fourth Industrial Revolution: speed, depth, and systemic impact (Malvezzi, 2013; Sutherland, 2013; Schmidt & Cohen, 2013).

Schmidt and Cohen (2013) point to the issue of broad connectivity as the most relevant characteristic of this new digital era. Digital connectivity is contributing to renew habits and everyday relationships in the most varied social spaces and institutions, which can lead to productivity gains, improvements in health, education, and quality of life. For these authors, the expansion of connectivity will even have important repercussions on how citizens and states will behave in the coming decades.

Evidently, it is necessary to pay attention to the offer of simple explanations and definitive solutions for the understanding of this new reality, made by the written and spoken media. However, I believe that the dimension of connectivity alone, by modifying the patterns of social interaction, has already been causing important changes in the world of work, although it is still not clear what its definitive scope will be.

In this scenario, Falzon (2014) suggests that the ergonomics of the activity should focus above all on the development of empowering work environments, going beyond promoting the adaptation of work, environments, and machines to man. An empowering environment is fundamentally characterized by preserving and promoting the health of workers, being inclusive, and continuously cooperating in the creation and accumulation of new knowledge and skills. The point already stated strengthens professional education processes, especially the ones focused on the development of actions that encourage the construction and accumulation of information, knowledge, and skills, on the part of workers, and on reflective actions, open to the innovation capabilities of the workers themselves, on the part of organizations and institutions. These actions will only be possible as far as the work environments provide room for maneuver and freedom for workers to act on their own activity and to continuously build the rules of work.

3 The Theory of Economic Complexity

Searching for answers to some of the previously mentioned questions and betting on the connectivity dimension of the digital age (Hidalgo et al., 2007; Hidalgo & Hausmann; 2009; Hausmann et al., 2014; Hidalgo, 2017), as already mentioned in the introduction, were likewise responsible for the development of the theory of economic complexity. Within this regard, based on studies of cybernetic movement and complex systems (Bateson, 1998; Shannon & Weaver, 1972; Wiener, 1961), they built a database on economic indicators and created the Economic Complexity Index (ECI). It is a method to measure the productive sophistication or "economic complexity" of countries (http://atlas.media.mit.edu).[2] This method ensured quick access to thousands of data on international trade since the 1960s. When comparing indicators of economic complexity across countries, they found a strong correlation between the degree of complexity and the development of economies (Hausmann et al., 2014, 2017).

The Economic Complexity Index (ECI) combines data on the complexity and ubiquity of the goods a country produces. As countries usually do not have data on all products produced, Hausmann et al. (2014) use foreign trade data as a kind of server for the accumulation of know-how (tacit knowledge). The economic complexity of a country increases, as there is more added value in the products it exports. For these authors, the composition of a country's exports can reveal, as we will see below, information about the knowledge and know-how incorporated in a given population.

The theory of economic complexity suggests that economic growth is related to the growth of information, which, in turn, defines the degree of complexity of an economy. In this case, the adopted concept of information comes from physics. As Hidalgo (2017) reports, our universe is composed of energy, matter, and information, but while energy and matter are here because they are physical, information needs to find ways to manifest itself and it does so through objects. That is, information by itself is not tangible, but when it appears ordered in objects, it is physical.

For Hidalgo (2017), "the physical accumulation of information is the blood of our society." Objects and messages connect and enable individuals to further elevate information growth. In his work, the author seeks to build bridges between the physical, biological, social, and economic factors that contribute to the growth of information and increase our ability to process it.

In favor of the information growing in each society, Hidalgo (2017) presents two fundamental capacities: storing and processing knowledge (materialized in the object) and know-how (tacit knowledge). It is worth clarifying that knowledge and know-how are not the same thing. Know-how is different from knowledge because it consists of the ability to perform actions, even without being able to explain them. It is tacit knowledge. Knowledge, on the other hand, refers to an expertise that can be represented before action. Both, as already explained above, are considered as the great engine of economic development.

Hidalgo (2017, p.326) further explains that:

> in a similar way to what happens with information, which is accumulated in objects, knowledge and know-how always need a physical embodiment. However, unlike information, knowledge and know-how are accumulate in human beings and human networks, whose capacity is finite. This finitude of humans and the networks we form limit our ability to accumulate and transmit knowledge and know-how, which gives rise to accumulations of knowledge and know-how that result in global inequalities. Therefore, the need for knowledge

[2]There is a Brazilian version of this database, "DataViva," the result of a partnership between MIT and the government of the State of Minas Gerais (http://www.dataviva.info/pt/).

> and know-how to accumulate in human beings and human networks can help explain the disparities in our world. (Hidalgo, 2017, p.326)

In such circumstances, the level of information growth depends on the place where it will be processed, that is, it will depend on the capacity of the city, a company or even a work team. This condition means that, in order to understand economic complexity, it is necessary to explore and understand the social and economic processes that allow groups of people to interact and produce information. Among these processes is the formation of social and professional networks in which reside the ability to process information socially, which in turn implies the accumulation of knowledge and know-how. This is the importance of the connectivity dimension of the digital age. In this way, the accumulation of knowledge and know-how in human networks becomes a major economic and social problem that needs to be solved. The sizes, shapes, and evolution of these networks are limited by historical and institutional factors (Hausmann et al., 2015; Hidalgo, 2017).

For Hausmann et al. (2014), the practical and social nature of learning introduces a bias in the accumulation of knowledge and know-how that favors what is already available in the places where these individuals reside. This means that the accumulation of knowledge and know-how has a geographic bias. Thus, according to Hidalgo and Hausmann (2009), the products produced by a given region portray the stock of knowledge and know-how available in that location, which in turn portray the economic complexity. Measuring the accumulation of knowledge and know-how from different locations was the challenge that the construction of the database, mentioned above, sought to face.

As seen, social institutions play a fundamental role in our ability to accumulate knowledge and know-how, as well as in the mechanisms that limit the production capacity people have. Consequently, it is necessary to develop professional education strategies in work environments, communities, and society as a whole, in order to contribute to the growth of information, which involves our ability to build large networks and establish more accessible connections. It is worth clarifying that economic complexity concerns not only the production of more complex products but also the possibility that knowledge and know-how are organically distributed among different industries in the same region.

4 Economic Complexity and Professional Education: The Initiatives of the Basque Country and Estonia

In 2006, an initiative by the University of Deusto gave rise to the "Orkestra – Basque Institute for Competitiveness," in the Basque Country. The point was to build a model of competitiveness toward inclusive and sustainable well-being of the region in a collaborative way. One of Orkestra's main lines of action is the promotion of productive diversification and the development of highly complex sectors. To this end, the institute carries out studies on the strategic sectors of the Basque economy and identifies opportunities for innovation and the creation of added value. In addition, Orkestra collaborates with companies and public institutions to develop innovation projects and stimulate the creation of new businesses in highly complex areas (Alcalde et al., 2017).

Orkestra's mission, according to Alcalde et al. (2017) is to support, through research, professional education, technical assistance, promotion of debates, and participation in networks of excellence. These activities carried out by public administration, socioeconomic agents, and by the university itself, in fields related to competitiveness, looking forward high levels of quality of life and socioeconomic growth of Basque citizens. This whole proposal is made through the development of policies that guarantee equal opportunities and freedom of choice.

As reported by Wolf (2021), after 15 years of existence, this initiative contributed to the fall of unemployment in the Basque Country to far below the Spanish average, raising the GDP (gross domestic product) nominal per capita and also the purchasing power, so that the average income per capita in the Basque Country in 2019 approached that of Germany; and, finally, to the improvement of the indexes referring to social well-being, which, according to the OECD, is at a similar level to the richest regions of Spain.

Currently, the Basque Country, like the other regions of the world, is experiencing a moment of turbulence in the global economy, associated with the COVID-19 pandemic and the invasion of Ukraine, which provoked or accelerated a series of instabilities in the energy markets, raw materials, logistics, and other critical supply chains. Such conditions, besides favoring uncontrolled inflation, have produced an environment of high uncertainty for companies, families, and governments, precisely at a time when "profound transitions related to the energy-environmental, technological-digital, and demographic-social domains" are taking place. (Orkestra, 2022, p.3).

While facing these challenges, the 2022 Basque Country Competitiveness Report, produced by Orkestra (2022, p.9), proposes five (5) transversal lines of action to guide the actions of companies, institutions, and other actors: promote leadership for a new sustainable industrial competitiveness by fostering multilevel collaborative governance; reinforce, through professional education actions, people's abilities to contribute to competitiveness and well-being; strengthen the culture, capacity, and orientation toward innovation in companies, administration, and society as a whole; contribute to the increase in the complexity of the economy, stimulating the development of a sophisticated economic and technological fabric; and, to intensify the international connections between the economy and society in its pursuit of offering new opportunities, financing, and talent attraction.

In this way, Orkestra's actions demonstrate how knowledge can be used to foster the economic complexity of a region, promoting innovation, productive diversification, and the development of highly complex sectors. Orkestra's example can be inspiring for other regions whose aspiration is to promote sustainable economic development based on knowledge and innovation.

Another relevant initiative is the case of Estonia, which shows how the adoption of information and communication technologies can generate added value and promote the economic complexity of a country. The rapid development of information technology played an essential role in the reconstruction of the state of Estonia after the restoration of its independence in 1991. Today, Estonia is one of the most digitized societies in the world, with a large number of public services available to citizens online and broadband Internet coverage in most parts of the country. Investments in basic and professional education were one of the main pillars of this process of change and modernization (Mets, 2017).

This is a successful experience, the result of partnerships established between the public and private sectors in the last two decades. The starting point was the launch of the Tiger Leap program in Estonian schools in 1996. The program fully equipped schools with computers, Internet access, and other ICT services. Computer classes were taught in 84% of schools over the following 8 years. Soon after, there were the electronic government and public services and the expansion of e-commerce, which has placed Estonia among Europe's leading innovation- and entrepreneurship-oriented knowledge-based societies. Furthermore, Estonia has become one of the leading startup ecosystems, the result of a successful combination of educational and entrepreneurial ecosystems. The promotion of a favorable environment for innovation and entrepreneurship stimulates the creation of new companies and products, generating high-quality jobs and increasing the complexity and competitiveness of the economy (Kraav & Mets, 2019).

5 Challenges for Professional Education in the Information Age

There are countless questions and reflections still without answers that we will need to face if the purpose is to contribute to the development of inclusive, sustainable, and dignified work environments. In this way, professional education actions that strengthen the experience of workers and the management of the corresponding tacit knowledge, may result, in addition to their formal recognition by organizations, in better working conditions and operational, social, and even financial results (Ribeiro, 2013).

To overcome these challenges, the US National Academies of Science Engineering and Medicine (2017) suggest conducting multidisciplinary research, where theoretical perspectives can complement each other, overcoming some of their limits. The idea is that micro and macro disciplines can be integrated, such as, social sciences, economics, psychology, informatics, and engineering, to verify and elucidate how digital technologies are advancing and impacting society and, in particular, the world of work.

In this scenario, the challenge is to find ways to organize work and promote economic development that takes this reality into account and promotes a more sustainable, unequal, and inclusive work environment. As Falzon (2014) suggests, work situations to be considered "empowering environments" must contain "manageable" challenges and demands, meaning that, they are work situations in which, while tasks are designed, it is possible to guarantee the availability of necessary social, psychic, cognitive, and technical resources. Empowering work environments is therefore the opportunity for the development of new knowledge and know-how. It is about the development of competences in an explicit and conscious learning process, based on reflective practices, both individual and collective.

References

Abrahão, J., & Pinho, D. L. M. (1999). Teoria e prática ergonômica: seus limites e possibilidades. In M. G. T. Paz & A. Tamayo (Eds.), *Escola, saúde e trabalho: estudos psicológicos* (pp. 229–240). Editora Universidade de Brasília.

Abrahão, J., & Pinho, D. L. M. (2002). As transformações do trabalho e desafios teóricos- metodológicos na Ergonomia. *Estudos de Psicologia, Natal, 7*(n. esp), 45–52.

Alcalde, H., Aranguren, M. J., & Wilson, J. (2017). Culture and organisational change in academic organisations: A reflective case, EKONOMIAZ. *Revista vasca de Economía, 92*(02), 300–321.

Barbosa, A. C. Q., & Cintra, L. P. (2012). Inovação, competências e desempenho organizacional – articulando construtos e sua operacionalidade. *Future Studies Research Journal, São Paulo, 4*(1), 31–60.

Bateson, G. (1998). *Pasos hacia una ecología de la mente: Una aproximación revolucionaria a la autocomprensión del hombre*. Ediciones Lohlé-Lumen.

Baumgartl, V. O., & Primi, R. (2006). Evidências de validade da bateria de provas de raciocínio (BPR-5) para seleção de pessoal. *Psicol. Reflex. Crit., Porto Alegre, 19*(2), 246–251.

Baumgartl, V. O., Pagano, A. P., & Lacerda, J. (2010). A utilização de testes psicológicos em organizações de Minas Gerais. *Arq. Bras. psicol., Rio de Janeiro, 62*(2), 178–186.

Bourdieu, P. (2001). Os três estados do capital cultural. In M. A. Nogueira & A. Catani (Eds.), *Escritos de educação* (pp. 73–79). Vozes.

Clot, Y. (2010). A psicologia do trabalho na França e a perspectiva da clínica da atividade. *Fractal, Rev. Psicol., Rio de Janeiro, 22*(1), 207–234.

Daniellou, F. (Ed.). (2004). *A Ergonomia em busca de seus princípios: debates epistemológicos*. Ed. Blucher.

Daniellou, F., Laville, A., & Teiger, C. (1989). Ficção e realidade do trabalho operário. *Revista Brasileira de Saúde Ocupacional, São Paulo, 17*(68), 7–13.

Durrive, L. (2011). A atividade humana, simultaneamente intelectual e vital: esclarecimentos complementares de Pierre Pastré e Yves Schwartz. *Trabalho, Educação e Saúde, 9*(Suppl. 1), 47–67.

Engeström, Y. (1987). *Learning by expanding. An activity-theoretical approach to developmental research*. Orienta-Konsultit Oy.

Engeström, Y. (2013). Aprendizagem expansiva no trabalho: em direção a uma reconceituação da teoria da atividade. In Â. P. Simonelli & D. S. Rodrigues (Eds.), *Saúde e Trabalho em debate: velhas questões, novas perspectivas*. Paralelo 15.

Engeström, Y., & Sannino, A. (2010). Studies of expansive learning: Foundations, findings, and future challenges. *Educational Research Review, 5*(1), 1–24.

Falzon, P. (Ed.). (2014). *Constructive ergonomics*. CRC Press - Taylor & Francis Group.

Fernández, G., & Clot, Y. (2007). Entrevistas de autoconfrontación: um método em clínica de La actividad. *Laboreal, Porto, 3*(1), 15–19.

Guérin, F., et al. (2001). *Compreender o trabalho para transformá-lo: a prática da Ergonomia*. Ed. Blücher.

Guile, D. (2008). O que distingue a economia do conhecimento? Implicações para a educação. *Cadernos de Pesquisa, 38*(135), 611–636.

Hartmann, D., et al. (2017). Linking economic complexity, institutions, and income inequality. *World Development, 93*, 75–93.

Hausmann, R., et al. (2014). *The atlas of economic complexity: Mapping paths to prosperity*. MIT Press.

Hausmann, R., Espinoza, L., & Santos, M. A. (2015). *Diagnostico de crecimiento de Chiapas: la trampa de la baja productividad*. Center for International Development, Harvard University.
Hidalgo, C. (2017). *El triunfo de la información: la evolución del orden, de los átomos a las economías*. Penguin Random House Grupo Editorial España.
Hidalgo, C. A., & Hausmann, R. (2009). The building blocks of economic complexity. *Proceedings of the National Academy of Sciences, 106*(26), 10570–10575.
Hidalgo, C., et al. (2007). The product space conditions the development of nations. *Science, 317*(5837), 482–487.
Kraav, M., & Mets, T. (2019). ICT Infrastructure Development as Political Entrepreneurship: The Case of the Estonian Education and Research Network (EENet). In *17th IEEE International Conference on Emerging eLearning Technologies and Applications, ICETA 2019*, Starý Smokovec, pp. 453–459.
Lacomblez, M., & Teiger, C. (2007). Ergonomia, formações e transformações. In P. Falzon (Ed.), *Ergonomia* (pp. 587–602). Edgard Blücher.
Leplat, J., & Cuny, X. (1977). *Introdução à psicologia do trabalho*. Fundação Calouste Gulbenkion.
Malvezzi, S. (2013). Gestão de pessoas no contexto da estrutura de redes: desafios para sociedade, empresas e indivíduos. *Perspectivas em Gestão & Conhecimento, 3*, 6–17.
Mets, T. (2017). From the Educational Tiger Leap Program to the ICT Startup Booming in Estonia. In *15th IEEE International Conference on Emerging eLearning Technologies and Applications, ICETA 2017*. Stary Smokovec, pp. 289–292.
Montmollin, M. (1974). *A psicotécnica na berlinda. Uma autocrítica da psicologia industrial*. Agir.
National Academies of Sciences, Engineering, and Medicine. (2017). *Information Technology and the U.S. Workforce: Where are we and where do we go from here?* The National Academies Press.
Orkestra. (2022). *2022 Basque Country competitiveness report. Foundations of competitiveness in a time of uncertainty*. Donostia-San Sebastián, Orkestra - Basque Institute of Competitiveness.
Pastré, P. (2017). A análise do trabalho em didática profissional. *Revista Brasileira de Estudos Pedagógicos, 98*(250), 624–637.
Ribeiro, R. (2013). Tacit knowledge management. Phenomenology and the cognitive sciences. *Phenomenology and the Cognitive Sciences, 12*(2), 337–366.
Ryan, A. M., & Ployhart, R. E. (2014). A century of selection. *Annual Review of Psychology, 65*, 693–717.
Santos, M., & Lacomblez, M. (2016). Do artefacto ao instrumento: um modelo de avaliação das relações entre trabalho e saberes numa ação de formação. In *Trabalho & saber: questões e proposições na interface entre trabalho e formação* (1st ed., pp. 13–32). Mercado de Letras.
Schmidt, E., & Cohen, J. (2013). *The new digital age: Reshaping the future of people, nations and business*. John Murray.
Shannon, C. E., & Weaver, W. (1972). *The mathematical theory of communication*. University of Illinois Press.
Spink, P. K. (1996). Organização como fenômeno psicossocial: notas para uma redefinição da psicologia do trabalho. *Psicologia & Sociedade, Porto Alegre, 8*(1), 174–192.
Sutherland, T. (2013). Liquid networks and the metaphysics of flux. Ontologies of flow in an age of speed and mobility. *Theory Culture & Society, 30*(5), 3–23.
Trinquet, P. (2010). Trabalho e educação: o método ergológico. *Revista HISTEDBR On-line, Campinas*, n. esp., p. 93–112.
Virkkunen, J., & Newnham, D. S. (2013). *The change laboratory. A tool for collaborative development of work and education*. Sense Publishers.
Wiener, N. (1961). *Cybernetics: Or control and communication in the animal and the machine*. M.I.T. Press.
Wisner, A. (1987). *Por dentro do trabalho: ergonomia, método e técnica*. FTD.
Wolf, M. (2021). *Lições do País Basco para o desenvolvimento econômico*. Folha de São Paulo (2021-11-30).

Luiz Gonzaga Chiavegato Filho has a Bachelor's degree in Psychology from the Federal University of Minas Gerais (1996); is specialized in Ergonomics from SENAC/SP (2005); and has a Master's degree in Collective Health from the State University of São Paulo (2002), a PhD in Psychology from the University of São Paulo (2011) and a postdoctoral degree in Work Psychology from the University of Porto/Portugal (2018/19). Currently, he is an associate professor at the Department of Psychology at the Federal University of São João Del-Rei (UFSJ), with experience in the area of work psychology, ergonomics and workers' health.

Educational Inclusion of Students with Disabilities in Situations of Social Vulnerability

17

Elifas Trindade de Paula and Maria Candida Soares Del-Masso

Abstract

The main objective of this research was to understand the factors that interfere in the activities of teachers dealing with special education service, concerning areas of social vulnerability. To delineate the social vulnerability condition, we have considered the concepts of social class and crisis, taking into account classical works of Marx and Engels. Considering this referential, it becomes clear that the social vulnerability is a contradiction of the capitalist society. Aside from its huge productive ability, it is not able to offer worthy conditions of life for most of the population to read the main goal. Hence, we have chosen to hear the professionals that work as teachers of special education service, with half-structured interviews. To know better the work conditions, we realized the interviews "in loco," taking notes as observations that allowed us to consider the material and human conditions of the places where the attendance happens. After collecting the data, we have organized a category analysis, considering the main factors that interfere on the work of experts that attend special education target students and performed a qualitative analysis. First, we have detached work conditions, family situations, and vulnerability. After, considering the public policy, we analyzed the support offered by equipment available in the community. The results have shown that even with equipment's and furniture available multifunctional classes and resources, the structural conditions and vulnerability conditions interfere negatively on the teacher's performance; the public health equipment is insufficient for demand fulfillment, and there is few informational and articulation with leisure, sport, and culture. Concerning the public policy, the data reveal that many times the schoolwork in isolation in the conclusion process. Finally, this research has shown that special education target students have the school as the only resource since other different services are lacking in socially vulnerable areas.

E. T. de Paula
Municipal Education Network of São Paulo,
São Paulo, Brazil
e-mail: e.paula@unesp.br

M. C. S. Del-Masso (✉)
São Paulo State University (UNESP),
São Paulo, Brazil
e-mail: del.masso@unesp.br

1 Between the Right to Education and Reality

The redemocratization process of Brazil, whose landmark is the Constitution of 1988 (Brasil, 1988), presented to all Brazilian people the possibility of accessing a set of rights, defined in the

A. Pereira Jr, F. Sousa (eds.), *Principles for Governance*, Sustainable Development Goals Series,
https://doi.org/10.1007/978-3-031-40025-4_17

body of the text as social and political, that would be fundamental for the exercise of citizenship and full participation in the democratic society reopened after 25 years of authoritarian governments. Among social rights, we highlight access to education, which, according to the constitutional text, aims at the full development of individuals, their preparation for citizenship and for work. In the principles established by the Charter, we highlight equality of conditions, freedom to teach and learn, guarantee of quality standards, and appreciation of professionals involved in the educational process. Looking through this prism, education appears in its idealized form, whose function would be to guarantee all the people access to the knowledge produced by humanity, necessary for a full life in accordance with the demands of their historical time.

However, it is necessary to verify how the law is applied and prescribed to put into practice. Saviani (2008) proposes a critical analysis of education, denying both noncritical theories and critical-productivist theories. The researcher recognizes that education, in class society, is organized to maintain domination, but he also sees it as a disputed space, and as such, its directions are conditioned to the relationship of forces in society. In this sense, the process of struggles for redemocratization that guaranteed access to education for all the people can be considered the first moment of the dispute, but for the ruling class it is enough to offer a minimum of knowledge and prepare the individual only for work. Further on, action by the dominated class is necessary, in the sense of continuing the movement and "bend the stick to the other side," through a "revolutionary pedagogy" (Saviani, 2013, p. 236).

In recent decades, we have experienced the intensification of debates about inclusive education, assuming the guarantee of access and permanence in education for all people, in different teaching modalities, for admission and/or continuation of studies, whether in education infant, basic, superior, and/or postgraduate. In this group are people with disabilities, autistic spectrum disorders (ASD) and high abilities/highly gifted, a group considered in accordance with official legislation as a special education target students (Brasil, 2011). This expansion, which is consistent with advances in science, has led to paradigm shifts in the way of seeing and acting, both by public authorities and by society, in relation to students eligible for special education.

Signatory of international conventions, treaties, and conferences on the guarantee of human rights, Brazil assumes the commitment to implement public policies in the perspective of inclusive education to fulfill the commitments to the international and national communities. Thus, the construction of the necessary conditions to guarantee quality education for special education target students cannot be understood as a gift, a concession from society; on the contrary, it is an obligation of the State, which should even evaluate its implementation and execution.

This information is not restricted to the recognition of the right and legal guarantees, but it leads to a change in the social paradigm, as it reveals that the problems are in the barriers that people with disabilities, autism spectrum disorder (ASD) and high abilities/highly gifted to access services, social facilities, and public spaces. They are attitudinal, physical, mobility, communication, and technological barriers, and, regarding education, we can highlight that are barriers to prevent or limit access to knowledge.

During this paradigm, special education conquers a new configuration and adopts the concept of transversality that permeates all levels and teaching modalities, ceasing to be a substitute to assume the role of specialized support, with the aim of guaranteeing the schooling of the special education target students. With this comes a set of student support services, including the multifunctional resource class, spaces that operate in schools under the responsibility of specialists in special education, inclusive education, or in some specific area related to the characteristics of the special education target students (De-Paula, 2022).

For an analysis of the intervening factors in the performance of these professionals, we resorted to the Marxist framework, which presupposes multiple determinations of educational reality (Marx, 2013; Netto, 2011). The individual work of the professional in specialized educational services is also a social and collective work, resulting from a set of factors:

1. Material conditions for the execution of its action plan, such as equipment, furniture, spatial dimensions, classroom, and school location
2. Subjective issues, such as the perspective of families in relation to the learning of special education target students, the world and education conceptions of school professionals
3. External factors, such as the provision of public equipment, provision of complementary services for those students who need autonomy to work

Table 17.1 Student services at multifunctional resources classes

School	Number of students assisted at multifunctional resources classes	Number of students assisted by the teacher of specialized educational services
01	52	52 (accumulate positions)
02	21	21
03	37	20
04	20 from 4 school	20
05	67/54 from CEU + 13 surroundings	19
06	22	22
07	30	30
08	43	43 (accumulate positions)
09	47	24
10	44	20

2 Working Conditions: SRM, Families, and Vulnerability

Access and permanence in schools is a necessary condition for the educational inclusion of all students. For this reason, a set of policies has been implemented with a view to eliminating barriers that prevent or limit special education target students' access to the curriculum. However, it is necessary to look at the quality of space, furniture, and equipment available since different aspects can interfere with the quality of education. Added to this, the conditions faced by teachers in their professional routine often clash with the legal assumptions that point out to a perspective on the number of special education target students in multifunctional resource classes and, what happens is the excessive number of students under the responsibility of the teacher.

To understand this context, we point out some information about the equipment and furniture available at the multifunctional resources classes, as well as the number of students in attendance. Data referring to equipment and furniture were collected through observation by the researcher and noted in the logbook; the number of students attended at the special education target students was informed by the teachers when the interview was realized. The number of students assisted appears in Table 17.1, as we did not bring the data collection by separating them by service modality because in the Municipal Network of São Paulo, all special education target students, indicated for attendance, are under the responsibility of the teacher who attends at multifunctional resources classes. The definition regarding the type of service, and whether it will be after-hours, collaborative or itinerant, should be the result of a joint evaluation by the common class teachers, school management, and the multifunctional resources classes teacher, in addition to reflections from the families (São Paulo, 2016b). When we refer to the opinions and observations of the nine participating teachers, they are spontaneous expressions that were noted in the logbook after interview moments.

The classes are well equipped with computers, printers, literacy games, math games, sensory panels, sound and television equipment, and stationery, and some have typing machines and Braille printers for students with visual impairments. In addition, we observed furniture such as desks and chairs, cabinets, and shelves, some of which were adapted for people with physical disabilities. Despite some of the teachers' observations regarding the lack of some equipment, Internet access problems, equipment in poor condition, furniture under renovation, and games and toys missing parts, this part has a positive influence on the teachers' work, as adapted furniture, and adequate equipment, in some cases, is essential for the development of students and is part of the Specialized Educational Services implementation program (MEC, 2010).

As for the number of students in attendance, the majority (seven classes) had the number of students above that defined by the Official Law of the Municipal Secretary of Education of São Paulo, which organizes the Specialized Educational Service in the city (São Paulo, 2016b). According to this Official Law, the service can be organized in three ways, collaborative, when the Specialized Educational Services teacher accompanies the students in the common classes to guide the professionals involved in their educational process; after school, modality in which the school team together with the family understand that there is a need for complementary or supplementary support, extending the student's permanence time at school; and the itinerant service, which are students from surrounding schools, where there is no multifunctional resources classes for Specialized Educational Services, and students are referred for follow-up at a closer multifunctional resources classes. Official Law n° 8.764 (São Paulo, 2016b) also defines that each teacher must attend from 12 to 20 students, regardless of the form. However, six of the nine teachers interviewed served more than what was defined in the document; the others had the maximum number of students. A teacher who accumulated a position attended to more than 50 students, when they should have attended a maximum of 40, with 20 students per position.

About the number of students assisted, there are complaints of a negative influence on the part of the teachers, as it is often necessary to reduce attendance after shifts to cope with the demand. In addition, according to the Official Law, the form of assistance must be carried out according to the needs of the students, as stated in the text of the law:

> Article 24 The Specialized Educational Services Plan will be prepared and executed by the EU educators together with the AEE Teacher and/or with the support of the Inclusion Support and Monitoring Teacher and must be preceded by a pedagogical evaluation/case study, including the following:
>
> I. Identification of abilities, existing barriers, and specific educational needs of male and female students
> II. Definition and organization of pedagogical and accessibility strategies, services, and resources
> III. Type of service according to the specific educational needs of male and female students
> IV. Service schedule
> V. Workload (São Paulo, 2016b)

To attend the demand, the Specialized Educational Services' teachers reported that they needed to disregard the specific needs of the students, offering support 1 day a week for 2 hours/class, or offering collaborative assistance to students who would need the extra shift. Considering the size of the classes, seven of the ten schools surveyed are suitable for serving groups of five students, and three schools have small spaces, insufficient to serve groups of three or more students.

About the lighting, only one classroom, which was being renovated, did not have adequate conditions. The main problem presented by the participants and observed by the researcher was regarding the location of the classrooms inside the schools and external interference. The classes were, in general, located close to courtyards and other places of great movement of personnel, some even close to multisport courts. These classrooms receive external impact, especially about noise. The teachers argued that the losses are greater for students with ASD, with hypersensitivity to sound and those who have greater attention and concentration difficulties. The advantage of carrying out the interviews in the teachers' workplaces was the experience of the problems faced. For a few moments, we had to interrupt the interviews or must raise our voices, because of external interference.

Structural conditions and suitable environments are constitutive parts of the pedagogical process and have a direct influence on the quality of care offered and on the worker's quality of life. From the observations and reports made available, the equipment and furniture were adequate for carrying out the specialized pedagogical support, as a complement to guaranteeing the Special Education Target Student's learning, but the spatial dimensions and location of the multifunctional resources classes did not receive the same attention from the public power.

Aiming at analyzing the interviews with the nine teachers participating in this study, we organized the data into two classifications: (1) facilities and difficulties encountered in the professional context and (2) families and vulnerability.

2.1 Classification: Facilities and Difficulties

For a better visualization of the main difficulties and facilities presented by the research participants, we prepared a comparative table, in which we established a dialogue with the teachers' speeches (Table 17.2).

Concerning questions about the facilities and/or difficulties faced in the development of the activities of teachers in the multifunctional resources classes, we identified problems in accessing Internet and external noise coming from games among other things.

Table 17.2 Facilities and difficulties of the professional context

Difficulties	Facilities
Infrastructure	Communication and relationships with families
Attitudinal barriers	Internal communication in schools
Number of students served Absence of public facilities	Proximity to students Reduced number of students
Number of students to follow	Possibility of using other spaces in the school
Noise pollution	Management learn support
Absence/limitation of support human resources, AVE and trainee	

Another difficulty presented by teacher P1 was the number of students, which was above that specified by Official Law No. 8764 (São Paulo, 2016b). As the teacher accumulated positions at the same school, she should have attended 40 male and female students, but in the period in which the interview was carried out, the teacher was assisting 52 students. As facility, the teacher at this School 1 cited the communication with most families, which took by WhatsApp.

The fact that it serves a larger number of students than established by law revealed the difficulty of the education network in responding to the demands for vacancies in the region, with the absence of a school with multifunctional resources classes and/or professionals to assume the role. According to the investigation, in 2021, there were 1207 special education target students enrolled in the schools of the Guaianases Regional Education Board (DREG) and 33 classrooms for attendance, according to information provided by the Inclusion Training and Monitoring Center. Even considering that a percentage is monitored directly by this center, the numbers revealed demand is greater than supply.

The teacher from School 2 (P2) pointed out that infrastructure was the main difficulty in her work, especially regarding the size of the classroom, but she pointed out that there is free movement in the unit, and she can use other spaces available in the school. As a facilitator, this same teacher cited the possibility of being closer to the children, as this allowed them to better understand the barriers and work with them. The teacher at School 3 (P3) presented the attitudinal barriers of some teachers as the main difficulty. In his speech, he pointed out that there is still the old discourse of some pointing out that "I don't have the training to work with these students" (P3). However, she highlighted that this number is very low in her unit, and the research work and exchange of information between the teachers in the multifunctional resource class and those in the regular class are processes that have been fundamental to break down these barriers and, by your view, is a work facilitator.

This is the same complaint of the teacher at School 5 (P5) who cited that "The biggest difficulty is still the attitudinal barriers. There's always the obstacle of, 'but it can't, but it won't work, but it doesn't have time, but it doesn't have resources [...].'' For the teacher at School 4 (P4), one of the main problems is the number of schools in which she works, because her can't find out what's really going on at the schools.

The family can either be a facilitator or a hindrance, according to the speech of the teacher from School 6. In another part of the interview, the teacher observed that this family often encounters problems in the social protection network and is also a victim of the situation of vulnerability that affects the peripheral regions of the city of São Paulo.

The teacher at School 8 (P8) highlighted, among the facilitators of the work, the small number of students, as it makes it possible to form small groups to meet the specific needs of each one of them and the good relationship with the management team. The good relationship with the management team also appeared as a facilitator of the speeches of other teachers, especially P2, P3, P4, P9, and P10.

As difficulties, the teacher from School 8 (P8) highlighted the noise pollution in the school, as the classroom is located at the beginning of the main corridor of the building and faces an important avenue in the region. The attitudinal barriers of some teachers were also pointed out. In addition, the teacher highlighted the difficulty of working with teachers of Elementary School II, due to the dynamics of exchanging classes, since most public schools in Brazil work with the system of specialists by subjects with classes of 45 or 50 minutes long.

For the teacher at School 9 (P9), as previously highlighted, the facilitator was the relationship and support of the management team. As for the difficulties, she pointed out the resistance of some professionals and the difficulties of families in getting support in the protection network, analyzing that "Our work depends on other works for it to happen." Teacher P9 also highlighted financial and human resources as a hindrance.

According to this, the first speech it is recurrence, although it does not appear much in the question about facilities and difficulties for the development of the work, problems with human resources appeared in several parts of the interviews. In addition to the Specialized Educational Services Teacher, Official Law No. 8764 (São Paulo, 2016b) provides for support for Special Education in schools, trainee, Scholar Support Aids, and the presence of these supports in the school should be done by assessing the needs of students and the school unit. However, most of the time, this is not what happens, as the DRE allocates a fixed number of trainees, and the school needs to organize itself based on this available number of supports.

2.2 Classification: Families and Vulnerability

In this subsection, we analyze parts of the teachers' speeches realized during the interviews: The question investigates: How do you see that the social reality of the students' families interferes in the schooling process of special education target students?

Vulnerability is presented by the teachers' reports in different ways. We highlight the points related to these difficulties in face of vulnerability such as contact with families, lack or limitation of access to therapies that in some cases are essential to the schooling process, lack of vacancies in the territory where families live, financial condition of families to find therapeutic assistance or follow-up in central regions of Sao Paulo city, the condition of single mothers who assume the responsibility of raising their children, and the difficulty of some families in understanding the role of education in the development of special education target students.

The main point of the responses of the teachers from different schools shows a relationship between the role of the family, the condition of vulnerability, and the difficulties in accessing the social protection network. Even some teachers who initially criticized the families, at the time of this question recognized that even with the efforts of the families, the difficulties were due to the absence and/or insufficiency of public policies. An important highlight in P3's speech, which initially did not consider the families of the children assisted as vulnerable, analyzing that it could be a momentary condition due to temporary unemployment. But, in another part of the interview, in which she spoke of leisure spaces, she recognized that the only spaces used by children were the parking lots of buildings. When he evaluated the territory's public policies, he considered them unsatisfactory, pointing out that many families only have the space of the school.

This lack of equipment and public policies is part of the Social Vulnerability Index (IVS) (Costa & Marguti, 2015). Therefore, even if the situation is apparently temporary, the region is experiencing a history of abandonment by the public authorities. Teacher P5 showed how this absence of public policies particularly affected the poorest families, as only 4 of the 50 students attended at the multifunctional resources classes of its school unit were well assisted, because they had medical insurance. That is, less than ten students attended had the ideal conditions, from the teacher's point of view. Similar situation are observed evaluating teacher P7, refering that support could come from other community resources. The one that has better financial condition presents better results, as a set of factors is necessary for adequate life in society.

We highlight the report by P9 about the families that receive the benefit of continued provision (BPC). To be entitled to the benefit, families need to prove a per capita income of less than a quarter of the current minimum wage, which in August 2022 was stipulated at R$ 1302.00 This lack of equipment and public policies is part of the Social Vulnerability Index (IVS) (Costa & Marguti, 2015). Therefore, even if the situation is apparently temporary, the region is experiencing a history of abandonment by the public authorities. Teacher P5 showed how this absence of public policies particularly affected the poorest families, as only 4 of the 50 students attended at the SRM of her school unit were well assisted, as they had medical insurance. That is, less than ten students attended had the ideal conditions, from the teacher's point of view. A similar situation in the evaluation of teacher P7, for whom the difference between the family's financial conditions could be perceived by support outside the school.

In order to have this benefit (BPC), families need to prove a per capita income of less than a quarter of the current minimum salary, which in August 2022 was stipulated at R$ 1302.00[1] (DIEESE, 2023). For comparative purposes, this department indicates the amount of R$ 6571.52, from March 2023 as the necessary salary for a family of four people, that is less than necessary to live. In that way, we can see a double process of exclusion, of the disabled person who receives an amount infinitely lower than what is necessary to live and the family that is excluded from the formal labor market, as mentioned by Cury (2008) as a process of exclusionary inclusion.

3 Public Policies, Articulation and Evaluation

According to Secchi (2017, p. 04), "Analyzing public policies is to strive for public policies to be more adequate, have more long-term benefits and are technically consistent, socially sensitive, politically viable." We understand this suggestion from two perspectives: first from the side of the dominant social class, which always to consider to public policies as a way of preserving its economic model (Del Pino, 2002) and from the side of the dominated social class, necessary but insufficient.

Public policies for the educational inclusion of special education target students are based on the elimination of barriers that prevent or limit access to knowledge (Brasil, 2008; São Paulo, 2016a). The reports by the teachers who work in the multifunctional resources classes, fundamental agents of this process, allow us to observe the limits and possibilities of these policies in the territory and, also, to know if they fulfill the function determined by the public power.

The way to verify the effectiveness of a public policy can be through the existence and effective-

[1]The value of the minimum salary in Brazil, in May 2023, will be R$ 1320.00 Brazilian real – 1 US dollar equal to 4.99 Brazilian real (Vilela, 2023).

ness of the equipment that is available for a region and the capacity to find the demand, in addition to the ability to articulate between the different services.

In the interviews, we presented a question that aimed to verify which public facilities were known by the teachers and which services could be accessed to guarantee the schooling of Special Education students in the territories. All the teachers pointed out that they knew about health equipment, such as Basic Health Units (UBS) and Psychosocial Care Centers (CAPS). In addition, other equipment was mentioned by some teachers, including the Lucy Montoro Rehabilitation Service, Ambulatorial Medical Assistance (AMA), and one interviewee, mentioned the University of São Paulo City (UNICID).

With regard to culture, one teacher mentioned the Municipal Library, which is close to the school; the Culture Factories were mentioned by three teachers; one teacher cited a group called Pombas Urbanas; and another mentioned a Municipal Sports Center (CDM), which, according to her information, also brings cultural activities. As for sports equipment, in addition to the CDM, two teachers mentioned private soccer schools and one remembered a center in Itaquera that has a Down Futsal team. In the laser, two teachers mentioned Parque do Rodeio, located in the district of Cidade Tiradentes (Neighborhood named Tiradentes City). However, one of them said that she knows the park and that it is in a poor state of conservation, presenting accessibility problems inside, despite having accessible entrances. In addition, according to the speech of one of the teachers, the leisure spaces that appeared in the reports are the parking lots of the buildings where they live and the spaces where dances take place.

As for social assistance, although it did not appear in the answers about known equipment, it is a public policy that permeated all conversations, always related to the vulnerable situation of families and considered extremely relevant for the schooling process of a part of special education target students.

Finally, other institutions that appeared in the interviews were the Brazilian Association for Social Assistance and Development (ABADS), a nongovernmental organization for people with intellectual disability and autistic spectrum disorders (ASD), and the Support and Monitoring Center for Learning (NAAPA), an agency linked to the São Paulo Municipal Education Network, which prioritizes students with learning delays and in vulnerable situations. The teachers who cited the NAAPA reported difficulties in meeting the demand, as there are few employees for the number of students in this situation that can be illustrated by the speech of the teacher from School 1 (P1) "There is the NAAPA too, but not much there is an employee, because there are many cases that we refer there, we do not have answers [...]." (Verbal information). Also, we have some references to Unified Educational Centers (CEU), with spaces for sports and culture and which could offer activities for students with disabilities. But, according to the reports, the activities are offered to the public, without suitability to meet the needs of students with disabilities. In another interview, the teacher pointed out problems with Recreio nas Férias (Vacation Recreations), a municipal government program that offers recreational, cultural, and sports activities for students during school holidays.

When we described the equipment for the "Guia Regional de Politicas Públicas" (Regional Public Policy Guide), available at https://repositorio.unesp.br/handle/11449/239379, we found the teachers' reports about the shortage of leisure spaces, culture, and sports and, for the few we found, is not visible information for special education target students. In the cultural program of the CEU for September 2022 (São Paulo, 2022), for example, we did not observe any information about the presence of a Libras (Brazilian sign language) interpreter or audio description professional to accessible activities to deaf students or with visual impairments.

In Jacoby's (2014, p. 108) criticism of Piketty, we found the following expression "Limited social capital condemns exclusion as much as an empty bank account." For the dominant class, the material poverty of the working class is not enough; spiritual misery is necessary.

Sports, cultural, and leisure activities are also important for the development of special education target students than as school content. There will be no provision of services by the government without claiming action by social actors, in this case, the school community with all its segments, in a process of articulation with equipment and services to create demand, putting students' needs on the agenda.

3.1 Articulation Between the School and the Equipment

Regarding the articulation of the school and the multifunctional resources classes with public equipment, we observed in the interviews, in general, the management team of the school unit. We observe that the articulation exists, but there is no relation between the professionals involved; the follow-up do not translate into case studies, in the articulation for the development of actions with students and/or families.

This teacher related an interesting experience, which involved several meetings between CEFAI and UBS representations to discuss cases and referrals; however, it was an experience that did not last long. It will be an action that could be extended to other services and become a permanent forum for discussing cases and follow-up.

In P3's statement, we observed the articulation between the teacher in the common class, the teacher of Specialized Educational Services and the family, which resulted in referral to the health service. It also shows the relationship between the school and the health service. The point for our attention once again is the relationship by e-mail, without direct contact between the professionals involved, condition that reveals problems.

In addition to the difficulties, the teacher related problems in last years, because the access to the health service is more difficult. We observed the participation of a representative from social assistance, which plays a very important role in a vulnerable territory such as the region in question. However, even with this participation, until this moment, the case cited by P4 had not been resolved.

One of the possible answers to the problem may be found in this teacher's report, as the process of bureaucratization of relationships ends up making actions difficult. The Specialized Educational Services teacher cannot contact the health service directly, and it is necessary to pass the problem on to the management of the school unit, which, in turn, forwards it to the sectors of the Regional Board of Education, such as CEFAI and NAAPA. In the same way that teacher P2 reported, an attempt to contact appears, in the 2019, for the articulation between the different services. But there was no sequel, and, once again, health, sports, and leisure services were not considered.

In the previous report, we highlighted the articulation with cultural equipment, in which the school usually takes students to participate. Fábrica de Cultura (Cultural Factory) is a project of the state government of São Paulo, which offers courses and cultural presentations, and in the territory studied, there is one in the District of Tiradentes and another on the border of the District of Lajeado with the District of São Miguel Paulista. They are equipment that could be provoked to think about training actions for special education target students.

The highlight of P7 is in the relationship with the management team of the school unit, especially with the management, apparently, that it is not a question of the absence of autonomy of the Specialized Educational Services teacher, but of a partnership future case.

The first highlight in P8's interview is about the articulation with Municipal Sports Clubs (CDM), currently known as Community Sports Clubs (CDC) (São Paulo, 2016c), which are municipal equipment managed by nongovernmental organizations, which cultural and sports activities. The teacher also cited the relationship with the management team of the school unit, especially with the management that, apparently, is not about the lack of autonomy of the Specialized Educational Services teacher, but about a partnership for the referral of cases.

In the last appointment about articulation, the teacher mentioned health equipment, such as UBS and CAPS, cited that the monitoring is done by the coordination, but in a specific case, a teacher can participate in meetings with the CAPS to discuss the situation of a student. In the development of the conversation, the teacher reported that the relationship with the management team is very good, and there is a joint work to seek the necessary referrals to guarantee the education of the special education target students. The management team establishes contact with public facilities, but the resource class's teacher also has the autonomy to discuss cases with professionals from other public facilities.

The Official Law that institutes the operational guidelines for Specialized Educational Services in Basic Education (MEC, 2009) defines that establishing intersectoral partnerships for the elaboration of strategies for accessibility is one of the attributions of the Specialized Educational Services teacher. The official document (São Paulo, 2016b) mentions that in addition to the duties of the SES teacher, it defines duties for the members of the Unit's Management Team and the Regional Education Board. The cooperation between the different teams is fundamental for the productive articulation between the school and other public facilities.

3.2 Evaluation of Public Policies

From the set of attributions conferred by the legislation (São Paulo, 2016b), both schools and teachers of Specialized Educational Services are fundamental actors for the implementation of the São Paulo Special Education Policy from the Perspective of Inclusive Education. However, education is just one of the many actions of the public power necessary to exclude the barriers that prevent and/or limit the effective participation of people with disabilities in social spaces. Educational action can often only be effective if a set of services are mobilized. Therefore, it is essential that the voice of the professional responsible for the Specialized Educational Services is heard.

Regarding the issue of how the teachers evaluate the set of public policies for the educational inclusion of special education target students, the teachers freely expressed their opinions about the process of access to public services, the assistance in the multifunctional resources classes, the difficulties of the families, the different conceptions of municipal management, advances and setbacks in the process of schooling special education target students, the role of families and the school, and material and human resources, among others.

We observed from the teacher's speech that the time allocated to care is not planned according to the students' needs, but according to the existing demand in the school space. In this case, the students only participated in the activities after school, once a week, which is considered insufficient by the teacher. This situation happens, in some cases, due to the number of students served, which exceeds what is established by law (São Paulo, 2016b). In relation to the set of public services, the teachers assess them as insufficient, especially with regard to the provision of sports, leisure, and cultural services, considered essential for the full development of individuals (Mazotta & D'antino, 2011).

For this participant, there was a setback in the last 4 years; she recognized that it may be linked to the lack of investment in public services in that period. We know that since 2016 the country has lived under the aegis of a public spending ceiling with limited budget allocations for services considered essential. In the opinion of Oreiro and Ferreira-Filho (2021), these economic policies are wrong; they deepen the crisis to ensure profit for rentier capitalism. Another factor to consider is the process of privatization and outsourcing that affects all public services, equipment, and municipal public spaces. Pietro and Laczynski (2020) show that there was a privatization offensive in the city of São Paulo, starting in 2016, but studies such as Contreiras and Matta (2015) point out that what happens in the capital of São Paulo is the continuity of the privatization process initiated in 1990s. This period coincides with the intensification of the guidelines of international organizations in Brazil (Biondi, 2003; Melo, 2005).

Although the interviewee points out that families are held accountable for absenteeism in relation to referrals, we must consider the difficulties faced in accessing public services, which can often cause discouragement. This suggests that the school, in vulnerable territories, becomes a reference space for the community.

One important point that drew attention in this teacher's speech was the distance between what the set of legislation proposes and which the school is inserted. Silvestre (2019) portrays this situation as a dichotomy between what is prescribed and what is experienced, since rights are guaranteed within the scope of the law, but a daily battle is needed for its implementation. In recent decades, there have been important advances in terms of the theoretical collection and the legal and normative set, but the reality of public services is still insufficient to guarantee the basics for the process of inclusion of people with disabilities.

Participant P6 brings other elements that reinforce the evaluation of P5, pointing out the architectural barriers for people with reduced mobility, the lack of equipment in the territory, and the financial difficulties of families to travel to central regions in search of care. This consideration is important for anyone who lives on the outskirts of the city of São Paulo, as traveling to the city center is a complicated task for anyone and sometimes a medical appointment means spending the whole day away from home, implying expenses with transport and food.

Participant P7 presents the faces of an unfinished construction. When looking at school and social inclusion as a process, the teacher leads us to a reflection on the historical process of education for people with disabilities who are now considered eligible for special education. A history marked by exclusion and segregation (Aranha, 2001; Mantoan, 2003; Mendes, 2006) which, in the early 1990s, takes the first steps in the process of integration and, later, that of inclusion. The landmark of the inclusive process in Brazil is the "National Policy on Special Education in the Perspective of Inclusive Education" (Brasil, 2008). The participant points out that the Municipal Network of São Paulo has advanced a lot, but there is a lack of support professionals, such as Libras interpreter interns. At other times, some teachers complained about the lack of a scholar support aids, a professional who assists the teacher in the classroom. Also, there is a moment when the presence of the researcher in the field showed facts of concrete reality, which confirmed the testimonies. During the observation of the spaces in one of the schools, together to the Specialized Educational Services teacher, we met an professional of scholar support aids, who told us, with some emotion, that she heard about hiring another professional to help with the work. In the report, the worker said that she was no longer able to work alone.

Oliveira and Pietro (2020, p. 357) when studying the training of multifunctional resources classes teachers to deal with the diversity of Special education target students, appointed that "The risk that must be avoided is that the school inclusion policy remains provisional and precarious, always waiting to reach the quality levels required by the commitment with quality education for all." Making an analogy with official national documents (Brasil, 2008), the risk we must avoid is to always keep the perspective on the future, on an increasingly distant horizon.

One of the hallmarks of the unfinished process of schooling special education target students appeared in the interview with teacher P8, as the impression is that inclusion only happens through and at school. When we look at the role of the school in vulnerable territories and compare it with information about the difficulties faced by families in accessing health and social assistance services, we observe the absence of culture, sport and leisure equipment, and the impression given by the teacher is understandable. When not finding an accessible support network, families consider the school as the appropriate space to seek guidance, information, and referrals.

Bringing Oliveira and Pietro's (2020) notes back to the discussion, multifunctional resources classes' teachers need to deal with the diversity of the special education target students, for which 84% do not feel prepared to realize the activities; let's add to that the working conditions and the number of attendances above that established by law.

We noticed that the need for therapies appeared many times in the participants' speeches, and it is important to analyze this issue. The teachers are clear that not all students need therapies as support for schooling, but for some they are fundamental. This goes for complaints about the importance of support from interns and stroke professionals. Participants highlighted some problems related to interns, such as, the transfer of educational responsibility by some teachers to these interns. However, the DRE hires a fixed number of students to work as interns, and sometimes the number indicated by the school does not meet the needs of the students. Schools and families compete for this support, including calling on the services of the Public Prosecutor's Office. In addition to the students, many times those who need support therapy are the families themselves, so that they accept the situation of their child's disability, as well as to overcome the grief. Another issue that appeared in a significant way was the material condition of the families, as we are talking about one of the poorest regions of the city of São Paulo. Resuming a quote from the teacher, "We cannot close our eyes to this situation."

Silvestre (2019), referring to Ida Brandão, call our attention to the inefficiency of legal and normative documents, when there are no concrete actions to implement what is written, as the documents alone do not solve the problems of schooling of special education target students. Cury (2008) appoints of an exclusionary inclusion process and that this process perfectly serves the interests of capitalism. Agreeing with both authors and going further, because if it serves the interests of a class, the lack of conditions is not an accident, but an intention. For Engels (2010), when a class acts deliberately and condemns the other regarding the situation of deprivation, we are facing a crime, a murder.

4 Final Considerations

Returning to the initial condition that motivated the proposition of this text, that is, the factors that interfere in the work of teachers of Specialized Educational Assistance in territories considered to be socially vulnerable, we present below some considerations that explain a complex reality that is far from the ideals of inclusive education.

The fact of studying and researching known territory, in an area that we have been dedicated to for more than 20 years as worked, allowed us to avoid what Gewirtz and Cribb (2011, p. 123) call "criticism from above." However, when listening to the multifunctional resources classes' teachers, we realized that the situation is more worrying than it might seem. Thus, we point out the main factors that interfere negatively in the work of specialist teachers.

When we observed that seven of the nine research participants accumulated positions or another professional activity to supplement the family income, it was already a worrying sign, since, in addition to the health problem of the worker, carrying by the articulation of public equipment often requires time that did not fit into the regular working hours of these teachers. In addition, the reports showed that some Specialized Educational Services teachers have relative autonomy to articulate with the equipment available in their community, depending on the school management or even on the Regional Board of Education so that they can establish contacts with other sectors.

Regarding leisure, sport, and culture facilities, the information is even more discouraging. There are virtually no activities planned to meet the needs of students with disabilities. Even in the CEU, which have a better structure and could offer diversified activities, there are no actions in this regard, requiring pressure from families for some activities to take place. Public health equipment in general and mental health equipment are struggling to meet the demands of the community. Therapies, which are fundamental for the school and social inclusion of some PAEE students, or their families, are scarce in the researched territory, which was pointed out by most of the interviewees, reporting a lack of vacancies and, when they do, the services are irregular. We understand that therapies are auxiliary and complementary actions to educational action, such as physiotherapy, occupational therapy, speech therapy, music therapy, psychological, and/or psychiatric treatment, also serving family members. However, these treatments do not replace the role of the school, but some students may have a truncated, incomplete development if they do not have this therapeutic support.

There are facilities in other regions that could serve these people and/or their families, but mobility and financial difficulties hinder access for these families. This finding shows that the situation of vulnerability of families has a direct influence on the social and educational inclusion of PAEE students and that the set of public policies is insufficient to guarantee the necessary conditions for its implementation.

The legal and normative set of Brazilian education presupposes access to knowledge as a form of social life for all citizens, and Special Education from the Perspective of Inclusive Education is the guarantee for people with disabilities, spectrum disorders autistic, and with high skills or giftedness have this right guaranteed. However, we cannot separate special education from education in general; we cannot talk about quality education for PAEE students if we do not have quality education for everyone.

The inclusive perspective of special education can only become effective with concrete actions to improve education and other public services to meet the needs of all people. Otherwise, the perspective will always be renewed and never reached, an increasingly distant horizon, perpetuating exclusionary inclusion.

References

Aranha, M. S. F. (2001). Paradigmas da Relação entre a Sociedade e as Pessoas com Deficiência. *Revista do Ministério Público do Trabalho, Brasília, DF, Ano XI*, (21), 160–173, março, 2001. Available in: https://claudialopes.psc.br/wp-content/uploads/2021/08/Paradigmas.pdf. Accessed in em: 23 jan. 2023.

Biondi, A. (2003). *O Brasil privatizado: um balanço do desmonte do Estado*. Fundação Perseu Abramo. (Coleção Brasil Urgente).

Brasil. (1988) [Constituição (1988)]. *Constituição da República Federativa do Brasil*. Presidência da República, [2023]. Available in: https://www.planalto.gov.br/ccivil_03/constituicao/constituicao.htm. Aceessed in: 20 jan. 2023.

Brasil. (2008). Secretaria de Educação Especial. *Política Nacional de Educação Especial na Perspectiva da Educação Inclusiva*. Brasília, DF. Available in: http://portal.mec.gov.br/arquivos/pdf/politicaeducespecial.pdf. Accessed em: 22 jan. 2023.

Brasil. (2011). Presidência da República. Casa Civil. Subchefia para Assuntos Jurídicos. *Decreto n° 7.611, de 17 de novembro de 2011*. Dispõe sobre a educação especial, o atendimento educacional especializado e dá outras providências. Brasília, DF.

Contreiras, H., & Matta, G. C. (2015). Privatização da gestão do sistema municipal de saúde por meio de Organizações Sociais na cidade de São Paulo, Brasil: 91 caracterização e análise da regulação. *Cadernos de Saúde Pública, Rio de Janeiro, 31*(3), 285–297. Available in:: https://www.scielosp.org/pdf/csp/2015.v31n2/285-297/pt. Accessed in: 24 jan. 2023.

Costa, M. A., & Marguti, B. O. (2015). *Atlas da vulnerabilidade social nas regiões metropolitanas brasileiras*. IPEA. 240 p. Available in: http://ivs.ipea.gov.br/images/publicacoes/Ivs/publicacao_atlas_ivs_rm.pdf. Accessed in: 22 jan. 2023.

Cury, C. R. J. (2008). A educação escolar, a exclusão e seus destinatários. *Educação em Revista, Belo Horizonte*, (48), 205–222. Available in: https://www.scielo.br/j/edur/a/PHhyxsVmtHVxX6Hjtn5ZkZp/?format=pdf&lang=pt. Accessed in: 22 jan. 2023.

De Paula, E. T. (2022). *A inclusão escolar de estudantes com deficiência em situação de vulnerabilidade social: fatores intervenientes na atividade do professor de atendimento educacional especializado*. Maria Candida Soares Del-Masso. Dissertação (Mestrado em Educação) –Universidade Estadual Paulista, Presidente Prudente. Available in: https://repositorio.unesp.br/handle/11449/239379. Accessed in: 13 mar 2023.

Del Pino, M. (2002). *Política educacional, emprego e exclusão social. A cidadania negada: políticas de exclusão na educação e no trabalho* (3rd ed., pp. 65–88). Cortez.

Departamento Intersindical de Estatística e Estudos Socioeconômicos [DIEESE]. (2023). *Nota à imprensa: Custo da cesta diminuiu em 13 capitais em março*. São Paulo, 10 de abril de 2023. Available in: https://www.dieese.org.br/analisecestabasica/2023/202303cestabasica.pdf. Accessed in: 28 abr 2023.

Engels, F. (2010). *A situação da classe trabalhadora na Inglaterra*. Boitempo.

Gewirtz, S., & Cribb, A. (2011). Concepções plurais de justiça social: implicações para a sociologia das políticas. In S. J. Ball & J. Mainardes (Eds.), *Políticas educacionais: questões e dilemas* (pp. 123–142). Cortez.

Jacoby, R. (2014). Entender a desigualdade: reflexões sobre o capital no século XXI. In S. C. Bava (Ed.), *Thomas Piketty e o segredo dos ricos* (pp. 9–18). Veneta-Le Monde Diplomatique Brasil.

Mantoan, M. T. E. (2003). *Inclusão escolar: o que é? por quê? como fazer?* Moderna. (Coleção cotidiano escolar).

Marx, K. (2013). *O capital: crítica da economia política.* Boitempo. (Livro I:o processo de produção do capital).

Mazzotta, M. J. d. S., & D'antino, M. E. F. (2011). Inclusão Social de Pessoas com Deficiências e Necessidades Especiais: cultura, educação e lazer. *Saúde e Sociedade, São Paulo, 20*(2), 377–389. Available in: https://doi.org/10.1590/S0104-12902011000200010. Accessed in: 23 jan. 2023.

Melo, A. A. S. (2005). Organismos internacionais na condução de um novo bloco histórico. In L. M. V. Neves (Ed.), *A nova pedagogia da hegemonia: estratégias do capital para educar o consenso.* Xamã.

Mendes, E. G. (2006). A radicalização do debate sobre inclusão escolar no Brasil. *Revista Brasileira de Educação, Rio de Janeiro, 11*(33), 387–405. Available in: https://doi.org/10.1590/S1413-24782006000300002. Accessed in: 22 jan. 2023.

Ministério da Educação [MEC]. (2009). *Resolução n° 4, de 2 de outubro de 2009.* Institui Diretrizes Operacionais para o Atendimento Educacional Especializado na Educação Básica, modalidade Educação Especial. MEC. Available in: http://portal.mec.gov.br/dmdocuments/rceb004_09.pdf. Accessed in: 24 jan. 2023.

Ministério da Educação [MEC]. (2010). Secretaria de Educação Especial. *Nota Técnica n°11, de 07 de maio de 2010.* Dispõe sobre Orientações para a institucionalização da oferta do Atendimento Educacional Especializado – AEE em Salas de Recursos Multifuncionais, implantadas em escolas regulares. MEC. Available in: http://portal.mec.gov.br/index.php?option=com_docman&view=download&alias=5294-notatecnica-n112010&Itemid=30192. Accessed in 11 set 2021.

Netto, J. P. (2011). *Introdução ao estudo do método de Marx.* Expressão Popular.

Oliveira, A. A. S. d., & Prieto, R. G. (2020). Formação de Professores das Salas de Recursos Multifuncionais e Atuação com a Diversidade do Público-Alvo da Educação Especial. *Revista Brasileira de Educação Especial, Bauru, 26*(2), 343–360. Available in: https://www.scielo.br/j/rbee/a/tM5kkt47pwmgdjHytrDCKvf/abstract/?lang=pt. Accessed in: 24 jan. 2023.

Oreiro, J. L., & Ferreira-Filho, H. L. (2021). A PEC 32 da Reforma Administrativa: Uma análise crítica. *Brazilian Journal of Political Economy, São Paulo, 41*(3), 487–506. Available in: https://doi.org/10.1590/0101-31572021-3308. Accessed in: 24 jan. 2023.

Prieto, G. F. T., & Laczynski, P. (2020). São Paulo à venda: ultra-neoliberalismo urbano, privatização e acumulação de capital (2017–2020). *Geousp – Espaço e Tempo (On-line), São Paulo, 24*(2), 243–261, ISSN 2179-0892. Available in: https://www.revistas.usp.br/geousp/article/view/168529. Accessed in: 24 jan. 2023.

São Paulo (Município). (2016a). *Decreto n° 57.379, de 13 de outubro de 2016.* Institui, no âmbito da Secretaria Municipal de Educação, a Política Paulistana de Educação Especial, na Perspectiva da Educação Inclusiva. Casa Civil do Gabinete do Prefeito. Available in: https://legislacao.prefeitura.sp.gov.br/leis/decreto-57379-de-13-de-outubro-de-2016. Accessed in: 19 jan. 2023.

São Paulo (Município). (2016b). *Portaria n° 8.764, de 23 de dezembro de 2016.* Regulamenta o Decreto n° 57.379, de 13 de outubro de 2016, que "Institui no Sistema Municipal de Ensino a Política Paulistana de Educação Especial, na Perspectiva da Educação Inclusiva". São Paulo. Available in: https://www.sinesp.org.br/179-saiu-no-doc/1217-portaria-n-8-764-de-23-12-2016-regulamenta-o-decreto-n-57--379-2016-institui-no-sistema-municipal-de-ensino--a-politica-paulistana-de-educacao-especial-na-perspectiva-da-educacao-inclusiva. Accessed in: 19 jan. 2023.

São Paulo (Município). (2016c). *Decreto n° 57.260, de 26 de agosto de 2016.* Confere nova regulamentação à Lei n° 13.718, de 8 de janeiro de 2004, que dispõe sobre a organização dos clubes da comunidade. Câmara Municipal. Available in: http://documentacao.camara.sp.gov.br/iah/fulltext/decretos/D57260.pdf. Accessed in: 24 jan. 2023.

São Paulo (Município). (2022). *Circuito municipal de cultura 2022 – programação CEUs: Setembro.* São Paulo. Available in: https://www.prefeitura.sp.gov.br/cidade/secretarias/cultura/noticias/?p=32104. Accessed in: 24 jan. 2023.

Saviani, D. (2008). *Escola e Democracia.* Autores Associados. (Edição comemorativa).

Saviani, D. (2013). Escola e democracia: para além da "teoria da curvatura da vara". *Germinal: Marxismo e Educação em debate, Salvador, BA, 5*(2), 227–239. https://doi.org/10.9771/gmed.v5i2.9713. Available in: https://periodicos.ufba.br/index.php/revistagerminal/article/view/9713. Accessed in: 20 jan. 2023.

Secchi, L. (2017). *Análise de Políticas Públicas: diagnóstico de problemas, recomendação de soluções* (2nd ed.). Cengage.

Silvestre, R. C. (2019). *Fluxo de encaminhamento dos estudantes com deficiência intelectual: do prescrito ao vivenciado.* Orientadora: Anna Augusta Sampaio de Oliveira. Dissertação (Mestrado em Educação) – Faculdade de Filosofia e Ciências, Universidade Estadual Paulista, Marília. Available in: https://repositorio.unesp.br/bitstream/handle/11449/190947/silvestre_rc_me_mar_sub.pdf?sequence=7&isAllowed=y. Accessed in: 23 jan. 2023.

Vilela, P. R. (2023). *Lula assinará a MP do salário mínimo de R$ 1.320,00 até 1° de maio. Agência Brasil.* Brasília, 27 de abril de 2023. Available in: https://agenciabrasil.ebc.com.br/politica/noticia/2023-04/lula-assinara-mp-do-salario-minimo-de-r-1320-ate-1o-de-maio. Accessed in: 28 abr 2023.

Elifas Trindade de Paula has a Master in Special Education from Universidade Estadual Paulista Júlio de Mesquita Filho (UNESP), Brazil. He is a specialist in Special Education with emphasis on deafness/hearing impairment (UNESP) and Special Education with emphasis on multiple disabilities (Mackenzie University at São Paulo, Brazil). He has graduated in Pedagogy and History. He is a member of GEPIS – Study and Research Group on Social Inclusion (UNESP). He currently works as a History professor at the São Paulo Municipal Public Network.

Maria Candida Soares Del-Masso is a PhD in Special Education, Professor at the Universidade Estadual Paulista Julio de Mesquita Filho (UNESP), Brazil, retired, with degrees in Education and Special Education. She is working in the UNESP Graduate Professional Program in Physical Education and Graduate Professional Program in Special Education. Coordinator of GEPIS – Study and Research Group on Social Inclusion (UNESP).

MIX
Papier aus verantwortungsvollen Quellen
Paper from responsible sources
FSC® C105338

If you have any concerns about our products,
you can contact us on
ProductSafety@springernature.com

In case Publisher is established outside the EU,
the EU authorized representative is:
Springer Nature Customer Service Center GmbH
Europaplatz 3, 69115 Heidelberg, Germany

Printed by Libri Plureos GmbH
in Hamburg, Germany